Memorial Volume for

JACK STEINBERGER

With Selected Papers and a Commentary by W-D Schlatter

Memorial Volume for

JACK STEINBERGER

With Selected Papers and a Commentary by W-D Schlatter

Editors

Julia Steinberger
University of Lausanne, Switzerland

Weimin Wu
formerly of Fermilab, USA

K K Phua
Nanyang Technological University, Singapore

World Scientific

NEW JERSEY • LONDON • SINGAPORE • BEIJING • SHANGHAI • HONG KONG • TAIPEI • CHENNAI • TOKYO

Published by

World Scientific Publishing Co. Pte. Ltd.
5 Toh Tuck Link, Singapore 596224
USA office: 27 Warren Street, Suite 401-402, Hackensack, NJ 07601
UK office: 57 Shelton Street, Covent Garden, London WC2H 9HE

Library of Congress Control Number: 2022947673

British Library Cataloguing-in-Publication Data
A catalogue record for this book is available from the British Library.

The editors and publisher would like to thank the publishers of the various journals for their assistance and permission to reproduce the selected reprints found in this volume: American Physical Society (*Physical Review, Physical Review Letters*); Elsevier (*Physics Letters B*); Springer Nature (*Zeitschrift für Physik C*).

While every effort has been made to contact the publishers of reprinted papers prior to publication, we have not been successful in some cases. Where we could not contact the publishers, we have acknowledged the source of the material. Proper credit will be accorded to these publications in future editions of this work after permission is granted.

MEMORIAL VOLUME FOR JACK STEINBERGER
With Selected Papers and a Commentary by W-D Schlatter

ISBN 978-981-126-442-9 (hardcover)
ISBN 978-981-126-443-6 (ebook for institutions)
ISBN 978-981-126-444-3 (ebook for individuals)

For any available supplementary material, please visit
https://www.worldscientific.com/worldscibooks/10.1142/13088#t=suppl

https://doi.org/10.1142/9789811264436_fmatter

Contents

Introduction to the Memorial Volume ix
Julia Steinberger

Photographs xi

Pepper and Salt, Enrico Fermi and Neutrinos 1
Jack Steinberger

Jack in Free Space 5
Hallstein Høgåsen

My Memories of Jack Steinberger in ALEPH 9
Jacques Lefrancois

Jack Steinberger — Memories 15
S. Lokanathan

Jack Steinberger, my Thesis Advisor at CERN 17
Vera Lüth

Jack in Pisa and at CERN: His Very Positive Influence 23
Italo Mannelli

My Memories of Jack Steinberger 27
Gigi Rolandi

Memories of Jack Steinberger 33
David N. Schwartz

Celebrating Jack Steinberger's 100th Birthday Memorial 41
Weimin Wu

Reprints of Selected Papers

Comments of the Selected Papers of Jack Steinberger 47
Dieter Schlatter

Early Papers

Evidence for the Production of Neutral Mesons by Photons 51
[*Phys. Rev.* **78**, 802–805 (1950)]

Possible Detection of Parity Nonconservation in Hyperon Decay 55
[*Phys. Rev.* **106**, 1367–1369 (1957)]

Noble Prize Experiment

Observation of High-Energy Neutrino Reactions and the Existence of Two Kinds of Neutrinos 59
[*Phys. Rev. Lett.* **9**, 36–44 (1962)]

CP-violation in Kaon Decay at CERN

A New Determination of the $K^0 \to \pi^+\pi^-$ Decay Parameters 73
[*Phys. Lett. B* **48**, 487–491 (1974)]

A Measurement of the K_L-K_S Mass Difference from the Charge Asymmetry in Semi-Leptonic Kaon Decays 79
[*Phys. Lett. B* **52**, 113–118 (1974)]

First Evidence for Direct *CP* Violation 85
[*Phys. Lett. B* **206**, 169–176 (1988)]

CDHS Neutrino Experiment at CERN

Inclusive Interactions of High-Energy Neutrinos and Antineutrinos in Iron* 93
[*Z. Phys. C* **1**, 143 (1979)] [*Abstract only]

QCD Analysis of Charged-Current Structure Functions 95
[*Phys. Lett. B* **82**, 456–460 (1979)]

Determination of the Gluon Distribution in the Nucleon from Deep Inelastic Neutrino Scattering 101
[*Z. Phys. C* **12**, 289–295 (1982)]

Experimental Study of Opposite-Sign Dimuons Produced in Neutrino and Antineutrino Interactions 109
[*Z. Phys. C* **15**, 19–31 (1982)]

Neutrino and Antineutrinos Charged-Current Inclusive Scattering in Iron in the Energy Range $20 < E_\nu < 300$ GeV 123
[*Z. Phys. C* **17**, 283–307 (1983)]

ALEPH e^+e^- Experiment at LEP/CERN

Determination of the Number of Light Neutrino Species 149
[*Phys. Lett. B* **231**, 519–529 (1989)]

Improved Measurements of Electroweak Parameters from Z Decays into Fermion Pairs 161
[*Z. Phys. C* **53**, 1–20 (1992)]

A Measurement of R_b Using Mutually Exclusive Tags.................. 181
[*Phys. Lett. B* **401**, 163–175 (1997)]

https://doi.org/10.1142/9789811264436_fmatter

Introduction to the Memorial Volume

Julia Steinberger

This memorial volume for my father Jack Steinberger was a labour of love and respect by some of his dearest former colleagues and friends. We are extremely thankful for them to have taken such an undertaking, especially of course Professor Weimin Wu, who organised the whole volume from start to finish. Weimin was a close friend of Jack and the whole family, who participated in many wonderful events, hosting us in China at the occasion of Jack's 65th birthday, and accompanying us on trips to Burgundy or around Geneva.

I am not qualified in any way to write the introduction to a volume on particle physics, so will focus on other aspects. We were lucky to have Jack longer than most: he died peacefully on December 12, 2020, aged 99 years old, after a long but painless decline, taken care of by Cynthia and John at home. He was still able to communicate that he appreciated the care and company of his family — he looked forward daily to John's ice cream treats, Cynthia's spiegeleier, and on weekends, telephone calls with Ned and Joe and Julia's pancakes with Maine maple syrup. This long life meant that he had the sadness of losing many dear colleagues, mentors and friends. Effectively, he was the last of his generation. Many friends still visited or cared for him in his final years, and helped him feel welcome and part of the CERN community which meant so much to him.

Having Jack as a family member meant many things: perhaps most importantly, it meant never being bored. He was always excited and passionate about something or other: very often physics of one sort or another, of course. Even something as mundane as the shape of water when it comes out of the faucet was an opportunity for experiment and discussion. But just as often, it could be outings in the great outdoors, with sailing and mountaineering two large loves of his life, frequent tennis matches near CERN, fixing his bicycle in later years, food, wine, music, concerts, history, culture, outings to art museums or old churches... he had so many interests, and was unhesitating in pursuing them fully, as his large collection of pre-Columbian art will attest.

Just as keenly fascinated as he was by some things, he was utterly unable to become interested in others. Fictional writing he enjoyed in his younger years, but lost interest in somewhere along the way, although he was quite impressed by particular authors. We read some aloud to each other, like Heinrich Böll's short stories, on the long days of sailing, and he was often extremely moved. He also had two books he read us repeatedly as children: Pinocchio, his favourite, and the Green Book of Fairy Tales. Along the way, he also lost interest in cinema and movies. From time to time, we could drag him out, but he never enjoyed the outings enough to be easily persuaded. Perhaps his hesitation with fiction and movies was a reflection of his engagement with reality: physics of course, first and foremost, but also real life, real people. He was always interested in talking to new people, learning how they thought and lived, especially if they were working class or immigrants, if they could tell him something he didn't yet know about the human condition. He loved to discuss and debate, on any topic, but certainly politics, often willing to test the other person's arguments, which could be infuriating as well as entertaining.

He was always upset about injustice and abuses, small or big, like nuclear armament, which he tried for a while reasonably successfully to curb, via his long-term engagement with Pugwash. He was proud to tell us of his terrible track record as a soldier ("the worst he had ever met" according to Leon Ledermann, who was an officer at the time), and how he had responded to racist comments made by superiors during his time in the military, and been punished for it.

For Jack, being part of life was being engaged, interested, in the thick of big mysteries and efforts to help others. He loved his work, his friends and colleagues, his family, his hobbies, and being part of things. We hope this volume will help give some sense of this love and excitement. Even before his death, during his illness, we missed the excitement and engagement he brought to everything he did. In remembering him at his brightest, we hope that those who read these words and remember him carry that something of that curiosity and joy of life forward with them, as a lasting gift.

https://doi.org/10.1142/9789811264436_fmatter

Photographs

Jack and his brothers: from left Herbert, Jack and Rudolph. Bad Kissingen, Germany, circa 1927–8.

Jack climbing his favorite tree, Bad Kissingen, same era.

Jack and his two brothers, again similar era. From left Herbert, Rudolph, Jack. This one is taken outside the Synagogue in Bad Kissingen, where their father Ludwig worked as a Cantor. The Synagogue was destroyed after Kristallnacht.

Father, mother, Herbert, Rudolf and I, 1926?

Jack with his parents Bertha and Ludwig, and brothers, Bad Kissingen, Germany, 1926. From left: Jack, Bertha, Ludwig, Rudolph, Herbert.

Jack in America, with Barnett Faroll, the Jewish family who took him in as part of the Kindertransport. Mid-1930s.

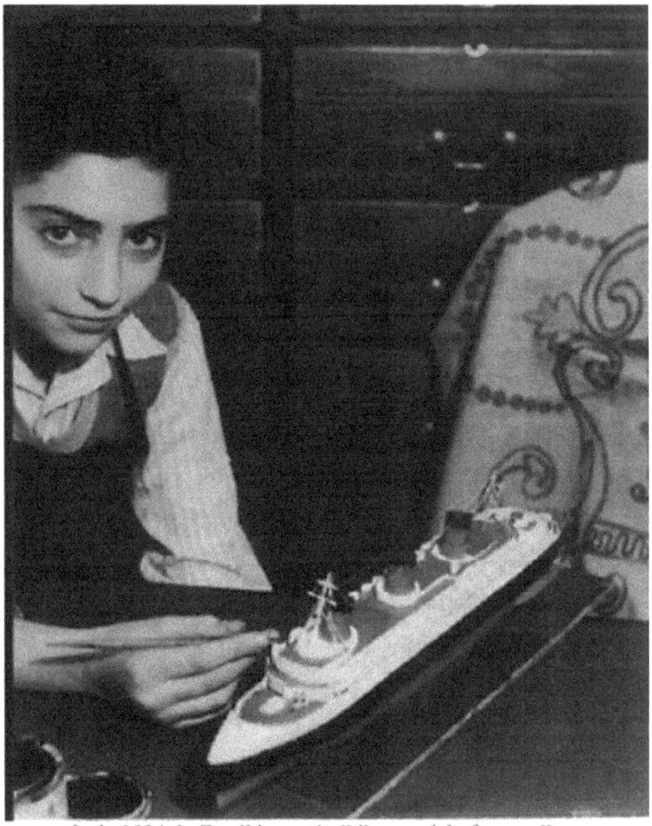

Jack, 1936, in Faroll home, building model of ocean liner.

Jack would have been 15 at the time, in America without his parents, or much hope of ever seeing them again.

The three brothers during WWII in the US armed services. By this point, Rudi (youngest) and their parents had been able to join Jack in the US thanks to sponsorship by Barnett Farroll. From left: Jack, Herbert, Rudolph. Jack was "the worst soldier he had ever seen" according to Leon Lederman, and got into trouble because of opposing racism in the US army, among other things.

Jack and his three sons, Maine, USA, late 1990s. From left Joe, Jack, Ned, John.

Jack and his first grandson, Abraham, the son of Ned and Denise.

Jack at his 80th birthday, 2001, Onex, Geneva. Julia in the background.

For dinner with the Chang's

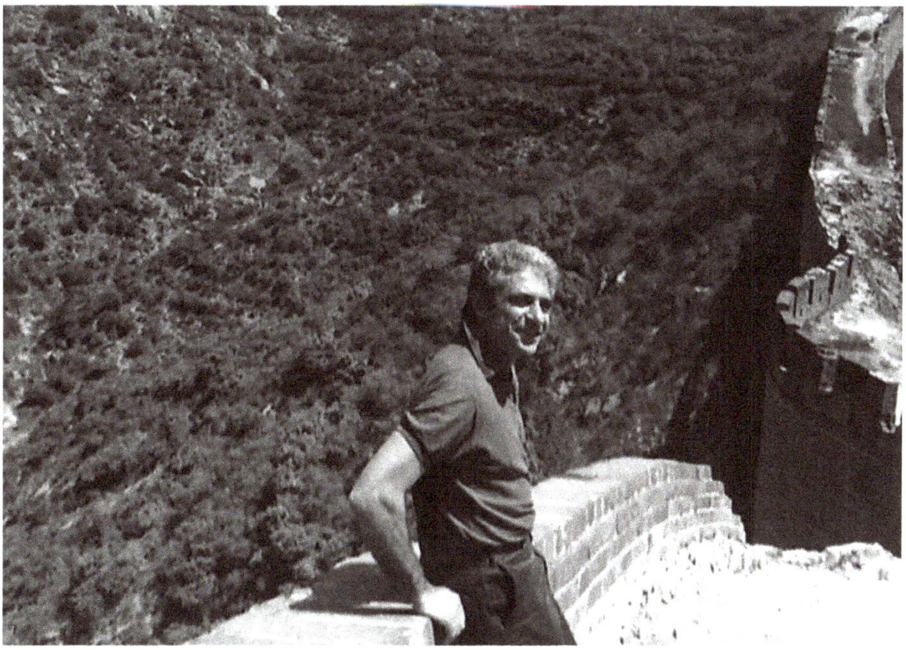

(Top) A trip to China with John Ellis, 1978. (Bottom) Same trip, 1978.

Jack with colleagues on the Neptune ship, lake Geneva, celebrating his 65th birthday. June 1986.

As director of the Physics Department, CERN, circa 1971.

Jack.

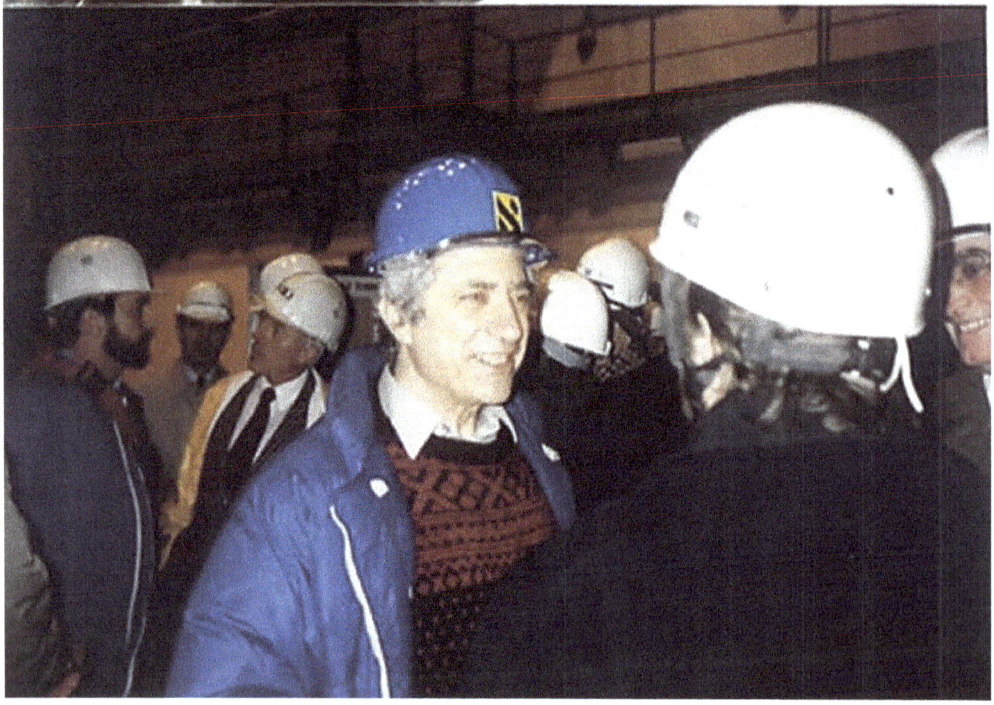

(Top) Transport of ALEPH coil 1987 with Pierre Lazeyras. (Bottom) Visit to ALEPH pit.

Visit of Elena Bonner and Andre Sakharov
Fall '89

With Picasso, Plass, Ellis
Wahl, Koulberg

Jack with colleagues and visitors.

Summer '76
With the Lee's and
the Wu's.

T. D. and Janet Lee, Julia, Jack, in Gex, France, above CERN.

On one of many sailboats.

A typical windblown family photograph. John, Julia, Jack, Cynthia.

One of my (Julia's) favorite photographs, taken by Konrad Kleinknecht, in Budapest.

https://doi.org/10.1142/9789811264436_0001

Pepper and Salt, Enrico Fermi and Neutrinos*

Jack Steinberger

When one mentions Fermilab, I remember Aurora, the town nearby. As you know, I escaped Hitler in Germany as a young man of 12 years of age and taken into an American foster home. My foster father had a farm, Farroll Farm, in Aurora which housed Pepper and Salt — two beautiful ponies for riding. So, my first impression of Chicago, Illinois was with the town of Aurora and with Pepper and Salt.

Coming back to Fermi now — after all a laboratory named after Fermi must be a great institution. I remember fondly my time at the University of Chicago, working under Fermi. Enrico Fermi is one of the greatest physicists of all time whom I have personally known and worked with. Fermilab should be very proud of being named after Fermi. It is a major laboratory for particle physics and I am honored to be asked to offer my thoughts. Fermi was extremely helpful to me, a fantastic mentor and a thoughtful physicist as everyone knows. My first recollection of him was a car trip to Mount Evans in Colorado, where Fermi's idea was to detect and then generate electricity or electric current from cosmic rays. I think Bernard Gregory, who later became the Director General of CERN, was also with us in that trip.

In the last years of his life, Fermi did something that nobody did; and this is about the Chicago cyclotron, the world's highest energy cyclotron at that time. If I remember correctly, Richard Garwin worked on that cyclotron as well. Everyone believed that cyclotron energy is fixed, but many experiments required the proton beam energy to have different values. Fermi found a completely new way to change the energy of the cyclotron beam on an experimental target . . . he attached the experimental target to a little cog which went around following the

*Transcribed by **Swapan Chattopadhyay** from his interview with Prof. Jack Steinberger at CERN on Thursday, April 27, 2017 and corrected and approved by Jack and Cynthia Steinberger.

magnetic field of the cyclotron. Coils were attached to the cog moving along the outer edge of the magnet. Thus, the cog carrying the target sees different fields and hence different energies of the protons.

During my experimental work at Columbia, several good students worked with us at that time... Marshall Rosenbluth, John Wheeler and many others. There was a paper published much afterwards, about the Fermi hierarchy of beta decays, which could also explain observations in cosmic rays. This eventually led to the universal theory of beta decay. But not everybody in this smart team saw it first. Being smart does not always mean that you can do things or observe things first — sometimes not. I have a son — he had lots of trouble in high school, but today he is the best-known Steinberger, for making musical instruments. Of course, I have another son, a mathematician, who is a professor at Tsinghua University in China.

About neutrinos now... my brain is far from being functional these days. Several types of neutrino experiments interest me. These are focused on decay of B-mesons into two muons or other forms of beta-decay, including double beta decay experiments. I do not know about the DUNE project you mention, but neutrinos are most interesting particles... any program of pursuing them must be a worthwhile exercise today.

You ask me about accelerator-based high energy physics. The field is dear to me but I am also troubled by it. It feels incredible to me, but as I cannot read anything anymore due to my vision, I also have not been able to give much thought to this topic lately. I am unhappy that I cannot think much anymore. But I believe that learning something about dark matter must be interesting. Who knows there maybe connection with neutrinos.

You ask me about my connection with Germany now. I am a wandering Jew as you know. In 1988, before I got the Nobel Prize, in the small town of my birth Bad Kissingen, Germany with a population of 20,000, people of the town decided to try to remember the Jewish experience and Jews like me. After I got the Nobel prize, I was invited back and I have made some great friends with people in the local high school there who I am very fond of. For my 95th birthday, there was a concert given by students of the local high school, which bears my name. This concert was a great pleasure for the students who gave it, and for me also.

You say T. D. Lee wrote something by hand... can you show it to me please, he is a very dear friend of mine.

Well, I have said much I think... you will of course write for me, knowing what I would have liked to have said, rather than what I am actually able to say now, for my memory makes it all jumbled up and it often feels empty. In any case, my best wishes to Fermilab on its 50th anniversary.

https://doi.org/10.1142/9789811264436_0002

Jack in Free Space

Hallstein Høgåsen
Department of Physics, University of Oslo, Norway
t.h.hogasen@fys.uio.no

These are some of the activities with Jack that I remember, which contributed nothing to his fame or wealth. Nevertheless they were important to him; This was his life outdoors. Nobody would say that he was a religious person, but often when surrounded by nature, I saw something pass over his face giving him a monk-like appearance.

I met Jack in 1965. I was a young fellow at the theory division at CERN. He was a world-famous experimentalist (with a beautiful wife almost my age: Cynthia). It came up that we both did some climbing. The summer before, I had been on a stay with the GUMS (Groupe Universitaire de Montagne et Ski) at Les Bossons in Chamonix under the guidance of a CAF guide (Bernezat?) who had found me competent enough to take people into the mountains. So I had some knowledge of that region and Jack and I went there together. First on easy tours and later on hikes where hands were necessary.

The first trip with some difficulty was in late August 65: the traverse of L'Aiguille de Chardonnet (3824 m). We started around three in the night from Refuge Albert Premier (2702 m) and crossed Glacier du Tour with flashlights under a clear sky. While mounting the snow slope towards the ridge, the sun rose, and it was with glee that the four companions reached the ridge leading to the summit. The view was fantastic! L'Aiguille Verte, Les Droites and Les Courtes just in front of us.

The traverse of the summit is what is called a 'mixed climb'. Stretches of snow/ice are separated by rock. And here our troubles began. One of us did not dare to climb on the rock with crampons on. Or in snow/ice without crampons. The operation crampons on-off-on-off took considerable time and we fell behind the planned time schedule. I was worried. In a way it was my fault, as I had agreed to bring along two people that I did not know. Jack behaved well. If he was worried (as I believe he was) he did not show it. Normally he could be rather crabby, but not up there.

When we reached the summit the sky had changed, promising incoming snow, further complicating matters. The descent has the reputation of being dangerous. Luckily that day there was little ice — mostly snow and rock. Still it was more tricky than we liked. The snow soon started to fall. Quite heavily too, when we reached the final slope above the glacier. Three of us passed down pretty fast, secured from above, and as the last I slid down on my boots to applause. Now we had the direction to the cabin on the compass so even with very bad visibility we felt safe. It was still very good to finally be inside!

After this Jack and I felt mutual confidence. That was as one could say “the beginning of a beautiful friendship”. No harm ever happened to us in the mountains, on foot nor on skis. Jack would often complain “I am miserable” but that was probably just a way of saying that it would be nice to sit down. Or perhaps stressing the triumph of his will as he pressed on.

My memory is weak and my pegasus is tired, so most of our trips are forgotten. Only L’Aiguille d’Argentière, L’Aiguille de l’M, la Nonne, le Petit Capucin and Zinalrothorn are clear in my memory. The last because Cynthia came with us to the Rothornhütte and surprised me with her stamina. (The cabin is at 3200 m and the summit at 4212 m.) I was thinking that Jack should have encouraged her to join the men in the sport. But he was somewhat old-fashioned in his opinion of what was for men and what was for married women. A real patriarch.

We spent much time outdoors together and I cannot resist telling how I came to share Jack’s darkest hours: In spring 69 we had a quite adventurous trip to the summit of the Petit Capucin. Happy and tired, we skied down La Mer de Glace to take the train from Montenvers down back to Chamonix. When we came up the ladders from the glacier, our worst fantasies had come true: The last train for the day had gone and the hotel had not yet opened for the summer. All doors on all buildings were locked. We sat down on the stairs of the hotel. The mood was morose when we finished the last of our provisions. We were not prepared for a bivouac. So we cheered ourselves up by saying that we would not have to pay for a hotel room.

We were not that tired anymore after the rest, and walking down to Chamonix did not seem impossible anymore. So we put the skis on our shoulders and trusted the horses of the apostles. It was now quite dark, and to avoid getting lost we walked along the railway tracks. We swept the boots along the rails. In the tunnels — there are many, and they are long — we joked that the photons in the visible spectrum could be counted on one finger. These were indeed our darkest hours! They passed without giving us too much pain, and we returned safely home.

As the years passed, the charm of keeping alive through the strength of one's fingers gradually passed away, and the blue sea became the road to the wide spaces where one best could meet nature's splendours. Jack told me how he had built a dinghy and bought a corn pipe when he was a young man. This was an example of how he believed a man should act. As a grown-up he was well off so Jack and Cynthia bought ocean-going sailboats.

I was lucky to be one of their many students. Their children also became sailors, and as far as I know none in the family were prone to seasickness. That is a gift among many they could enjoy. That Jack did not reach hundred years came as a big bad surprise to me. But to me and all my friends there are fond memories to exchange for many evenings in the coming years.

https://doi.org/10.1142/9789811264436_0003

My Memories of Jack Steinberger in ALEPH

Jacques Lefrancois
LAL, CNRS/IN2P3, Université Paris-Saclay, Orsay, France
lefranco@lal.in2p3.fr

It is very moving for me, as it is I am sure for many physicists, to remember the days in ALEPH with Jack, it was a wonderful time in our physicist life, I will try to cover the various period of ALEPH and our interactions.

1. The Apparatus Conception

It started in 1980, when a small group of physicists (17) were invited in Jack's office after the announcement by CERN of the LEP project. About half were collaborators in his previous experiments, the idea was to discuss a collaboration, for building a new instrument and preparing for the physics which was announced for 1987–88 at the time.

We were enthusiastic about collaborating with Jack in such a project and for the next two years he organized our collaboration as a kind of brainstorming period until we froze the design in 1982, to then start the construction until 1989.

Jack's leadership during this period was remarkable, he encouraged new ideas even unusual ones (he studied himself for a short time a central magnet in form of a sphere... probably to encourage our imagination!) He also succeeded in having a very open atmosphere: physicist who mainly worked on calorimeter ideas could for some time think and even write a note about tracking detectors and inversely. This was done without compromising on his very demanding standards of rigor and deep understanding. The atmosphere helped to create rapidly a team spirit where we were happy to meet together and exchange ideas.

Retrospectively it is clear that, even with this encouragement of new ideas, there was also his insistence on many key points which were already there in his own experience of building remarkable instruments (I remember my amazement when I saw the apparatus built for his neutrino experiment at Brookhaven in 1962).

To present two of the key points:

Jack insisted that the detector was going to be complicated to build, and to manage after, and therefore it was essential to minimize the numbers of different techniques: if one chooses the technique of streamer tube for the Hadronic calorimeter, then the muon detector should be built with the same technique; or if we decide to build the barrel electromagnetic calorimeter using a technique of proportional wire chamber and lead plates, then the End cap ECAL and even the small angle luminosity monitor would be built with the same technique. This was a very wise decision but it took Jack's strong personality to have it implemented (there are counter examples in the other LEP experiments) because of course the various laboratories and physicists in charge of constructing one of the instruments would prefer to each adopt their own technique.

Another important point, of Jack's influence, was the premium attached to the redundancy of a detector, this had a tremendous influence in helping us to understand (and decrease!) possible sources of systematic error on our data. For example, in the TPC, for the tracking of the particles, clearly there had to be some edges where there would be no detection, so the design was arranged such that no tracks would be lost some tracks would have only half the points but all had therefore enough points to be measured. There is a nice picture (Fig. 1) of Jack explaining the TPC concept in a LEPC meeting.

Similarly in the ECAL and HCAL where the sensitive elements were wire chambers the most important measurements were cathode pads defining towers but all wires were also readout and the consistency could be checked and noise in some elements therefore identified. The dead region in the ECAL and HCAL were offset again to recuperate particles missed in the first detector. Jack's role was certainly mainly to encourage discussions but when there were critical points it was clear he was taking the final decision, and because of this excellent team spirit this was accepted.

2. The Apparatus Construction

After all these discussions and decisions between 1982 and 1989 we concentrated in building the detectors. There were periodic general meetings about every 2 months but in between also dedicated meetings for the people working on a particular element, in my case the ECAL. Jack was coming to each of these "more specialised meetings" and had a major influence either by his challenging remark we often heard "but are you sure of that" usually spotting a weak point to be studied again for the next time but also for brilliant ideas. I remember for example that we had discovered that our ECAL was loosing accuracy when immersed in a magnetic field because low energy electron would progress as spiral along the field lines. Jack suggested that we used Xenon instead of Argon in our proportional chambers scattering these low energy electrons out of their spiral path, this idea was checked by simulation and worked beautifully, first in the simulation, and then in the ECAL small size prototype, and in the final detector.

Because of his charisma and the remarkable collaboration atmosphere he was also able to attract excellent specialists in fields where he was not an expert ,like data taking software with John Harvey, or offline software with Jugen Knobloch. He also attracted Pierre Lazeyras our technical coordinator, who was a remarkably competent person and Jack had a lot of respect for him. Our event displays were designed by Hans Drevermann, they were beautiful they contained all the information but for me they were also almost "artistic" compared to what I had seen before (Fig. 2).

When the apparatus was built, but also visible before the installation of electronics he decided to have a picture made (Fig. 3) which has become famous.

3. The First Analysis

After a few events in August 1989 , there was sizeable data taking during the beginning of the fall of 1989 and Jack was leading a small group of physicist using events as seen by the calorimeters to measure the Z^0 cross section and therefore the number of neutrino families. As often in further analysis there was another group obtaining the same results by observing the events in the trackers. The two measurements agreed (most events were in common) and we published our result on the number of neutrino types by averaging the two results which was an excellent procedure for encouraging to have more than one analysis for important measurements. This first data gave $N_\nu = 3.27 \pm 0.24_{\rm stat} \pm 0.16_{\rm sys} \pm 0.05_{\rm th}$. Jack was especially happy about this publication since he had been awarded the year before the Nobel prize (with Leon Lederman and Melvin Schwartz) for his experiment in 1961 at Brookhaven on the existence of two different neutrinos ν_μ and ν_e now it was proven that there were only 3 different types of neutrinos.

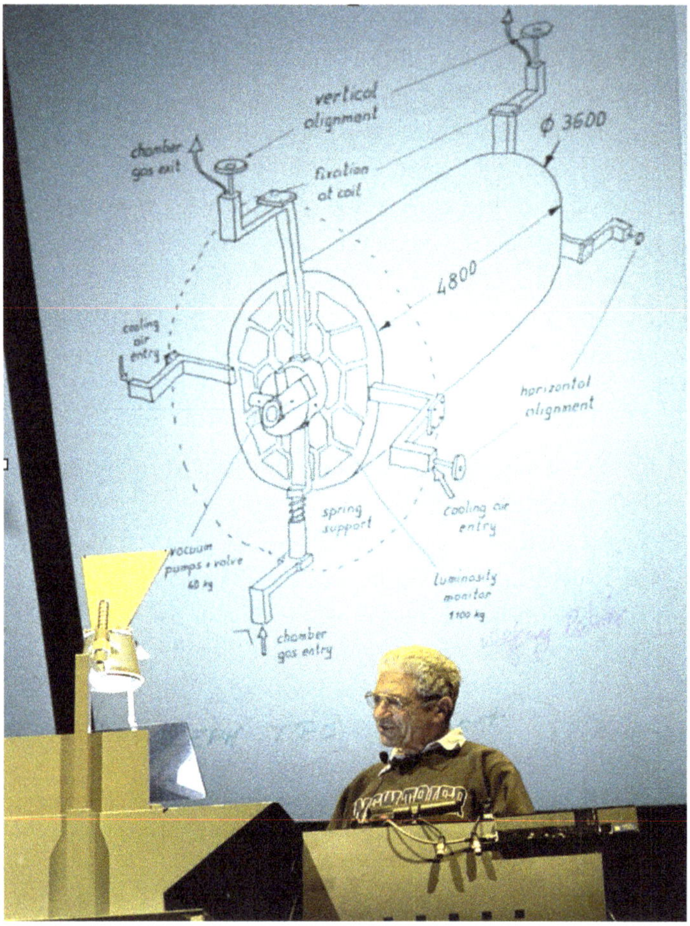

Fig. 1. Jack Steinberger at LEP Fest 2000 — Science Symposium. Source of this photo: CERN. © CERN.

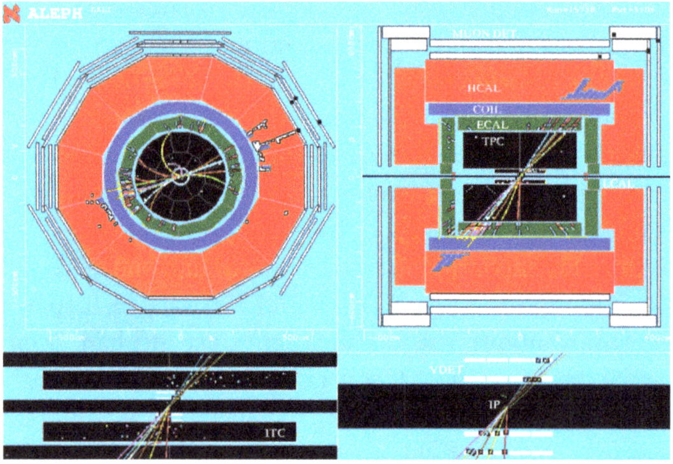

Fig. 2. ALEPH event displays designed by Hans Drevermann.

Fig. 3. Photo of the ALEPH apparatus in the cavern. From right to left one finds Pierre Lazeyras, Lorenzo Foa, Jack Steinberger, and myself. Source of this photo: CERN. © CERN.

Fig. 4. Jack, Lorenzo, and myself at a party. (Photographer unknown.)

4. The 90's

Jack had already reached the age of 65 in1986 and according to CERN rules had therefore reached the retirement age, he nevertheless continued to be ALEPH spokesman until February 1990 when he resigned and I became his successor as spokesman until 1993. He continued doing analysis with the ALEPH data and produced with a small group of physicists a beautiful measurement on Rb the fraction Z_0 decays to a bb pair. His idea was to use different quark type signature and therefore it was possible to control efficiency and systematic errors.

But in my opinion what was his most important achievement in this period was the collaboration of the 4 LEP experiments on doing joint publications for the averaging of the results. This was not a small feat because 4 experiments with the same physics program are usually more in a mood of competition than collaboration, but it worked very well after he contacted the 4 spokesperson and initiated joint discussions. This was an essential work because the physicist who have worked on the measurements are certainly the most competent to understand for example common systematic errors due to models.

A moment dear to my heart is in 1993 when I was replaced by Lorenzo Foa as ALEPH spokesman and there was the usual party. I remember I told the audience that initially I thought it would be an almost impossible task to succeed to Jack but that he had created such a remarkable spirit of collaboration in ALEPH that the job was then easier that I had been afraid. During the drink after we had an interesting discussion between Jack Lorenzo and myself for serious moments but also lighter ones as seen in the pictures (Fig. 4)!

In summary, I can say that physicist who collaborated in ALEPH remember this as a wonderful period. Of course, it had to do with the fact that the physics was interesting but also that the ALEPH detector worked very well and also that there was a very open atmosphere in the physics discussions and it is clear for all of us that the remarkable personality of Jack was the main reason for the last two points.

https://doi.org/10.1142/9789811264436_0004

Jack Steinberger — Memories

S. Lokanathan
(Retired) Professor of Physics, University of Rajasthan, Jaipur, India

I joined Columbia University as a graduate student in the fall of 1951. It was not just a New World geographically for one coming from a traditional Indian background. There were very few Indians to advise or share their experience.

In the first few semesters there were the required courses of study to go through, not always with ease. Later, I had a part time job as a Scanner, poring through a microscope looking at Nuclear Emulsions that had been exposed to meson beams from a cyclotron and recording 'events' of interest. It was a good education for me since hitherto Mesons and other particles had been just text book words. And I still had the ominous prospect of passing the qualifying examination for Ph.D. research.

It was then that Jack Steinberger dropped in one day to see what we were doing and chatting with us. A few days later he asked me if I would like to work with him at the Nevis Cyclotron Labs. Would I! Jack's reputation was already a byword in the student grapevine and I eagerly jumped at the chance.

Those were days before Solid state devices and we learnt to mount electronic tubes in a chassis, wiring and soldering and using an Oscilloscope to check and get them going. Jack was always encouraging and if he could have done it all in a fraction of the time I took, he did not show his impatience. Letting students learn was important. Indeed motivating students was to be his endearing trait.

In one of the experiments he put me on to, we had to try to distinguish higher energy electrons expected from possible direct decays of pions, from the lower energy ones from muons. In those times there were the crude ways of filtering them through absorbers. Jack put me on to do some Monte Carlo calculations to study the energy loss. I had never heard of it except as a place in the South of France! It was fun to learn without being embarrassed at my ignorance.

One morning in a hotel lobby just before a session of the American Physical Society in Chicago Jack was having a chat with Enrico Fermi and spotting me asked me to come and meet the great man. I was 'tongue tied'. Meeting a 'Grand Guru' was a rare privilege — Fermi had been Jack's supervisor, guru.

By 1957, I was in Clarendon Laboratory, at Oxford, Britain as a post-doc and had fewer opportunities of meeting Jack. One was at that lovely town of Sienna in northern Italy in 1963 at an international conference. By then the pioneering experiment of Schwartz, Lederman and Jack using neutrino beams from accelerators was shown feasible and Jack was increasingly involved in CERN experiments.

In 1969 I was in the Physics Department of Rajasthan University, Jaipur and would have an occasional telephone converstion with Jack. It was a great pleasure to hear that Jack was awarded the Physics Nobel Prize in 1987 along with Mel Schwartz and Leon Lederman. In 1989 when I had occasion to attend a meeting in Switzerland, I spent a delightful evening with Jack in his home in Geneva exchanging pleasantries. Typically he did not want to talk about his Nobel! Such was his friendship that he insisted on driving me next day to the airport for my return to India.

Jack's physics contributions are legion as others would recount. I want to share my remembrance of him as an unusually warm person. He was not yet thirty when he joined the staff in Columbia University and at that tender age he had shown a toughness of spirit to take a stand on principle, refusing to sign an oath to conform to a political straightjacket, which forced him to leave a job and seek another. Yet he hardly flaunted his beliefs and continued with his profession as a physicist. And in that calling he not merely reached great heights but showed a wonderful quality, kindness and empathy with students and colleagues of diverse cultures. In today's troubled world it is to remind us that Science IS always a Social activity. And for Jack, the two were inseparable.

https://doi.org/10.1142/9789811264436_0005

Jack Steinberger, my Thesis Advisor at CERN

Vera Lüth

SLAC National Accelerator Laboratory, USA

The four years I spent at CERN as a member of the small CERN-Heidelberg group building a beyond the state-of-the-art detector and producing high precision measurements of K^0 decays under Jack Steinberger's leadership were extremely important for my later career in experimental particle physics. Observing and interacting with Jack was a great experience in these early years, as well as during occasional visits to CERN in later years. He stood out as a great mentor and scientist and as a person with a very broad spectrum of knowledge and interests, and concerns about the world we live in!

I first met Jack in 1968 soon after he and Cynthia had moved to Europe. At that time, CERN was building the Intersecting Storage Rings (ISR), a proton collider at high energy. There were plans for a large general-purpose detector with a strong magnetic field to measure the momenta of the charged particles produced in the proton-proton collisions. Three designs for a large magnet structure had been proposed, and for each of them a scale model had been built and their magnetic fields had been measured.

In the summer of 1968, I had almost finished the analysis of bubble chamber data for my MSc thesis at the Institute of High Energy Physics at University of Heidelberg (IHEP), when Heinz Filthuth, the IHEP director, sent me to CERN for the summer. He had been contacted by Jack Steinberger who had recently joined CERN and was trying to assess the performance of the ISR detector for three different magnet designs and was looking for a young scientist to work with him on this study for a few months.

To evaluate the acceptance and precision achievable with each of these three proposed magnets I was asked by Jack to simulate and reconstruct tracks of different angles and energies produced in these high energy interactions. By the end of the year, we documented the results of this study and on the basis of the written report and various engineering studies the so-called split-field magnet, proposed by Adolf Minten and favored by Jack Steinberger, was chosen by the CERN management.

Just before I returned to Heidelberg to finish writing my master's thesis, Jack took me aside and told me that he had recently proposed an experiment to study CP violation in decays of neutral K mesons in collaboration with the Heidelberg group. He asked me if I would be interested in joining the group. Knowing Jack and having spent time at CERN, this was a very attractive offer I could not turn down. Jack had been nominated as a "Honorary Professor", a title that allowed him to be my advisor! A couple of weeks later I received a letter from IHEP in Heidelberg offering me support as a graduate student working at CERN!

A few years earlier in 1964, the discovery of the violation of CP invariance in a decay of long-lived neutral kaons, $K_L \to \pi^+\pi^-$, by Christensen, Cronin, Fitch, and Turley was a great surprise. In other words, a violation of the concept that the laws of physics should be the same if a particle is interchanged with its antiparticle and the spatial coordinates are inverted. The violation of this invariance in K^0 decays could not be explained by any theoretical model, a very puzzling phenomenon. Jack devoted about a decade to experimental studies of these decays.

In 1965, while on sabbatical leave at CERN, Jack teamed up with Carlo Rubbia and others on a new experiment which resulted in a variety of measurements of neutral kaon decays, specifically the interference between the K_L and K_S decays in the K^0 decay time dependence and the first measurements of their mass difference. The next result was the observation of the small, CP-violating, charge asymmetry in semi-leptonic decay in 1966. Upon his return to the US, Jack joined a new experiment at a higher energy beam at Brookhaven Laboratory which confirmed this result in 1967 with somewhat higher precision than the earlier measurements at CERN and also at SLAC.

After Jack moved to Europe in 1968, he started thinking about a high precision experiment to study K^0 decays to two oppositely charged particles, $K^0 \to \pi^+\pi^-$ or $K^0 \to \pi^+\lambda^-\nu$, here the charged lepton λ refers to an electron or muon and ν refers to the neutrino. His concept of a two-arm spectrometer relied in many ways on earlier experiments. It was to be placed close to the production target to enhance the detection of the K^0 decays with short decay times. The momenta of the two charged particles from the K^0 decays were to be measured using three very large multiwire chambers, one in front and two behind a dipole magnet, each equipped with a horizontal and a vertical signal wire planes. The novelty was the use of proportional wire chambers recently invented by George Charpak.

When I arrived at CERN in the summer of 1969 the proposed experiment was approved, and the CERN-Heidelberg collaboration had been formed. Apart from Jack, the group had ten members, among them two graduate students.

Jack had realized that the new proportional wire chambers would allow substantially higher data rates, cleaner trigger signals, and higher position accuracy. The biggest challenge was the design and fabrication of these large chambers, 2.6 m wide and 1 m high, with more than thousand signal wires each, installed with considerable tension to sustain the electrostatic forces between neighboring wires. Starting out with a few small chambers (10×10 cm^2) used for beam tests, it took us about two years to arrive at a suitable design and to build these chambers with assistance from Charpak's group and top engineers from the CERN mechanical shop.

Almost all detector components were beyond the state-of-the-art. These included the first microchip front-end electronics with full on-line calibrations and a very smart trigger system to select the two-track signal events. Each of the signal wires required their separate amplifier and readout circuits! The electronics were developed by Bill Sippach, a leading engineer at Nevis Laboratory, taking advantage of the high counting rate of these wire chambers and achieving high time resolution to minimize backgrounds. Jack was proud of the beautiful layout of the three sets of wire chambers and their technical coherence and powerful performance. All of us were involved in developing and building the detector, installing the associated electronics and developing the on-line and off-line software, truly enjoyed the opportunity to participate and face the challenges we encountered.

Once completed and installed, very large data samples were collected at the CERN PS accelerator. With a read-out system selecting events with two charged particles in each of the three wire planes, we recorded more than 1000 events per beam pulse, resulting in a total sample of more than 10^9 events (198 bits in length) stored on more than 1000 magnetic tapes, typically one tape every 20 minutes.

While a given data analysis was usually taken on by one scientist, we worked as a team. Jack insisted on thorough checks of the analysis methods and detailed studies of sources of systematic uncertainties. When I showed him the final result of my analysis of the time dependence of the charge asymmetry in semi-leptonic K^0 decays, he asked me for the data I had used and proceeded to write a program to verify my results. To my relief, his analysis confirmed my results!

The experiment produced a series of precise measurements of the decay time dependence of K^0 mesons to two pions or a single pion plus a charged lepton of opposite charge and an associated, but undetected, neutrino. This led to very precise measurements of the CP-violating parameters in these K^0 decays. Several of the measured values of these parameters differed significantly from earlier measurements, but all future measurements agreed with our values. For instance, our measured ratio of two complex decay amplitudes, referred to as η_{+-}, was $(2.30 \pm 0.035) \times 10^{-3}$, whereas the average of earlier measurements was $(1.91 \pm 0.09) \times 10^{-3}$, a difference of about four standard deviation! Measurements of two other parameters characterizing the K^0 decay also resulted in sizable differences from other experiments. Given our large data sample and the outstanding capabilities of the detector, we concluded that it was very unlikely that our results were wrong. Nevertheless, the graduate student working on two-pion decays started checking every detail of his analysis, and at one of our meetings he announced with a great smile on his face, "Jack, I am now getting the right result!!" Jack jumped up. "What do you mean by the right result? The right result is the one you understand. It doesn't matter what anybody else has measured." It took several days for Jack to calm down. This has been the most important of the many lessons I learned from Jack and will never forget! Nowadays, most experiments use blind analyses for precision measurements. They mask the measured value in the analysis by a certain numerical offset, so that the person performing the analysis cannot compare the result with the values expected from theoretical predictions or earlier measurements. The offset, which is kept secret by one person, will be removed after all uncertainties have been carefully studied and the results are ready for publication. Since then, we observe significantly larger spreads of measured values obtained from different measurements. The CERN-Heidelberg group was small, and our offices were all on the same floor, so we would meet frequently. Group meetings were ad-hoc, whenever there was a problem or some good news to report. Frequently we had lunch together in one of the CERN cafeterias. On occasions Jack would invite a colleague or friend to join us. I remember he introduced me to T.D. Lee, the famous theorist and former colleague of his at Columbia University. I tremendously enjoyed T.D.'s annual lectures at CERN.

Fig. 1. Jack Steinberger. Picture courtesy of Konrad Kleinknecht.

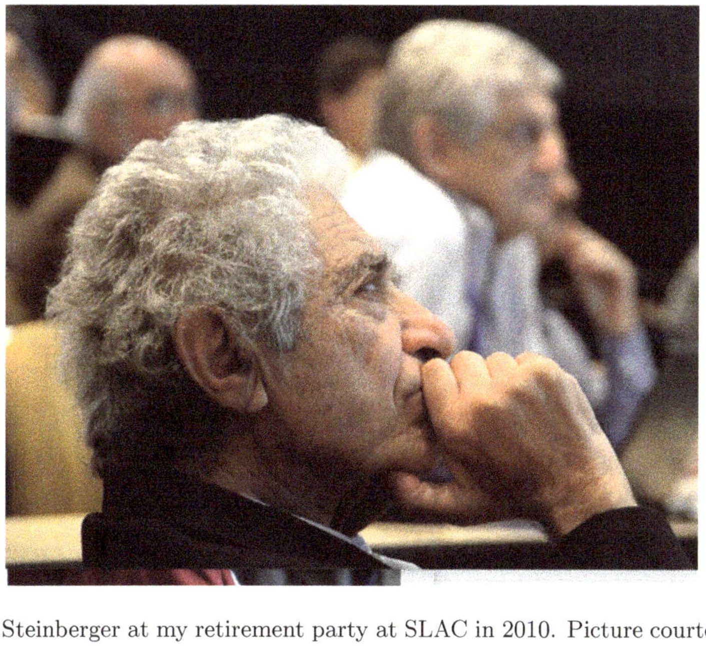

Fig. 2. Jack Steinberger at my retirement party at SLAC in 2010. Picture courtesy of Harvey Lynch.

Every year, the group arranged a number of outings in the beautiful areas around Geneva, family picnics in the summer, ski excursions in the winter, and chamber music concerts at Jack's home. Many of the friends he went skiing and hiking with were German, among them experienced mountaineers. Music was very important to Jack, the son of competent pianist and organist. He enjoyed music, especially baroque and classic chamber music, from J. S. Bach to G. Mahler.

I consider myself very fortunate to have had chance to work and interact with Jack Steinberger for four years at CERN as a graduate student. To answer technical questions he often resorted to explanations he had learned as a student from Enrico Fermi. I fondly remember many occasions talking to Jack on a variety of topics, recent scientific discoveries as well as the many concerns about the world we live in, on a local as well as global level.

On a personal level, Jack was always very helpful. For instance, when I first arrived in Geneva to work with him on the ISR detector studies, he invited me to stay at his house above Gex for a few days, until I could find a place to stay in Meyrin. I remember waking up every morning to Bach cantatas.

I was extremely pleased that Jack attended my retirement party at SLAC in 2010, and reported on the kaon experiments at CERN, 40 years earlier! I had offered him a business ticket for the very long flight from Geneva and tried to arrange this with the travel agency at CERN, but he refused to accept this personal luxury!

These are just a few of the many memories I cherish since those years at CERN.

https://doi.org/10.1142/9789811264436_0006

Jack in Pisa and at CERN: His Very Positive Influence

Italo Mannelli
Scuola Normale Superiore e INFN, Sezione di Pisa, I-56100 Pisa, Italy
italo.mannelli@pi.infn.it

In 1956–57, Jack was in Bologna and Pisa on leave from Columbia University. In the Autumn of 1956 Jack Steinberger came to Italy with the intent of establishing a scientific collaboration with Marcello Conversi, Director of the Physics Institute in Pisa, and with Giampietro Puppi, Director of the Physics Institute in Bologna. He had been very impressed by the Conversi–Pancini–Piccioni experiment, which had established the weak interacting property of the majority component of the cosmic rays at sea level, and by the theoretical work by Puppi who recognized the connection between different experimental facts as due to a single weak interaction with a unique coupling constant.

His proposal to put together a local activity, by building a projector and analyze Bubble Chamber pictures taken at the BNL Cosmotron was a wide window opening on a new exciting world.

At Brookhaven the Cosmotron 3 GeV proton syncrotron had come into operation and a large sample of negative ~ 1 GeV pion interactions in a H2 and in a Propane Bubble Chamber had been obtained. For the first time, events with production of so-called Strange Particles, previuosly seen rarely in cosmic rays and detected in cloud chambers or nuclear emulsions, had become available in fairly large number and with the possibility of accurate reconstruction of their kynematical properties. Jack offered the opportunity to two groups of very enthusiastic but totally unexperienced young physicists to enter a new extremely interesting field of physics, just waiting to be explored.

At the instigation by T.D. Lee an experimental challenge was to try answering the following question: Was the newly discovered Parity Violation in beta and in pion and muon decays, due to the presence of Neutrinos, with Zero mass and hence defined helicity, or was Parity Violation a general property of Weak Interactions?

In the strangeness conserving $K^0-\Lambda$ production from π^- proton interactions the Λ could be polarized but, due to Parity invariance, only in the direction of the normal to the $\pi^- - \Lambda$ production plane. In the subsequent Λ to π^- proton decay Parity violation could manifest itself as an UP–DOWN asymmetry. Concretely the task consisted in measuring, event by event, with the help only of pencil, paper and slide ruler, the cosine of the angle between the proton and the normal to the production plane.

After one year of intensive work by a collaboration of Bologna, Columbia, Michigan and Pisa University, the YES answer was published in a very short Letter to Physical Review, following the presentation by Jack at the 1957 (Padova–Venezia) International Conference on High Energy Physics. It was the subject of my Thesis in December 1957. Not bad to have your name on the list of less than 20 authors, four of whom (Glaser, Perl, Steinberger and Schwartz) later got the Nobel Prize!

In the period 1960–1980, I was working on other experiments, not involving strange particles. Jack with Lederman, Schwartz and others at Brookhaven discovered in 1962 the neutrino flavour, for which they got the Nobel Prize, and then he moved permanently to CERN. Following the discovery of CP violation by Fitch and Cronin in 1964, Jack promoted some very accurate experiments both at CERN and at BNL, mostly on Kaon decays. Later he was the leader of a large scale collaboration to study high energy neutrino interaction and subsequently the leader of the great ALEPH experiment at LEP. But he had not abandoned his deep interest in Kaon physics.

In 1976 John Ellis, with M.K. Gaillard and others, first realized that, in allowing in the Standard Model the possibility of CP violation as due to the complex phase allowed in the CKM mixing matrix, a not yet discovered contribution to the two-pion decays of the K_S, K_L was to be expected. To check this expectation, in early 1980 Jack decided that the challenge was worth the effort of inventing a necessary new experimental strategy and promoted discussions among some CERN colleagues. I was attracted by the idea and tried to contribute on the basis of the experience I had acquired working with liquified noble gas ionization calorimeter and high energy photon detection.

After debating the pros and cons of different proposals, Jack came up with a basic idea which convinced the potential participants to the experiment, later approved as NA31, on the feasibility of the enterprise. The experiment implied the accurate comparison of the so-called K_S and K_L in both their $\pi^+\pi^-$ $\pi^0\pi^0$ decays. The K_S decay close to where they are produced while the K_L persist for a time of order 500 times longer than the K_S. Jack proposed to house the beam transport, magnets and collimators, to produce the K_S, on a train inside the vacuum tube, along which the $K > \pi\pi$ were observed, which would stop at 40 stations each at 1.2 m along the 48 m length of the tube.

Together with a Liquid Argon calorimeter for detecting the photons from the $\pi^0\pi^0$ decays and wire chambers followed by hadronic calorimeters for the $\pi^+\pi^-$ the apparatus allowed to obtain a result, limited by statistics rather than systematic, with a first evidence of the new effect predicted by the Standard Model. The statistics limitation was later overcome by the NA48 experiment, to which Jack continued to collaborate, and by a concurrent KTeV Fermilab experiment.

During the 80s and 90s, Jack spent several weeks in Pisa as a guest of his long time friend Luigi Radicati and gave lectures on experimental particle physics as Galileo Galilei Chair of the Scuola Normale Superiore. Wanting to honor Gian Carlo Wick, to whom he had been an assistant at Berkeley, immediately after his PhD with Enrico Fermi in Chicago, Jack commisioned to the Danish sculptress Hilde Maehlum a bronze bust of Gian Carlo, which he donated to the SNS.

Fig. 1. Bronze bust of Prof. Gian Carlo Wick: Work commisioned by Jack Steinberger to Hilde Maehlum, donated to SNS-Pisa. Photo courtesy by SNS-Pisa.

Jack last visit to Pisa was in 2013, when he came to attend the ceremony in celebration of Bruno Pontecorvo 100 years from the his birth. During the last 40 years, I have had the privilege of having an office at CERN in the same corridor as Jack's office, which he frequented daily coming first by car, then by bicycle and only since ~ 2015 by tram. It was natural for me, coming to the laboratory in the morning, to stop and have, a some time very long, chat with Jack.

He was concerned and preoccupied about the global problems facing our society and our entire world. They ranged from the question of controlling the construction of the arms of mass destruction, the viability of the production, storage and distribution of energy, the injustice and poverty which prevails in some continents. He wanted to be concrete in approaching these problems and I definetly believe he felt he had at least the duty of dedicating some time and effort in thinking and analyzing them. But of course he realized the complexity of geopolitical factors that seem to render inapplicable most, if not all, technical potentially workable solution.

The first encounter with Jack in 1956–57 has had long lasting consequences and has contributed essentially to my choice of working in experimental high energy particle physics. Interacting with him over a period of more than 60 years has had an important influence on my activity and actually on the way I have approached life, not only professionally.

Fig. 2. Pisa, 14/05/1999. Jack at SNS. Photo courtesy by SNS-Pisa.

https://doi.org/10.1142/9789811264436_0007

My Memories of Jack Steinberger

Gigi Rolandi
Scuola Normale Superiore - Pisa, Italy and CERN - Geneva, Switzerland
gigi.rolandi@sns.it

I have several memories related to Jack mostly in the years of the ALEPH preparation. I was a young post-doc. Together with Francesco Ragusa we were working on the ALEPH Time Projection Chamber (TPC) project from Trieste, in close collaboration with the CERN-TPC group led by Jurgen May.

This picture (Fig. 1) was taken in the TPC90 hall, and Jack is in the back on the left of the picture. I am on the top of the magnet in the middle. The group TPC90 group was 30 people from CERN, Mainz, Glasgow, Imperial, MPI Munich, Pisa, Trieste and Wisconsin. Jack was following closely the progress of the group and contributing to it with new ideas. In the early 80's, the TPC was a very new detector and there were only two working large TPCs: the PEP-4 TPC at SLAC, which was suffering distortions of the trajectories of the drifting electrons, and the TRIUMF TPC, which could not reach the design resolution. There was a lot of attention on our progress from the LEP Experiment Committee, who was asking how our TPC could guarantee the promised performance.

At that time Francesco and I were trying to understand the details of the trajectories of electrons in gas in presence of magnetic and electric fields. The e-mail did not exist yet and Jack was calling us often in Trieste on the phone and we had to rush to the phone boot in the corridor to tell him about the progress of our work. Eventually our TPC big prototype, the TPC90, was ready and equipped with a laser producing ionization as straight tracks at three different drift lengths. In three days, we could test our theories with experimental data. The picture that we had developed on paper was proven correct and we demonstrated that the ALEPH TPC design, with atmospheric pressure and large magnetic field, was robust against the distortions seen in the SLAC TPC.

Fig. 1. TPC90 Group around the TPC90 Magnet, probably 1983.

Jack wrote in the ALEPH memory book: "I think that the happiest moment of my life in ALEPH was listening, at our plenary meeting in November 1983, to the report by Julia Sedgbeer on the measurements with TPC90, in which one after the other of the predictions on the functioning of a TPC with a large magnetic field were quantitatively confirmed." The theory developed by Francesco and myself could also explain why the TRIUMF TPC could not reach the design resolution: while the electrons were approaching the sense wire from the edge of the cell they were experiencing an ExB force that was spreading them along the wire.

The work done in Trieste proved to be very productive, but the communication was difficult. Jack convinced me to go to CERN every week and I was doing so, taking the train on Sunday evening and going back on Tuesday evening to teach in Trieste on Wednesday morning. Every Tuesday morning, at 11 am,

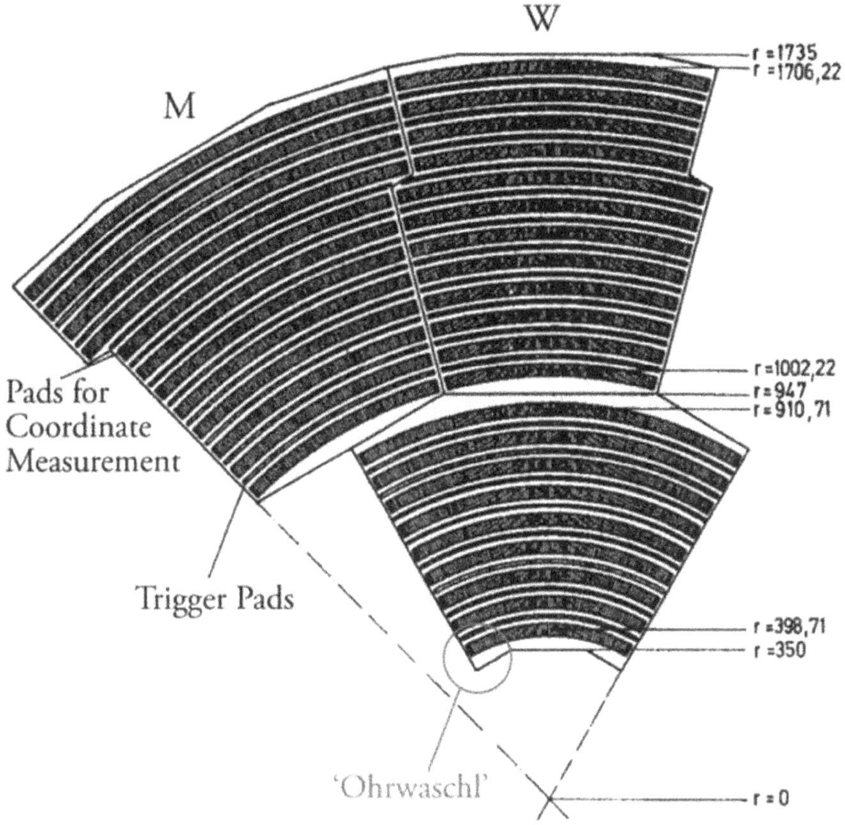

Fig. 2. Portion of the end-plate of the ALEPH TPC. See ALEPH-handbook page 55.

I met Jack in his office at CERN. Jack was always very quick to understand the progress and ask the right questions! I was always coming out of his office having learnt something new. I remember very well his drawing on his blackboard when he proposed to mill long radial pads. (See Fig. 2.) This was the right way to mitigate the ExB effect near the wire and improve the resolution for high momentum tracks. This was a real breakthrough, which once seen was obvious... like Columbus egg.

When the data-tacking was approaching Jack understood that we had to make a transition from "subdetector groups" to "one collaboration" and he asked me to call a meeting of the whole ALEPH collaboration to discuss the readiness of the data-tacking with a physics-oriented agenda. This was called "Tuesday Meeting" because in Jack's vision "Physics" is the expression of the individual and cannot be coordinated. In ALEPH we had the role of "Tuesday meeting organizer", which was what in the other Collaborations was the Physics Coordinator. And amazingly the name was kept when the meeting was moved

DETERMINATION OF THE NUMBER OF LIGHT NEUTRINO SPECIES

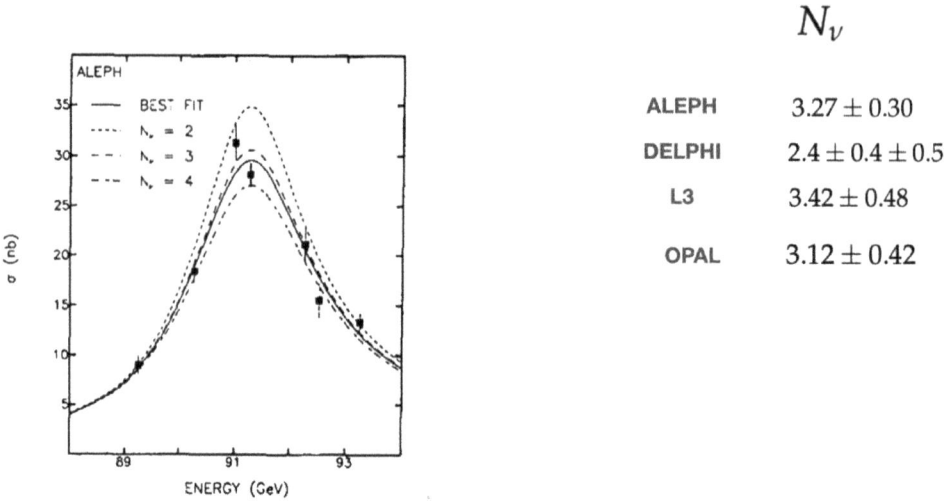

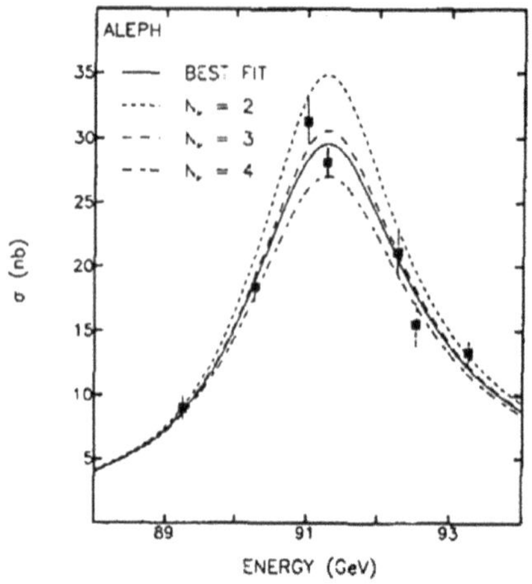

	N_ν
ALEPH	3.27 ± 0.30
DELPHI	$2.4 \pm 0.4 \pm 0.5$
L3	3.42 ± 0.48
OPAL	3.12 ± 0.42

Received 12 October 1989

Fig. 3. (Left) Z boson lineshape measured by ALEPH. Figure taken from Phys. Lett. B Vol. 231 (1989) 519–529. (Right) Number of neutrinos measured by the four LEP collaboration in the first LEP run.

to Thursday. ALEPH was ready at startup and the first results were really exciting. We could measure very quickly the number of neutrinos to be three with an uncertainty that was smaller than the other three competing experiments. (See Fig. 3.)

Let me finish with a vivid image that I have of Jack from when I was a young post-doc. At that time, we had to decide on the design of ALEPH and we had several intense meetings in the PS auditorium in building 6. During these meetings, Jack was silent most of the time. Sometimes looking at him one could think even that he was sleeping. However, near the end of the meeting, when the audience was still divided with many opposite and strong arguments, Jack used to take the lead of the meeting, summarizing the essence of a few-hours long discussion with the two to three most important statements and showing the road where to go with simple and convincing arguments. And magically everybody agreed. At that time, I was dreaming to become like him and now, after 40 years, I still think I would like.

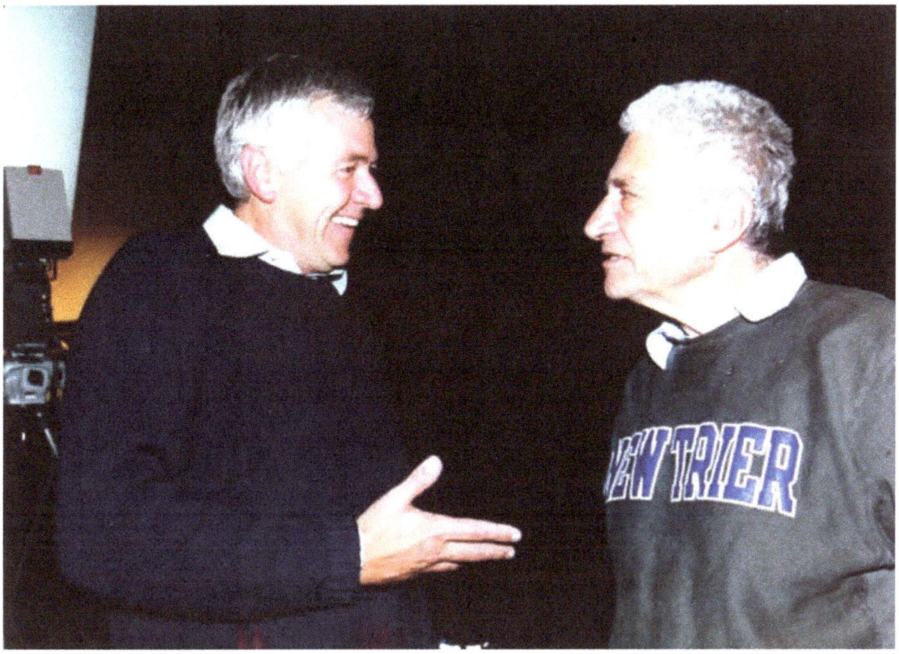

Fig. 4. Jack and Gigi, early '90.

https://doi.org/10.1142/9789811264436_0008

Memories of Jack Steinberger

David N. Schwartz

I knew Jack my entire life. My dad was Mel Schwartz, and Jack was his Ph.D. advisor; when I was born my dad was a bit over two years into his graduate studies. In fact I am sure that I "knew" Jack way before I remember him!

I would like to divide this short essay into two parts. The first is an appreciation of Jack through the eyes of my late father, who worked with him for the better part of a decade at Columbia. The second is my own perspective, reflecting on the relationship I developed with him over the years.

When my dad was alive, if you were to have asked him how he became a physicist, he would have told you an anecdote from his undergraduate years at Columbia. He had signed up for a physics course taught by Jack. The first day of class, Jack came in looking incredibly disheveled. Jack had been up all night running an experiment at Nevis Lab's cyclotron — one of the most powerful in the world at that time. Instead of covering the material for the class, Jack began by explaining the experiment he had been running. By the time Jack was finished, Mel Schwartz had decided he wanted to become an experimental physicist.

A few years later, he approached Jack to announce that he wanted to pursue a doctorate in physics at Columbia, and he wanted Jack to be his advisor. Jack had gotten to know my dad, and was receptive to the idea, but suggested that my father leave Columbia and go to Chicago to study with Enrico Fermi, Jack's graduate advisor. Jack adored Fermi — as I was to learn later — and thought not only that the change would do my father good, but that Fermi would be a spectacular mentor for him. With all the callowness of youth, my father rejected Jack's advice. He had accumulated a fair number of graduate credits, all of which would have been lost if he had moved to Chicago. So he insisted on studying with Jack, and Jack relented.

My dad began his graduate studies alongside Jack Leitner and Nick Samios. Leitner passed away tragically young, but Samios went on to a distinguished career as an experimentalist, the discoverer of the omega minus particle, and eventually director of Brookhaven National Lab. My dad worked at Brookhaven for a year after his doctorate, and came back to the Columbia faculty in 1959, working alongside Jack, Leon Lederman, II Rabi, TD Lee, and many other distinguished colleagues.

In 1959 he had the idea for what became the muon neutrino experiment and worked closely with Jack and Leon on the project. In 1962 the experiment was a success, and for the next year the three principal investigators made the rounds of conferences, seminars, and summer schools to discuss the results. I attach a photo of my dad and Jack on a boat on Lake Como during this period (Fig. 1). For me it perfectly sums up the relationship between the two of them. (If anyone can identify the third member of this happy group, please let me know!)

Fig. 1. Schwartz (middle) and Steinberger (right). Photo from the Schwartz Family Archives; photographer unknown.

My dad left Columbia for Stanford in 1966, Jack left for CERN in 1969, and the two never worked together again. Of course they ran into each other at physics meetings and enjoyed each other's company — here is a photo of the two of them with Tini Veltman and Jack Sandweiss looking on (Fig. 2). I know my dad carried enormous respect for Jack for the rest of his life. They enjoyed the recognition of the Nobel Prize committee in 1988, and the photo of Jack, Leon, and my dad from that event shows how much affection they had for each other (Fig. 3).

Fig. 2. From left: Schwartz, Sandweiss, Veltman, Steinberger. Photo from the Schwartz Family Archives; photographer unknown.

Fig. 3. From left: Steinberger, Schwartz, and Lederman. Reproduced courtesy of Brookhaven National Laboratory.

Years later, if you were to ask my father who was the best experimental physicist he ever worked with he would say, without hesitation, that Jack Steinberger was. If you asked him why, he would say, very simply, that Jack had exquisite taste in physics. By this he meant that Jack was always working on the most interesting problems in the field, but I think his respect for Jack transcended the issue of taste; he had enormous respect for Jack's honesty and integrity, and an understanding of experimental practice that enabled Jack to continue to be a pioneer in the field for decades.

After my father passed away, Jack would always begin a conversation with me by talking about him. He would say things like, "Your dad was more than a student to me — he was a teacher," or "Your dad was incredibly unlucky" — referring to the health problems that cut short my father's life — or "Your dad was too much of a gambler!" referring to my father's penchant for trading in the stock market, something that Jack had little time for! Jack also claimed that my father made an important improvement in the design of the spark chambers that were used on the muon neutrino experiment. I have been unable to find out exactly what he was talking about – perhaps someone out there might be able to add to Jack's comment.

My own relationship with Jack began when I was a child — he was a presence at Nevis, and when my mom would drive to pick my dad up at Nevis, in the early 1960s, I would sometimes explore the halls of that institution and run into Jack. He would look me up and down, seriously, and then smile. I remember a time when Jack and my father entertained me by doing a "calculation" on one of the mechanical calculators at the lab. As the machine crunched away it made a rhythmic pattern that was almost like music! The two of them took enormous delight in entertaining me this way.

As an adult, though, I got to know Jack on my own. The experience is one of my most cherished. When I first encountered Jack as an adult, I was assigned to the US delegation to the arms control talks with the Soviet Union in Geneva in the spring of 1982. Knowing that Jack and Cynthia lived there, I called them and immediately got an invitation to dinner at their lovely home. Those invitations were repeated several times during my stay in Geneva. Of course, Jack was intensely interested in what I was doing, and to the extent I was able, within the confines of security clearances, to discuss what was going on I did. He was an incredibly intelligent and well-informed interlocutor, a severe critic of US policy but civil and never rude to me. (I hasten to add that I was a junior member of the US team in Geneva and had almost no input into US policy as such.) We also talked about life in Geneva, food, and culture; he gave me great advice on restaurants to dine at, and things to do and see while I was there. We also talked about his family — he was enormously proud of all his children and spoke about them with love and affection. During these dinners he was almost always hosting other physicists, including my dad's close colleague Stan Wojcicki and good friend Burt Richter. The conversations about physics, about CERN and Stanford gossip, were thoroughly enjoyable for someone like me, who had grown up listening to this around our dining room table.

As I explained, my dad died in 2006. From that point onward I was in touch with Jack on a fairly regular basis, mainly because I viewed him as someone who could explain developments in physics in a way that I could understand. One pithy example: when the OPERA experiment in 2012 suggested that neutrinos might move faster than light, I asked Jack what he thought, and he replied, characteristically, "If this is true, we really know nothing about physics." Note that he did not completely dismiss the findings, although he certainly was skeptical, as he was with all physics results! But while the results were being checked he was unwilling to say whether he believed them or not; he was, however, willing to comment on the implications if the results were to be confirmed.

We became closer, however, during my research for my biography of Jack's great mentor, Enrico Fermi. Jack was perhaps the first graduate student that Fermi took on after the war, an experience about which Jack wrote movingly in his memoir, "Learning About Particles." In January 2014, I emailed him to describe the project and to ask if I could interview him. He wrote back to say he would be delighted to discuss his time with Fermi, but he noted that he was an "old man" and that we should hurry! I proposed to come to CERN the next month, but he explained that he was travelling to the US to see his kids there and to attend a symposium at Berkeley in honor of another of his students, David Nygren. So much for being an old man! We arranged to meet up in Berkeley. My wife joined us, and she was just as charmed as I was when he sat down with us to relive his days in Chicago.

He started off by apologizing that his memory was poor, but as he relaxed and began to reminisce it was clear that his memory was quite sharp. He loved talking about Fermi. The great scientist was obviously a key mentor in Jack's life, and he spoke with real emotion about how Fermi influenced him. The honesty and skepticism that Fermi displayed in his work was something that Jack admired and tried to live by. He also was moved by Fermi's kindness — for example, noticing that Jack was struggling with a theory dissertation, he suggested that Jack move instead to an experimental dissertation looking at cosmic ray decay. (The experiment became an important indicator that what was then known as the Fermi Interaction governing beta-decay was a more universal interaction associated with other decay processes.) But Jack was also quick to note the kindness of others when he was at Chicago, particularly department chair William Zachariasen, who gave a green light for Jack to retake the qualifying exams he failed the first time around. He made a mental note of the people who were kind to him or gave him help in one way or another when he needed it. He still remembered these people and shared those experiences with us. I am sure that their examples shaped the way he dealt with younger physicists as he grew in stature.

Jack was particularly delighted when I told him that New Trier Township High School, his alma mater on Chicago's north shore, had awarded my son Alex the Jack Steinberger Prize for scientific research. I explained that Jack's Nobel Prize display in the library of the school was something that Alex passed every day in school, serving as an inspiration for him and presumably many other New Trier students. My fully grown son works as an industrial chemist, which I believe is largely due to the example of my father and Jack.

One final memory from the meeting at Berkeley: Jack insisted that while he was pleased to receive the Nobel Prize it was inherently a bad thing. I pressed him on this, and he explained that the Nobel Prize by its very nature tends to celebrate the individuals honored, neglecting the many other important people involved in any great scientific discovery. This was not a case of false modesty; as he said this I could see that he really meant it.

We met with Jack one last time, in 2016, when we traveled to CERN for more Fermi research. He joined us for a lively meeting with CERN Director General Fabiola Gianotti and afterward had lunch with us in the CERN cafeteria. Jack was 95 years old by that time, but still biking to the lab on a regular basis, still climbing stairs to get around the main office complex, and still very much himself. While my wife and I had lunch, Jack opted for a Dove Bar. We chatted about our project again, and he reiterated how much he owed to Fermi. He expressed skepticism about a lot of recent physics results — he was particularly skeptical about the discovery of gravity waves by LIGO the previous September and wanted to see more data before he made up his mind. As we spoke, we certainly got the feeling that the CERN staff members who crowded the cafeteria that morning were wondering who those people were who were having lunch with the most distinguished physicist at CERN!

I continued to correspond with Jack by email from time to time, asking about recent physics news, but by 2020 I was not getting replies. I spoke with his wife

Cynthia, who explained that his health had declined, and that he was mostly bed-ridden. The news saddened both me and my wife, who had come to have great affection for this giant of physics. When Cynthia called me with the news of Jack's passing, I made sure to alert a few others here in the US who might not otherwise have heard, and of course was honored to be invited to the Zoom memorial event in his honor.

I still think about Jack often, and not simply when some news comes along about particle physics. I think about what it was like when Jack and my dad and Leon Lederman and others were working together during the Golden Age of high energy particle physics; when they worked in small teams and built much of the equipment themselves, in contrast to today's experiments involving thousands of physicists; when they knew literally everything that was going on in an experiment, which is simply impossible given the magnitude of the experimental setups today; and when an experiment might take a year or so to complete, as opposed to a decade or so now. Those days must have been fun, and thrilling, in a way that young physicists today cannot appreciate. With Jack's passing we move one step closer to a time when the giant figures of that Golden Age are no longer with us. The world will be a sadder place for their passing. It certainly is for Jack's.

https://doi.org/10.1142/9789811264436_0009

Celebrating Jack Steinberger's 100th Birthday Memorial

Weimin Wu
Fermilab Retiree

I met Jack Steinberger for the first time on November 3, 1979, in his office, and since then, he has become my professor, mentor, role model and friend.

After the terrible Cultural Revolution, China began to join the international research community of high energy physics. In Spring 1979, Prof. T.D. Lee gave a lecture on QCD in Beijing and selected 33 scientists from the audience to join various experiments in the USA and CERN as visitor scholars. I was lucky to be selected to join the CDHS experiment that was led by Steinberger.

The first thing I learned from Jack was about working on shifts in the CDHS experiment. The shift schedule was generated by computers randomly, and one night I was on shift together with Jack. It surprised me that a famous professor like him is required to work on shift like any other ordinary person, especially on the night shift for a senior professor, and Jack had no problem with the duty. He even climbed up to the top of the CDHS detector to check the gas tubes for any bubbles which could be due to a gas leakage. Jack insisted that he should do the checking by himself even though I offered him my help. This impressed me and I will remember it forever.

Jack is a true humanitarian. I remember there was an engineer who came from IHEP to join ALEPH for the muon chamber construction. Unfortunately, he got a heart attack and his medical insurance with an Italian company was not accepted by the Swiss hospital. Jack spent lots of time helping him to solve this problem, and he even visited him at the hospital almost every day, talked to doctors, nurses, and everyone was deeply touched by Jack's warm spirit and deep care for him.

The ALEPH collaboration was the largest experiment collaboration between China and the international community. It is truly created by Jack. He once told me that collaborating with his Chinese counterpart on ALEPH was one of the happiest events in his life. The picture in Fig. 1 shows the establishment of the ALEPH collaboration with the IHEP China at CERN.

Jack told me that he has a deep appreciation of the collaboration with Chinese scientists, especially in the 1940s, with T.D. Lee and C.N. Yang. He also shared his thoughts with me several times and believed that Mdm. C.S. Wu deserved a Nobel Prize but was unfortunately unable to due to the discrimination of women at that time.

Due to the well-known reason, Jack has had less contact with C.N. Yang lately. In 1986, when Jack visited China, I arranged a meeting for Jack and C.N. Yang who was also in Beijing at that time. It was such a great opportunity for both Steinberger and C.N. Yang to meet in person and they were invited to a dinner organized by Prof. Guangzhao Zhou, the then president of Chinese Academy. The picture in Fig. 2 was taken with my camera at the meeting. It was a nice meeting, and both had a very good conversation. Jack was very grateful for my initiative in arranging the meeting.

After IHEP completed the construction for the outer layer of ALEPH muon detector, I was informed by Jack that the final purpose of building an ALEPH detector is to do physics research in Beijing. Thus, a computer network between CERN and IHEP is of great importance. He proposed a task force which was led by both Paolo Palazzi at CERN and myself at IHEP. After a month of hard work, we created the computer network that realized the first-ever email from China, sent out by me in Beijing to Jack at CERN on August 25, 1986 (Fig. 3) — this has been recorded officially as the first event in China's internet history.

In 1998, during Jack's visit to Fermilab, I thought it would be a great opportunity for both Nobel laureates Prof. Leon Lederman and Prof. Jack Steinberger to meet up. Both agreed with my suggestion of meeting up at my home for a private gathering, and we had a great dinner at my home. The picture in Fig. 4 captured this historical moment.

Jack's 100th birthday memorial gave us an opportunity to celebrate his wonderful life. He lives in my heart forever.

Fig. 1. The establishment of the ALEPH collaboration with the IHEP China at CERN. A copy of this picture was given to Weimin Wu as an event participant by CERN media services.

Fig. 2. At a dinner organized by Prof. Guangzhao Zhou. From left: Weimin Wu, Jack Steinberger, Zhou and C.N. Yang. Photo courtesy of Weimin Wu.

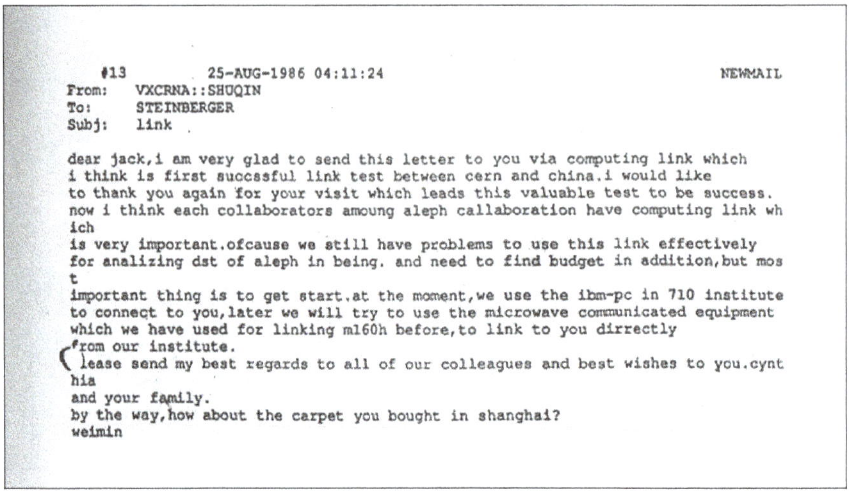

```
#13          25-AUG-1986 04:11:24                                   NEWMAIL
From:   VXCRNA::SHUQIN
To:     STEINBERGER
Subj:   link

dear jack,i am very glad to send this letter to you via computing link which
i think is first succssful link test between cern and china.i would like
to thank you again for your visit which leads this valuable test to be success.
now i think each collaborators amoung aleph callaboration have computing link wh
ich
is very important.ofcause we still have problems to use this link effectively
for analizing dst of aleph in being. and need to find budget in addition,but mos
t
important thing is to get start.at the moment,we use the ibm-pc in 710 institute
to connect to you,later we will try to use the microwave communicated equipment
which we have used for linking m160h before,to link to you dirrectly
from our institute.
lease send my best regards to all of our colleagues and best wishes to you.cynt
hia
and your family.
by the way,how about the carpet you bought in shanghai?
weimin
```

Fig. 3. Original email of August 25, 1986. Picture courtesy of Weimin Wu.

Fig. 4. Jack Steinberger and Leon Lederman. Photo courtesy of Weimin Wu.

Reprints of Selected Papers

https://doi.org/10.1142/9789811264436_0010

Comments of the Selected Papers of Jack Steinberger

Dieter Schlatter
CERN
dieter.schlatter@cern.ch

The selection of a dozen of the most important publications from the approximately 350 papers signed by Jack Steinberger is inherently difficult and suffers from personal bias. The number of times a paper is cited is certainly helpful in this selection. However, over the 50 years of Jack's scientific career, the field of particle physics has grown considerably. The number of experimental physicists has probably increased by a factor of at least 100, and with it the mean number of citations per article has increased accordingly. Furthermore, in the 1950s the experiments were performed by a handful of physicists, while at the end of the century particle physics experiments were performed by large collaborations of a few hundred physicists.

Jack became spokesman of the big neutrino experiment CDHS at CERN and later for the even bigger ALEPH experiment at the e^+e^- collider LEP. These experiments resulted in many important articles over the years. Therefore, my selection of the significant papers was guided by the personal interest Jack had taken in a particular publication.

The first publication in the list is from the early days of particle physics in the 1950s. It reported results of a beautiful experiment with the X-ray beam of the 300 MeV Berkeley synchrotron. Together with Panofsky and Steller, Steinberger was able to show that the observed photon spectrum agreed with the assumption that the two photons arose from the in flight decay of a neutral pion, the π^0 (Paper 1, pp. 51–54).

The second paper dealt more with theoretical ideas, along with T.D. Lee, C.N. Yang and others. After discovering parity violation in beta decays, pion and muon decays, they proposed to check whether parity is violated in hyperon decays as well. Bubble chamber experiments that Jack performed at that time were well suited for such a test (Paper 2, pp. 55–57).

The 1962 paper is about Jack's most important experiment, the discovery of the muon neutrino. A new proton synchrotron, the Brookhaven AGS, became available and with its 15 GeV proton beam sufficiently high neutrino flux intensities for high-energy neutrino interactions could be achieved. The experiment showed that there is a second type of neutrino associated with the muon (Paper 3, pp. 59–72). Today we know that there are three such families. Twenty-six years later, this experiment was rewarded with the Nobel Prize.

After discovering CP violation in 1964 in the decay of the neutral K meson, Jack began a series of neutral kaon beam experiments at CERN in the first half of the 1970s to learn more about CP violation. An improved beam design and a new detector technology (multi-wire proportional chambers) increased the achievable precision. One such result is the measurement of the proper time dependence of the decay $K^0 \to \pi^+\pi^-$ (Paper 4, pp. 73–77). Another result is the measurement of the mass difference $K_L - K_S$ (Paper 5, pp. 79–84).

The ultimate experimental goal on CP violation in K^0 decay was the measurement of the two decay amplitudes in $2\pi^0$ and $\pi^+\pi^-$. However, an order of magnitude improvement was needed. New, higher-energy accelerators enabled better detection of the neutral decay. The increased capacity of electronics and computer technology allowed for much higher event rates to achieve the precision required. Finally, in 1987, the NA31 experiment at CERN, in which Jack participated, showed that direct CP violation is real (Paper 6, pp. 85–92).

From 1979 to 1983, Jack was spokesperson for the CDHS high-energy neutrino experiment at CERN, which involved about 40 scientists. It provided a large amount of data on the charged and neutral current reactions in iron (Paper 7, p. 93). First precise measurements of the structure functions provided quantitative confirmation of the QCD prediction of the Q^2 evolution (Papers 7 and 8, pp. 93, 95–99). The study of multi-muon events quantitatively supported the GIM model through its prediction on charm production (Paper 9, pp. 101–107). Analysis of the full data set gave an estimate of the strong coupling constant and showed remarkably good agreement with perturbative QCD (Paper 10, pp. 109–121).

With the event of the e^+e^- collider LEP at CERN, Jack led the ALEPH experiment during construction and initial data collection from 1982 to 1997. The first achievement was the precise measurement of the number of neutrino families (Paper 11, pp. 123–147). Certainly a great satisfaction for Jack, 27 years after establishing the second family by the discovery the muon neutrino.

A more general precision test of the Standard Model at the 1% level became possible with the Z-resonance data set from the first year's running (Paper 12, pp. 149–159). Measurements of important parameters of the model such as the Z mass and width, production cross sections for quarks and leptons, the "Weinberg" weak mixing angle and the strong coupling constant at the Z mass were determined.

The last topic that Jack was actively involved with concerns the b-quark production in Z-decay. A more than doubling of the statistics collected by the LEP experiments required a deeper understanding of the systematics of these decays. Together with a small group of young scientists, new algorithms were developed and the resulting ratio of b-quark to hadronic Z-decays was once more in excellent agreement with the Standard Model prediction (Paper 13, pp. 161–193).

PHYSICAL REVIEW VOLUME 78, NUMBER 6 JUNE 15, 1950

Evidence for the Production of Neutral Mesons by Photons*

J. STEINBERGER, W. K. H. PANOFSKY, AND J. STELLER
Radiation Laboratory, Department of Physics, University of California, Berkeley, California

(Received April 28, 1950)

In the bombardment of nuclei by 330-Mev x-rays, multiple gamma-rays are emitted. From their angular correlation it is deduced that they are emitted in pairs in the disintegration of neutral particles moving with relativistic velocities and therefore of intermediate mass. The neutral mesons are produced with cross sections similar to those for the charged mesons and with an angular distribution peaked more in the forward direction. The production cross section in hydrogen and the production cross section per nucleon in C and Be are comparable.

I. INTRODUCTION

NEUTRAL mesons which are coupled strongly to nuclei must be expected to be unstable against decay into two or more gamma-rays. The modes of decay, and expected lifetimes, have been discussed extensively.[1] These gamma-rays are then supposed to be responsible for the soft showers which often accompany energetic cosmic-ray nuclear events.[2] The evidence in favor of the existence of the neutral meson has recently been greatly strengthened by the discovery at Berkeley[3] of gamma-rays which behave in all ways as if they were due to the disintegration of a neutral meson. They are produced by proton bombardment of various nuclei and have a production cross section which depends on proton energy much like that of charged mesons. Their energy is approximately 70 Mev on the average, half that of the charged π-meson, and the energy spread is in agreement with the Doppler shift due to the velocity of the parent mesons. The lifetime of the mesons is less than 10^{-13} sec., which is in agreement with the theoretical expectations.

The evidence is therefore already much in favor of the existence of a gamma-unstable neutral meson. However, until now, coincidences between the two gamma-rays have never been observed. We report here the detection of such coincidences, produced by the bombardment of various nuclei in the x-ray beam of the Berkeley synchrotron. This must be regarded as strong additional evidence supporting the existence of the neutral meson.

II. EXPERIMENTAL ARRANGEMENT

The apparatus is sketched in Fig. 1. The synchrotron x-ray beam of 330-Mev maximum energy is collimated in two successive collimators. The second collimator serves only to intercept some of the electrons produced at the edge of the first collimator. The beam then strikes a target, which, for most of the experiment, is a cylinder of beryllium, $1\frac{1}{2}$ inches long and 2 inches in diameter. The particles produced in the target are detected in two telescopes, each consisting of three scintillation counters. A converter, usually $\frac{1}{4}$ inch of lead, is inserted between the two crystals nearest the target in each telescope. An event is recorded if simultaneous (resolving time 10^{-7} sec.) pulses are recorded in the outer four crystals, but none in the two near the target. That is, we require that there be two particles, one in each telescope, neutral at first which are converted into charged particles by the lead, and which penetrate one crystal and enter the next. With a beam intensity of about 10^{11} Mev/min. the counting rate for such coincidences at favorable orientations of the telescopes is about 10 counts/min.

III. NATURE OF THE COINCIDENCES

Let us first describe the experiments which identify the particles as gamma-rays, indicate their energy and show that their origin is the nuclear rather than the Coulomb field. In Table I we list the relative detection

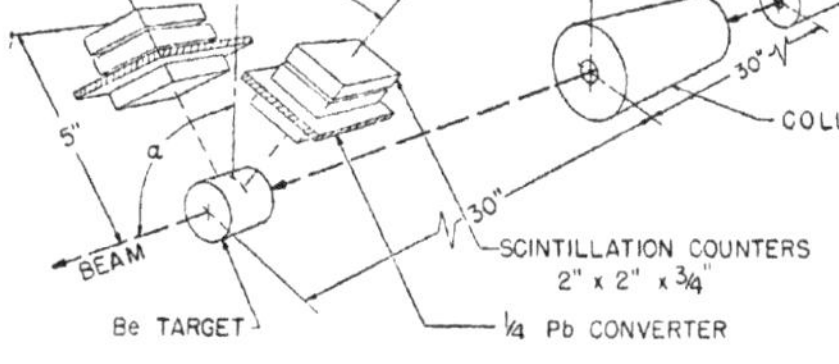

FIG. 1. Experimental arrangement.

* This work was performed under the auspices of the AEC.

[1] Y. Tanikawa, Proc. Phys. Math. Soc. Japan 24, 610 (1940). R. J. Finkelstein, Phys. Rev. 72, 414 (1947). H. Fukuda and Y. Miamoto, Prog. Theor. Phys. 4, 347 (1949). Ozaki, Oneda, and Sasaki, Prog. Theor. Phys. 4, 524 (1949). J. Steinberger, Phys. Rev. 76, 1180 (1949). C. N. Yang, Phys. Rev. 77, 243 (1950).

[2] The implications of the gamma-decay of neutral mesons for the soft component in the cosmic radiation were pointed out by J. R. Oppenheimer (Phys. Rev. 71, 462 (T) (1947). It was assumed that in high energy nuclear events neutral mesons are emitted with multiplicities similar to those for charged mesons. The neutral mesons decay into photons and account for the early development of extensive showers, as well as the large total amount of soft radiation. These bursts of soft radiation accompanying energetic nuclear events were then actually observed in the cloud chamber by W. Fretter, Phys. Rev. 73, 41 (1948), 76, 511 (1949); C. Y. Chao, Phys. Rev. 75, 581 (1949); Gregory, Rossi, and Tinlot, Phys. Rev. 77, 299 (1949); and J. Green, Thesis, University of California, 1950. They were found in photographic plates by Kaplan, Peters, and Bradt, Phys. Rev. 76, 1735 (1949). Both the cloud-chamber pictures and the photographic star show that the showers begin with gamma-rays rather than electrons.

[3] Bjorklund, Crandall, Moyer, and York, Phys. Rev. 77, 213 (1950).

efficiency for various converter materials and thicknesses. Without converters the counting rate is almost zero, then increases as the converter thickness in each arm is increased to $\frac{1}{4}$ inch of lead, and only slightly from $\frac{1}{4}$ inch to $\frac{1}{2}$ inch. This is as expected from shower theory for about 100-Mev photons. Copper of $\frac{1}{4}$ inch thickness has approximately the same conversion efficiency as has $\frac{1}{16}$ inch of lead, again in agreement with shower theory, since the number of shower units is the same for these thicknesses.

The coincidences are attenuated by a factor of four when $\frac{1}{4}$ inch of lead is inserted between the target and the anticoincidence crystals. This again is as expected for photons. Furthermore, it can be seen from Table I that both telescopes require converters, so that both particles must be photons.

TABLE I. Relative detection efficiency as a function of absorber material and thickness.

Converter in telescope 1	Converter in telescope 2	Relative counting rate $\alpha=\beta=90°$
none	none	0.01±0.005
$\frac{1}{32}$-in. Pb	$\frac{1}{32}$-in. Pb	0.17±0.013
$\frac{1}{16}$-in. Pb	$\frac{1}{16}$-in. Pb	0.3 ±0.02
$\frac{1}{8}$-in. Pb	$\frac{1}{8}$-in. Pb	0.67±0.08
$\frac{1}{4}$-in. Pb	$\frac{1}{4}$-in. Pb	1.00±0.06
$\frac{1}{4}$-in. Cu	$\frac{1}{4}$-in. Cu	0.39±0.03
none	$\frac{1}{4}$-in. Pb	0.15±0.05
$\frac{1}{16}$-in. Pb	$\frac{1}{4}$-in. Pb	0.62±0.07
$\frac{1}{2}$-in. Pb	$\frac{1}{4}$-in. Pb	1.07±0.1
$\frac{1}{4}$-in. Pb	$\frac{1}{4}$-in. Pb	0.28±0.05

$\frac{1}{4}$-in. Pb absorbers placed in front of both telescopes.

To measure the energy of the conversion electrons, aluminum absorbers were inserted between the last two crystals of one of the telescopes. Unfortunately, at these energies the radiation losses are important, and therefore the straggling large. We have plotted in Fig. 2 the coincidence counting rate as a function of the average energy required to traverse the telescope. Because the photons originate in moving mesons, the average gamma-ray energy is expected to be approximately 100 Mev, and the average electron energy 50 Mev, quite in agreement with the observed attenuation.

The nuclear origin of the photons is demonstrated by the fact that the cross section for these coincidences is only six times as big for a lead nucleus as for beryllium, which is less than the ratio of the nuclear areas. On the other hand, ordinary shower cross sections increase by a factor of 400.

Finally, we have looked for coincidences with the beam energy reduced to about 175 Mev with angles α and β of the telescope both 90°. The cross section per Q (the number Q for a bremsstrahlung beam is equal to the total energy divided by the maximum energy of the spectrum) is at least 50 times smaller here than at 330 Mev. This steep excitation function is also observed for charged meson production.

We believe, therefore, that it is demonstrated that the observed coincidences are caused by gamma-rays of about 100-Mev average energy, of non-Coulombic origin, and with a threshold similar to that for charged mesons.

IV. ANGULAR CORRELATION AND DISTRIBUTION OF THE GAMMA-RAYS

To study further the properties of these coincidences, we have measured their rate as a function of the angle, α, between the beam direction and the plane of the telescopes and of the correlation angle β (see Fig. 1). Consider first the variation with β at a fixed α, say 90°. 180° coincidences are rare. The counting rate increases with decreasing β to a maximum at 90°, and then drops sharply. This behavior must actually be expected of gamma-rays which are the decay products of neutral mesons, because of the motion of the decaying mesons. A meson at rest decaying into two gamma-rays, emits them in opposite direction. But when this is seen from a system in which the meson has a total energy E, then the included angle β varies between π and $2\sin^{-1}(1/E)$ with a probability which favors the small angles tremendously. The median angle is $2\sin^{-1}[2/(3E^2+1)^{\frac{1}{2}}]$. E is the total meson energy in units of its rest energy.

For 70-Mev mesons the minimum angle of β is 84° and the median angle 92°. Since the distribution is so heavily peaked, not much error is introduced if one assumes, as is done in the following, that to an angle β corresponds a unique energy, that of the median angle. Therefore a measurement of the distribution in β is a measure of the distribution in energy of the neutral mesons, although the angular resolution of our telescopes is insufficient to give more than a glimpse of the energy distribution. We have included in Fig. 3 curves in which the observed[4] energy distributions of the π^+-meson made by the same x-rays on hydrogen are transformed into distributions in β and arbitrarily normalized. All corrections due to scattering and angular resolution are omitted. The general shape of the curves is certainly well reproduced by the experiment. It is

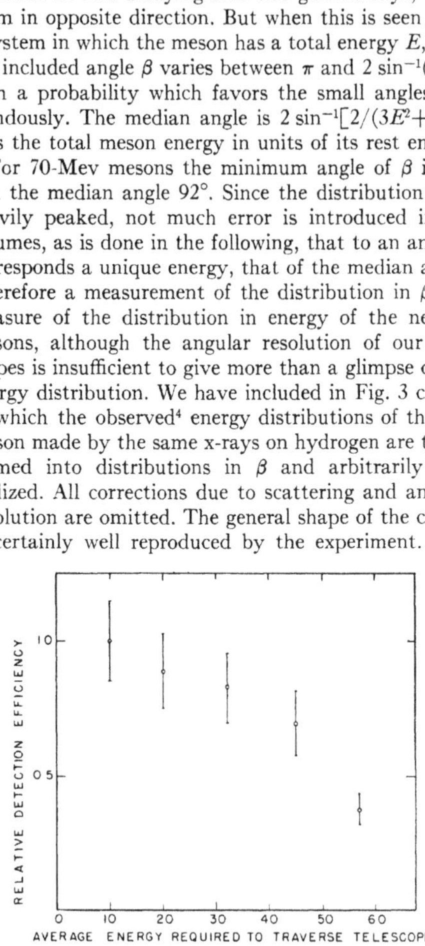

FIG. 2. Absorption of conversion electrons in aluminum. The energy includes the average radiation loss.

[4] Bishop, Cook, and Steinberger (to be published).

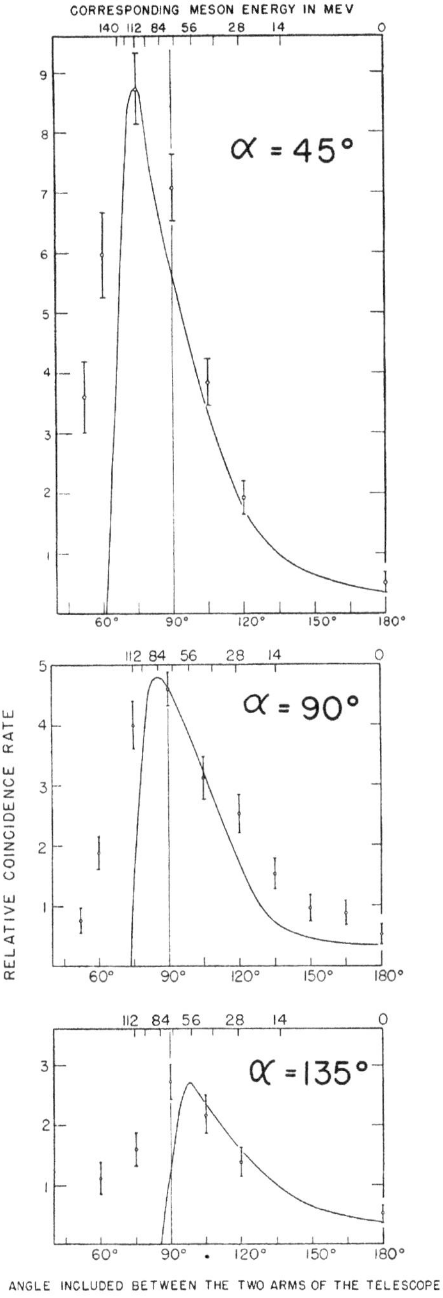

FIG. 3. Variation of coincidence rate with the included angle β between the two arms of the telescope. The curves are those expected on the assumption that the gamma-rays are the decay products of a neutral meson, emitted with the same energy distribution as are π^+-mesons from hydrogen. The curves are arbitrarily normalized for each angle α.

therefore clear that if the gamma-rays are the decay product of intermediate particles, these particles must move with velocities of the order of $v/c \simeq 0.8$. Excited nucleons of this velocity cannot be produced by x-rays of 330 Mev; the particles must therefore have an intermediate mass. Furthermore, it is possible to see that the decay must be into only 2 photons, since the expected angular distributions for a decay into more than two photons would not show a valley for small angles β.

The distribution in the angle α of the beam with the plane of the telescope shown in Fig. 4, is interesting chiefly because of the difference between this distribution and the angular distribution of π^+-photo-mesons[5] from either carbon or hydrogen targets. This is not particularly surprising, since various theories also give quite different results for charged and neutral mesons.

V. HYDROGEN CROSS SECTION AND TOTAL CROSS SECTION

At one setting of the telescopic angles, $\alpha=\beta=90°$, we have compared the cross sections of hydrogen and carbon. This was done by comparing the count from a polyethylene (CH_2) block and a perforated carbon block of the same size and carbon content as the CH_2. The result is: $\sigma H_{\pi 0}/\sigma C_{\pi 0}=0.12\pm 0.03$. This again differs from the results for positive mesons, where $\sigma H_{\pi^+}/\sigma C_{e^+} \simeq 0.55$. The difference is probably in part caused by the fact that both neutrons and protons can contribute to neutral meson production, but only protons to π^+-production. In part, it may also be possible to ascribe this to the same phenomenon which, according to Chew,[6] is responsible for the large hydrogen-carbon ratio for the positive mesons. In the case of π^+-production, the reaction is inhibited by the fact that, when the proton is changed into a neutron, there is an oversupply of neutrons in the immediate neighborhood and the number of states available to it is small because of the Pauli principle. This is not significant in the neutral case because the nucleon's charge does not change.

The curves in Fig. 3 can be integrated to yield a total cross section for beryllium. $\sigma_{Be}=7.5\times 10^{-28}$ cm^2 per Q, while for carbon and hydrogen, assuming the same angular distribution, $\sigma_C=10\times 10^{-28}$ and $\sigma_H=1.3\times 10^{-28}$ cm^2 per Q. The absolute x-ray intensity is known[7] to about 10 percent, but the efficiency of the detecting system only to within a factor of two, so that there is a corresponding error in the above cross sections. The hydrogen cross section is approximately the same as that for π^+-production;[5] those for carbon and beryllium are somewhat higher.[8]

One might assume that the charge of the meson would play an important role in the production of mesons in the electromagnetic field of the photon. This

[5] J. Steinberger and A. S. Bishop, Phys. Rev. **78**, 493 (1950).
[6] G. Chew (private communication).
[7] Blocker, Kenney, and Panofsky (to be published).
[8] McMillan, Peterson, and White, Science **110**, 579 (1949).

is contradicted by the observed angular distribution of π^+-mesons produced by photons in H_2. The angular distribution indicates that the principal process responsible for charged meson production is the interaction of the photon with the spin of the nucleon. If the neutral meson has the same transformation properties as the charged, it then appears plausible that the production cross sections in hydrogen should be comparable, as seems to be the case. However, actual calculations on the basis of pseudoscalar theory, both in the classical and in the perturbation theory approximation, which give a reasonable angular distribution for the π^+-production, give smaller values for neutral meson production. Whether or not this is a new difficulty in a theory which already has several, is not clear. From a less restricted point of view it is not a surprising result.

VI. SUMMARY

In the bombardment of various nuclei by 330-Mev x-rays, photons with the following properties are emitted:

(1) At least two are emitted in coincidence.
(2) They each have an average energy of about 100 Mev.
(3) The Z dependence of the production indicates that they have their origin in a nuclear interaction, and not in the Coulomb field.
(4) The threshold for their production is at least 150 Mev.
(5) The angular correlation of the photons shows that they are emitted in pairs as the only decay products of particles moving with velocities of the order of $v/c=0.8$, and therefore of intermediate mass.
(6) The total cross section for production from hydrogen is about the same as that for production of π^+-mesons; other light nuclei cross sections are somewhat higher than those for the positive mesons.

It is clear from these properties that the gamma-rays are the decay products of neutral mesons. Since spin $\frac{1}{2}$,

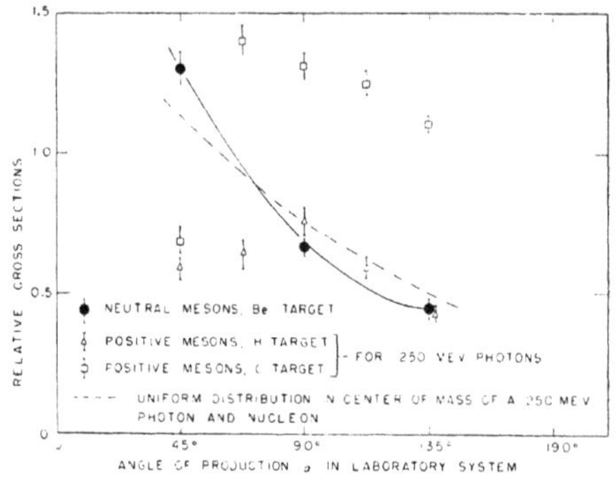

FIG. 4. Variation with the angle α between the plane of the telescope and the beam. Each point represents an integral over the angle β.

and spin 1 mesons are forbidden to decay into two photons,[1] the spin must be zero, excluding the possibility of very high intrinsic angular momenta. It seems reasonable, and it is in good agreement with all observations, to assume that both charged and neutral mesons are of the same type. It then follows from the angular distribution of the x-ray produced π^+-mesons, and the high cross sections for making neutral mesons by x-rays, that the π-meson is a pseudoscalar. This remark applies, of course, only to the character of the meson, and not to any particular field theory for the interaction of mesons with nucleons.

All phases of this experiment have been discussed with Professor Edwin McMillan and his advice has been of great help. The bombardments were carried out by the synchrotron crew under the direction of W. Gibbons.

Possible Detection of Parity Nonconservation in Hyperon Decay*

T. D. LEE, J. STEINBERGER,
Columbia University, New York, New York

AND

G. FEINBERG, P. K. KABIR, AND C. N. YANG,
Institute for Advanced Study, Princeton, New Jersey

(Received May 2, 1957)

RECENTLY, various experiments[1–3] established the nonconservation of parity in β decay, π decay, and μ decay. The purpose of this note is to emphasize that, in view of these developments, experiments on hyperon production and decay in $(\pi+p)$ collisions of the type done by various groups using bubble chambers,[4] seem now to be especially important for a clarification of the following related questions: (i) whether parity conservation is violated in hyperon decays[5] and (ii) whether parity doublets exist.[6]

A detailed analysis concerning the possible detection of parity doublets exists in the literature.[6] In the following we shall make a phenomenological study of the problem of possible detection of parity nonconservation in hyperon decay under the assumption that there exist no parity doublets for either K mesons or hyperons.[7]

To make the analysis unambiguous and to draw conclusions that are relatively definite, it is necessary that one knows something about the polarization of the hyperons produced. It seems that a good plan is to study hyperon production and decay near threshold.

Production and decay of Σ^-. For example, let us consider the production of Σ^- from (π^-+p) collisions:

$$\pi^-+p\rightarrow\Sigma^-+K^+. \qquad (1)$$

It is perhaps worthwhile to try to do the experiments at laboratory kinetic energies of the pion of, say, 955 Mev and 1 Bev, corresponding to center-of-mass total kinetic energies of the Σ^-+K^+ system of 30 Mev and 60 Mev. At these energies one hopes that only s and p waves are produced in the Σ^-+K^+ system.

It is then easy to see that the differential production cross section per unit solid angle $d\Omega$ (in the center-of-mass system of production) of the Σ^- produced is given by

$$I(\theta)=|a+b\cos\theta|^2+|c|^2\sin^2\theta, \tag{2}$$

where a can be chosen as real and b and c are complex numbers. We use the following notations:

$$\begin{aligned}\mathbf{p}_{\text{in}}&=\text{momentum of the incoming }\pi^-,\\ \mathbf{p}_\Sigma&=\text{momentum of the }\Sigma^-\text{ produced,}\\ \theta&=\text{angle between }\mathbf{p}_{\text{in}}\text{ and }\mathbf{p}_\Sigma.\end{aligned} \tag{3}$$

In (3) both $\mathbf{p}_{\text{in}}$ and $\mathbf{p}_\Sigma$ are measured in the center-of-mass system of production. The polarization of the Σ^- produced at the angle θ is always in the direction of $\mathbf{p}_{\text{in}}\times\mathbf{p}_\Sigma$ and has the magnitude

$$P(\theta)=[I(\theta)]^{-1}2\sin\theta\times\text{Im}[c^*(a+b\cos\theta)], \tag{4}$$

where $P(\theta)$ is defined to be the average spin of the Σ^- in units of $\frac{1}{2}\hbar$. In Eq. (4) the *assumptions have been made that the spin of* Σ^- *is* $\frac{1}{2}$ *and that the spin of* K^+ *is* 0.

If parity is not conserved in the decay of Σ^-, the polarization $P(\theta)$ can be measured by using the decay process of Σ^-,

$$\Sigma^-\rightarrow n+\pi^-, \tag{5}$$

as an analyzer. Let R be the projection of the momentum of the decay pion in the direction of $\mathbf{p}_{\text{in}}\times\mathbf{p}_\Sigma$. The distribution function for R at an angle θ of production is given by

$$W(\theta,\xi)d\Omega d\xi=I(\theta)d\Omega\times\tfrac{1}{2}[1+\alpha p(\theta)\xi]\alpha\xi, \tag{6}$$

where

$$\xi=R/(\text{maximum value of }R)\cong R/(100\text{ Mev}/c).$$

In terms of the coefficients a, b, and c, defined in Eq. (2), $W(\theta,\xi)$ can be written as

$$\begin{aligned}W(\theta,\xi)d\Omega d\xi=&[|a+b\cos\theta|^2+|c|^2\sin^2\theta]d\Omega\times\tfrac{1}{2}d\xi\\ &+\alpha\sin\theta\,\text{Im}[c^*(a+b\cos\theta)]d\Omega\times\xi d\xi.\end{aligned} \tag{7}$$

The existence of a nonvanishing α would constitute an unambiguous proof of parity nonconservation in Σ^- decay. In such a case the final state of $(n+\pi^-)$ in process (5) would be a mixture of $s_{\frac{1}{2}}$ and $p_{\frac{1}{2}}$ states with amplitudes, say, A and B respectively. The asymmetry parameter α is related to these amplitudes by

$$\alpha=2\,\text{Re}(A^*B)/(|A|^2+|B|^2). \tag{8}$$

If time reversal leaves invariant the decay process of Σ^-, then[8]

$$\alpha=\pm\frac{2|A|\times|B|}{|A|^2+|B|^2}\cos(\delta_p-\delta_s), \tag{9}$$

where δ_p and δ_s are, respectively, the phase shifts of $(n+\pi^-)$ scattering in the $p_{\frac{1}{2}}$ and $s_{\frac{1}{2}}$ states at about 117 Mev in their center-of-mass system. If the decay interaction is invariant under charge conjugation, then[8]

$$\alpha=\pm\frac{2|A|\times|B|}{|A|^2+|B|^2}\sin(\delta_p-\delta_s). \tag{10}$$

The following remarks are useful concerning the measurements of α and $p(\theta)$.

1. The polarization $P(\theta)$ may sometimes be very small. E.g., if (1) gives

$$I(\theta)=(1+\cos\theta)^2,\quad\text{or}\quad I(\theta)=(1-\cos\theta)^2, \tag{11}$$

then $P(\theta)=0$ identically.

2. At production energies *near the threshold*, the variations of the quantities a, b, and c, introduced in Eq. (2), with respect to p_Σ are given by

$$\begin{aligned}a&=a_0(p_\Sigma)^{\frac{1}{2}},\\ b&=b_0(p_\Sigma)^{\frac{3}{2}}\exp(i\chi_b),\\ c&=c_0(p_\Sigma)^{\frac{3}{2}}\exp(i\chi_c),\end{aligned} \tag{12}$$

where a_0, b_0, c_0, χ_b, and χ_c are all *real* constants independent of p_Σ. Thus by selecting two or three energy values near threshold, it is possible to determine a_0, b_0, c_0, and χ_b from the angular and energy dependence of $I(\theta)$ alone.

If the energy dependence of the cross section should be not representable by (12), one would have an indication that resonance effects might be important in the π^-+p system near the threshold for Σ^- production.

3. If $\chi_b\neq0$, then by comparing the coefficients of the $\sin\theta$ and $\sin\theta\cos\theta$ terms in $W(\theta,\xi)$, the phase χ_c cal also be determined.

4. From the values of these five real constants, a_0, b_0, c_0, χ_b, and χ_c, the asymmetry parameter α can then be deduced from $W(\theta,\xi)$ [Eq. (7)].

5. If

$$|\alpha|>|\sin(\delta_p-\delta_s)|,$$

then from Eq. (10) both invariance under charge conjugation and conservation of parity do not hold in the decay of Σ^-.

Since the phase shifts in the $J=\frac{1}{2}$ states are all small, the conclusion is essentially that any appreciable asymmetry with respect to the sign of ξ in $W(\theta,\xi)$ is an indication that conservation of parity and invariance under charge conjugation do not hold in the decay of Σ^-.

Production and decays of other hyperons. The foregoing analysis can also be applied to the productions and decays of other hyperons. We consider, for definiteness, the following processes concerning Λ^0:

$$\pi+p\rightarrow\Lambda^0+K^0, \tag{13}$$

and

$$\Lambda^0 \rightarrow p + \pi^-. \quad (14)$$

All the previous formulas for $I(\theta)$, $p(\theta)$, $W(\theta,\xi)$, and α [i.e., Eqs. (2), (4), (6), (7), and (8)] remain unchanged. The only difference is that in Eq. (8) the amplitudes A and B of s- and p-wave final states in the decay process of Λ^0 are now each a mixture of two isotopic spin states. These amplitudes can be written as

$$A = (\tfrac{2}{3})^{\frac{1}{2}} A_{\frac{1}{2}} + (\tfrac{1}{3})^{\frac{1}{2}} A_{\frac{3}{2}},$$
$$B = (\tfrac{2}{3})^{\frac{1}{2}} B_{\frac{1}{2}} + (\tfrac{1}{3})^{\frac{1}{2}} B_{\frac{3}{2}}, \quad (15)$$

where $A_{\frac{1}{2}}$, $B_{\frac{1}{2}}$ are, respectively, the s- and p-wave amplitudes for final states with the total isotopic spin value $I=\frac{1}{2}$, and $A_{\frac{3}{2}}$, $B_{\frac{3}{2}}$ the corresponding amplitudes for states with $I=\frac{3}{2}$. In place of Eqs. (9) and (10) we have now the following conditions for invariance under time reversal and charge conjugation:

If the decay process is invariant under time reversal, then we can choose[8]

$$A_{\frac{1}{2}} = |A_{\frac{1}{2}}| e^{i\delta_1},$$
$$A_{\frac{3}{2}} = \pm |A_{\frac{3}{2}}| e^{i\delta_3},$$
$$B_{\frac{1}{2}} = \pm |B_{\frac{1}{2}}| e^{i\delta_{11}}, \quad (16)$$
$$B_{\frac{3}{2}} = \pm |B_{\frac{3}{2}}| e^{i\delta_{31}}.$$

On the other hand, if the decay process is invariant under charge conjugation operation, then these amplitudes are[8]

$$A_{\frac{1}{2}} = |A_{\frac{1}{2}}| e^{i\delta_1},$$
$$A_{\frac{3}{2}} = \pm |A_{\frac{3}{2}}| e^{i\delta_3}, \quad (17)$$
$$B_{\frac{1}{2}} = \pm i |B_{\frac{1}{2}}| e^{i\delta_{1,1}},$$
$$B_{\frac{3}{2}} = \pm i |B_{\frac{3}{2}}| e^{i\delta_{3,1}}.$$

The phase shifts δ are the usual pion-nucleon scattering phase shifts at 37-Mev total kinetic energy:

δ_1 = phase shift for s waves, $I=\frac{1}{2}$, $J=\frac{1}{2}$,

δ_3 = phase shift for s waves, $I=\frac{3}{2}$, $J=\frac{1}{2}$,

$\delta_{\lambda\mu}$ = phase shift for p waves, $I=\lambda/2$, $J=\mu/2$.

All these phase shifts are small at 37-Mev total kinetic energy. Therefore any appreciable asymmetry in $W(\theta,\xi)$ with respect to the sign of ξ is an indication that conservation of parity and invariance under charge conjugation do not hold in the decay of Λ^0.

A measurement of the branching ratio in the decay processes

$$\Lambda^0 \rightarrow p + \pi^-, \quad (18)$$

$$\Lambda^0 \rightarrow n + \pi^0, \quad (19)$$

and a measurement of the distribution function $W(\theta,\xi)$ for process (19) would lead to additional information concerning the amplitudes $A_{\frac{1}{2}}$, $B_{\frac{1}{2}}$, $A_{\frac{3}{2}}$, and $B_{\frac{3}{2}}$.

* Work supported in part by the U. S. Atomic Energy Commission.

[1] Wu, Ambler, Hayward, Hoppes, and Hudson, Phys. Rev. **105**, 1413 (1957).

[2] Garwin, Lederman, and Weinrich, Phys. Rev. **105**, 1415 (1957).

[3] J. Friedman and V. Telegdi, Phys. Rev. **105**, 1681 (1957).

[4] Budde, Chretien, Leitner, Samios, Schwartz, and Steinberger, Phys. Rev. **103**, 1827 (1956); D. A. Glaser (private communication); R. Adair (private communication).

[5] T. D. Lee and C. N. Yang, Phys. Rev. **104**, 254 (1956).

[6] T. D. Lee and C. N. Yang, Phys. Rev. **104**, 822 (1956).

[7] In view of the recent experimental developments (references 1, 2, and 3) there appears to be at present no *theoretical* necessity to introduce the complication of parity doublets. (See footnote 8 in reference 5.)

[8] Lee, Oehme, and Yang, Phys. Rev. **106**, 340 (1957). See the proof of Theorem 2.

OBSERVATION OF HIGH-ENERGY NEUTRINO REACTIONS AND THE EXISTENCE OF TWO KINDS OF NEUTRINOS*

G. Danby, J-M. Gaillard, K. Goulianos, L. M. Lederman, N. Mistry, M. Schwartz,† and J. Steinberger†

Columbia University, New York, New York and Brookhaven National Laboratory, Upton, New York
(Received June 15, 1962)

In the course of an experiment at the Brookhaven AGS, we have observed the interaction of high-energy neutrinos with matter. These neutrinos were produced primarily as the result of the decay of the pion:

$$\pi^{\pm} \rightarrow \mu^{\pm} + (\nu/\bar{\nu}). \tag{1}$$

It is the purpose of this Letter to report some of the results of this experiment including (1) demonstration that the neutrinos we have used produce μ mesons but do not produce electrons, and hence are very likely different from the neutrinos involved in β decay and (2) approximate cross sections.

Behavior of cross section as a function of energy. The Fermi theory of weak interactions which works well at low energies implies a cross section for weak interactions which increases as phase space. Calculation indicates that weak interacting cross sections should be in the neighborhood of 10^{-38} cm^2 at about 1 BeV. Lee and Yang[1] first calculated the detailed cross sections for

$$\nu+n\rightarrow p+e^-,$$

$$\bar{\nu}+p\rightarrow n+e^+, \qquad (2)$$

$$\nu+n\rightarrow p+\mu^-,$$

$$\bar{\nu}+p\rightarrow n+\mu^+, \qquad (3)$$

using the vector form factor deduced from electron scattering results and assuming the axial vector form factor to be the same as the vector form factor. Subsequent work has been done by Yamaguchi[2] and Cabbibo and Gatto.[3] These calculations have been used as standards for comparison with experiments.

Unitarity and the absence of the decay $\mu\rightarrow e+\gamma$. A major difficulty of the Fermi theory at high energies is the necessity that it break down before the cross section reaches $\pi\lambda\!\!\!^{-2}$, violating unitarity. This breakdown must occur below 300 BeV in the center of mass. This difficulty may be avoided if an intermediate boson mediates the weak interactions. Feinberg[4] pointed out, however, that such a boson implies a branching ratio $(\mu\rightarrow e+\gamma)/(\mu\rightarrow e+\nu+\bar{\nu})$ of the order of 10^{-4}, unless the neutrinos associated with muons are different from those associated with electrons.[5] Lee and Yang[6] have subsequently noted that any general mechanism which would preserve unitarity should lead to a $\mu\rightarrow e+\gamma$ branching ratio not too different from the above. Inasmuch as the branching ratio is measured to be $\lesssim 10^{-8}$,[7] the hypothesis that the two neutrinos may be different has found some favor. It is expected that if there is only one type of neutrino, then neutrino interactions should produce muons and electrons in equal abundance. In the event that there are two neutrinos, there is no reason to expect any electrons at all.

The feasibility of doing neutrino experiments at accelerators was proposed independently by Pontecorvo[8] and Schwartz.[9] It was shown that the fluxes of neutrinos available from accelerators should produce of the order of several events per day per 10 tons of detector.

The essential scheme of the experiment is as follows: A neutrino "beam" is generated by decay in flight of pions according to reaction (1). The pions are produced by 15-BeV protons striking a beryllium target at one end of a 10-ft long straight section. The resulting entire flux of particles moving in the general direction of the detector strikes a 13.5-m thick iron shield wall at a distance of 21 m from the target. Neutrino interactions are observed in a 10-ton aluminum spark chamber located behind this shield.

The line of flight of the beam from target to detector makes an angle of 7.5° with respect to the internal proton direction (see Fig. 1). The operating energy of 15 BeV is chosen to keep the muons penetrating the shield to a tolerable level.

The number and energy spectrum of neutrinos from reaction (1) can be rather well calculated, on the basis of measured pion-production rates[10] and the geometry. The expected neutrino flux from π decay is shown in Fig. 2. Also shown is

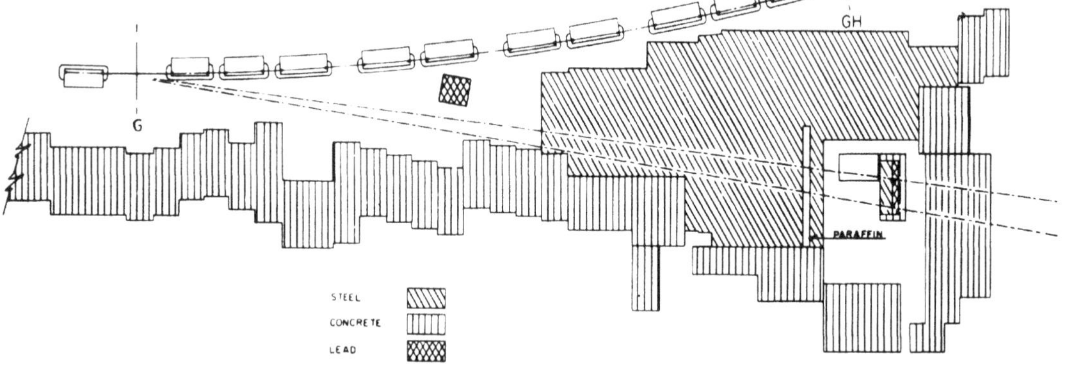

FIG. 1. Plan view of AGS neutrino experiment.

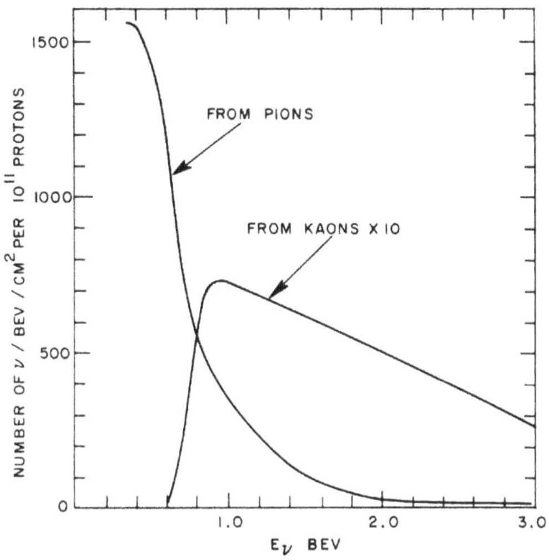

FIG. 2. Energy spectrum of neutrinos expected in the arrangement of Fig. 1 for 15-BeV protons on Be.

an estimate of neutrinos from the decay $K^{\pm} \rightarrow \mu^{\pm} + \nu(\bar{\nu})$. Various checks were performed to compare the targeting efficiency (fraction of circulating beam that interacts in the target) during the neutrino run with the efficiency during the beam survey run. (We believe this efficiency to be close to 70%.) The pion-neutrino flux is considered reliable to approximately 30% down to 300 MeV/c, but the flux below this momentum does not contribute to the results we wish to present.

The main shielding wall thickness, 13.5 m for most of the run, absorbs strongly interacting particles by nuclear interaction and muons up to 17 BeV by ionization loss. The absorption mean free path in iron for pions of 3, 6, and 9 BeV has been measured to be less than 0.24 m.[11] Thus the shield provides an attenuation of the order of 10^{-24} for strongly interacting particles. This attenuation is more than sufficient to reduce these particles to a level compatible with this experiment. The background of strongly interacting particles within the detector shield probably enters through the concrete floor and roof of the 5.5-m thick side wall. Indications of such leaks were, in fact, obtained during the early phases of the experiment and the shielding subsequently improved. The argument that our observations are not induced by strongly interacting particles will also be made on the basis of the detailed structure of the data.

The spark chamber detector consists of an array of 10 one-ton modules. Each unit has 9 aluminum plates 44 in. × 44 in. × 1 in. thick, separated by $\frac{3}{8}$-in. Lucite spacers. Each module is driven by a specially designed high-pressure spark gap and the entire assembly triggered as described below. The chamber will be more fully described elsewhere. Figure 3 illustrates the arrangement of coincidence and anticoincidence counters. Top, back, and front anticoincidence sheets (a total of 50 counters, each 48 in. × 11 in. × $\frac{1}{2}$ in.) are provided to reduce the effect of cosmic rays and AGS-produced muons which penetrate the shield. The top slab is shielded against neutrino events by 6 in. of steel and the back slab by 3 ft of steel and lead.

Triggering counters were inserted between adjacent chambers and at the end (see Fig. 3). These consist of pairs of counters, 48 in. × 11 in. × $\frac{1}{2}$ in., separated by $\frac{3}{4}$ in. of aluminum, and in fast coincidence. Four such pairs cover a chamber; 40 are employed in all.

The AGS at 15 BeV operates with a repetition period of 1.2 sec. A rapid beam deflector drives the protons onto the 3-in. thick Be target over a period of 20-30 μsec. The radiation during this interval has rf structure, the individual bursts being 20 nsec wide, the separation 220 nsec. This structure is employed to reduce the total "on" time and thus minimize cosmic-ray background. A Čerenkov counter exposed

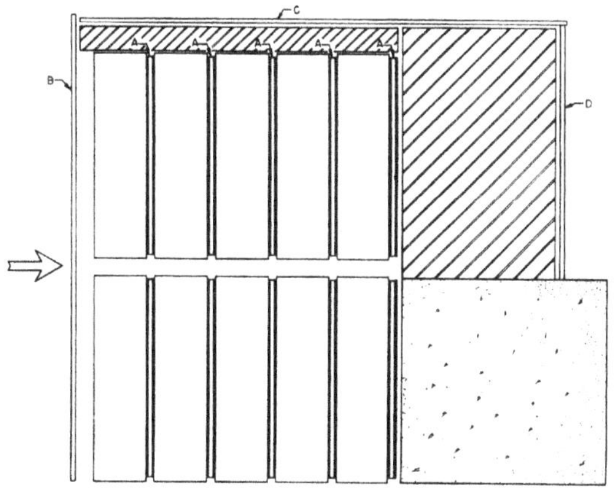

FIG. 3. Spark chamber and counter arrangement. A are the triggering slabs; B, C, and D are anticoincidence slabs. This is the front view seen by the four-camera stereo system.

to the pions in the neutrino "beam" provides a train of 30-nsec gates, which is placed in coincidence with the triggering events. The correct phasing is verified by raising the machine energy to 25 BeV and counting the high-energy muons which now penetrate the shield. The tight timing also serves the useful function of reducing sensitivity to low-energy neutrons which diffuse into the detector room. The trigger consists of a fast twofold coincidence in any of the 40 coincidence pairs in anticoincidence with the anticoincidence shield. Typical operation yields about 10 triggers per hour. Half the photographs are blank, the remainder consist of AGS muons entering unprotected faces of the chamber, cosmic rays, and "events." In order to verify the operation of circuits and the gap efficiency of the chamber, cosmic-ray test runs are conducted every four hours. These consist of triggering on almost horizontal cosmic-ray muons and recording the results both on film and on Land prints for rapid inspection (see Fig. 4).

A convenient monitor for this experiment is the number of circulating protons in the AGS machine. Typically, the AGS operates at a level of $2-4\times10^{11}$ protons per pulse, and 3000 pulses per hour. In an exposure of 3.48×10^{17} protons, we have counted 113 events satisfying the following geometric criteria: The event originates within a fiducial volume whose boundaries lie 4 in. from the front and back walls of the chamber and 2 in. from the top and bottom walls. The first two gaps must not fire, in order to exclude events whose origins lie outside the chambers. In addition, in the case of events consisting of a single track, an extrapolation of the track backwards (towards the neutrino source) for two gaps must also remain within the fiducial volume. The production angle of these single tracks relative to the neutrino line of flight must be less than 60°.

These 113 events may be classified further as follows:

(a) 49 short single tracks. These are single tracks whose visible momentum, if interpreted as muons, is less than 300 MeV/c. These presumably include some energetic muons which leave the chamber. They also include low-energy neutrino events and the bulk of the neutron produced background. Of these, 19 have 4 sparks or less. The second half of the run (1.7×10^{17} protons) with improved shielding yielded only three tracks in this category. We will not consider these as acceptable "events."

(b) 34 "single muons" of more than 300 MeV/c. These include tracks which, if interpreted as muons, have a visible range in the chambers such that their momentum is at least 300 MeV/c. The origin of these events must not be accompanied by more than two extraneous sparks. The latter requirement means that we include among "single tracks" events showing a small recoil. The 34 events are tabulated as a function of momentum in Table I. Figure 5 illustrates 3 "single muon" events.

(c) 22 "vertex" events. A vertex event is one whose origin is characterized by more than one track. All of these events show a substantial energy release. Figure 6 illustrates some of these.

(d) 8 "showers." These are all the remaining events. They are in general single tracks, too irregular in structure to be typical of μ mesons, and more typical of electron or photon showers. From these 8 "showers," for purposes of comparison with (b), we may select a group of 6 which are so located that their potential range within the chamber corresponds to μ mesons in excess of 300 MeV/c.

In the following, only the 56 energetic events of type (b) (long μ's) and type (c) (vertex events) will be referred to as "events."

Arguments on the neutrino origin of the ob-

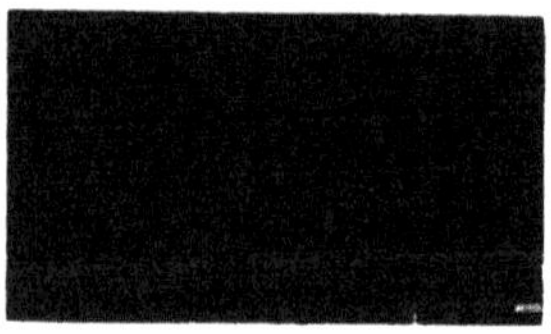

FIG. 4. Land print of Cosmic-ray muons integrated over many incoming tracks.

Table I. Classification of "events."

Single tracks			
$p_\mu < 300$ MeV/c [a]	49	$p_\mu > 500$	8
$p_\mu > 300$	34	$p_\mu > 600$	3
$p_\mu > 400$	19	$p_\mu > 700$	2
Total "events" 34			
Vertex events			
Visible energy released < 1 BeV	15		
Visible energy released > 1 BeV	7		

[a] These are not included in the "event" count (see text).

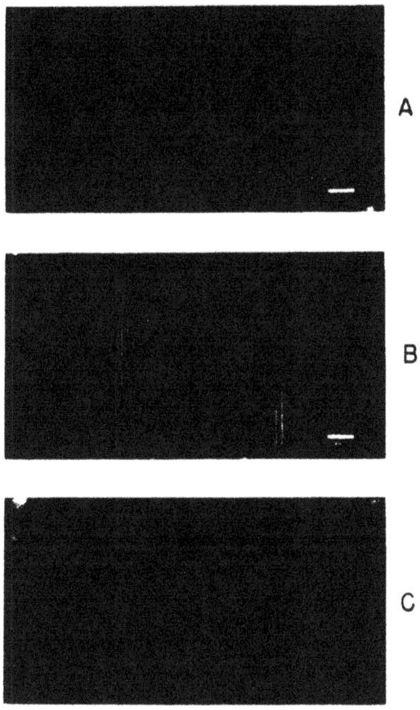

FIG. 5. Single muon events. (A) $p_\mu > 540$ MeV and δ ray indicating direction of motion (neutrino beam incident from left); (B) $p_\mu > 700$ MeV/c; (C) $p_\mu > 440$ with δ ray.

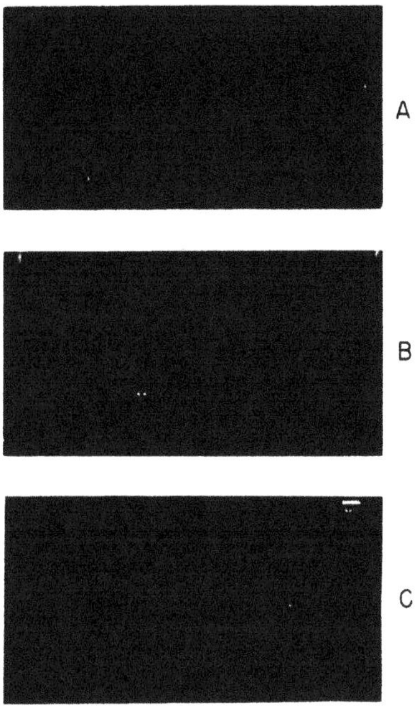

FIG. 6. Vertex events. (A) Single muon of $p_\mu > 500$ MeV and electron-type track; (B) possible example of two muons, both leave chamber; (C) four prong star with one long track of $p_\mu > 600$ MeV/c.

served "events."

1. The "events" are not produced by cosmic rays. Muons from cosmic rays which stop in the chamber can and do simulate neutrino events. This background is measured experimentally by running with the AGS machine off on the same triggering arrangement except for the Čerenkov gating requirement. The actual triggering rate then rises from 10 per hour to 80 per second (a dead-time circuit prevents jamming of the spark chamber). In 1800 cosmic-ray photographs thus obtained, 21 would be accepted as neutrino events. Thus 1 in 90 cosmic-ray events is neutrino-like. Čerenkov gating and the short AGS pulse effect a reduction by a factor of $\sim 10^{-6}$ since the circuits are "on" for only 3.5 μsec per pulse. In fact, for the body of data represented by Table I, a total of 1.6×10^6 pulses were counted. The equipment was therefore sensitive for a total time of 5.5 sec. This should lead to $5.5 \times 80 = 440$ cosmic-ray tracks which is consistent with observation. Among these, there should be 5 ± 1 cosmic-ray induced "events." These are almost evident in the small asymmetry seen in the angular distributions of Fig. 7. The remaining 51 events cannot be the result of cosmic rays.

2. The "events" are not neutron produced. Several observations contribute to this conclusion.

(a) The origins of all the observed events are uniformly distributed over the fiduciary volume, with the obvious bias against the last chamber induced by the $p_\mu > 300$ MeV/c requirement. Thus there is no evidence for attenuation, although the mean free path for nuclear interaction in aluminum is 40 cm and for electromagnetic interaction 9 cm.

(b) The front iron shield is so thick that we can expect less than 10^{-4} neutron induced reactions in the entire run from neutrons which have penetrated this shield. This was checked by removing 4 ft of iron from the front of the thick shield. If our events were due to neutrons in line with the target, the event rate would have increased by a factor of one hundred. No such effect was observed (see Table II). If neutrons penetrate the shield, it must be from other di-

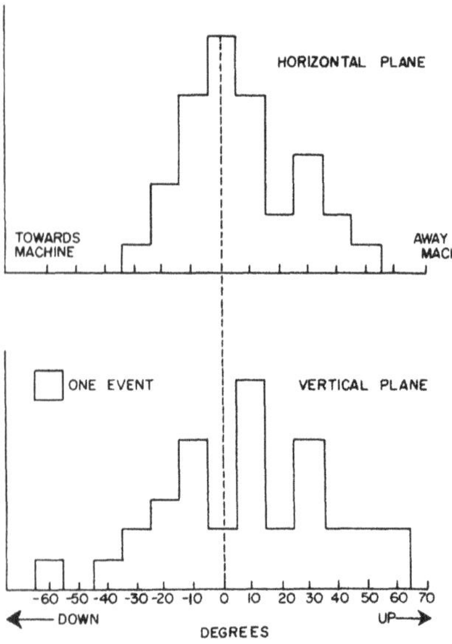

FIG. 7. Projected angular distributions of single track events. Zero degree is defined as the neutrino direction.

rections. The secondaries would reflect this directionality. The observed angular distribution of single track events is shown in Fig. 7. Except for the small cosmic-ray contribution to the vertical plane projection, both projections are peaked about the line of flight to the target.

(c) If our 29 single track events (excluding cosmic-ray background) were pions produced by neutrons, we would have expected, on the basis of known production cross sections, of the order of 15 single π^0's to have been produced. No cases of unaccompanied π^0's have been observed.

Table II. Event rates for normal and background conditions.

	Circulating protons $\times 10^{16}$	No. of Events	Calculated cosmic-ray[c] contribution	Net rate per 10^{16}
Normal run	34.8	56	5	1.46
Background I[a]	3.0	2	0.5	0.5
Background II[b]	8.6	4	1.5	0.3

[a]4 ft of Fe removed from main shielding wall.
[b]As above, but 4 ft of Pb placed within 6 ft of Be target and subtending a horizontal angular interval from 4° to 11° with respect to the internal proton beam.
[c]These should be subtracted from the "single muon" category.

3. The single particles produced show little or no nuclear interaction and are therefore presumed to be muons. For the purpose of this argument, it is convenient to first discuss the second half of our data, obtained after some shielding improvements were effected. A total traversal of 820 cm of aluminum by single tracks was observed, but no "clear" case of nuclear interaction such as large angle or charge exchange scattering was seen. In a spark chamber calibration experiment at the Cosmotron, it was found that for 400-MeV pions the mean free path for "clear" nuclear interactions in the chamber (as distinguished from stoppings) is no more than 100 cm of aluminum. We should, therefore, have observed of the order of 8 "clear" interactions; instead we observed none. The mean free path for the observed single tracks is then more than 8 times the nuclear mean free path.

Included in the count are 5 tracks which stop in the chamber. Certainly a fraction of the neutrino secondaries must be expected to be produced with such small momentum that they would stop in the chamber. Thus, none of these stoppings may, in fact, be nuclear interactions. But even if all stopping tracks are considered to represent nuclear interactions, the mean free path of the observed single tracks must be 4 nuclear mean free paths.

The situation in the case of the earlier data is more complicated. We suspect that a fair fraction of the short single tracks then observed are, in fact, protons produced in neutron collisions. However, similar arguments can be made also for these data which convince us that the energetic single track events observed then are also noninteracting.[12]

It is concluded that the observed single track events are muons, as expected from neutrino interactions.

4. The observed reactions are due to the decay products of pions and K mesons. In a second background run, 4 ft of iron were removed from the main shield and replaced by a similar quantity of lead placed as close to the target as feasible. Thus, the detector views the target through the same number of mean free paths of shielding material. However, the path available for pions to decay is reduced by a factor of 8. This is the closest we could come to "turning off" the neutrinos. The results of this run are given in terms of the number of events per 10^{16} circulating protons in Table II. The rate of "events" is reduced from 1.46 ± 0.2 to 0.3 ± 0.2 per 10^{16} in-

cident protons. This reduction is consistent with that which is expected for neutrinos which are the decay products of pions and K mesons.

Are there two kinds of neutrinos? The earlier discussion leads us to ask if the reactions (2) and (3) occur with the same rate. This would be expected if ν_μ, the neutrino coupled to the muon and produced in pion decay, is the same as ν_e, the neutrino coupled to the electron and produced in nuclear beta decay. We discuss only the single track events where the distinction between single muon tracks of $p_\mu > 300$ MeV/c and showers produced by high-energy single electrons is clear. See Figs. 8 and 4 which illustrate this difference.

We have observed 34 single muon events of which 5 are considered to be cosmic-ray background. If $\nu_\mu = \nu_e$, there should be of the order of 29 electron showers with a mean energy greater than 400 MeV/c. Instead, the only candidates which we have for such events are six "showers" of qualitatively different appearance from those of Fig. 8. To argue more precisely, we have exposed two of our one-ton spark chamber modules to electron beams at the Cosmotron. Runs were taken at various electron energies. From these we establish that the triggering efficiency for 400-MeV electrons is 67%. As a quantity characteristic of the calibration showers, we have taken the total number of observed sparks. The mean number is roughly linear with electron energy up to 400 MeV/c. Larger showers saturate the two chambers which were available. The spark distribution for 400 MeV/c showers is plotted in Fig. 9, normalized to the $\frac{2}{3} \times 29$ expected showers. The six "shower" events are also plotted. It is evident that these are not consistent with the prediction based on a universal theory with $\nu_\mu = \nu_e$. It can perhaps be argued that the absence of electron events could be understood in terms of the coupling of a single neutrino to the electron which is much weaker than that to the muon at higher momentum transfers, although at lower momentum transfers the results of β decay, μ capture, μ decay, and the ratio of $\pi \rightarrow \mu + \nu$ to $\pi \rightarrow e + \nu$ decay show that these couplings are equal.[13] However, the most plausible explanation for the absence of the electron showers, and the only one which preserves universality, is then that $\nu_\mu \neq \nu_e$; i.e., that there are at least two types of neutrinos. This also resolves the problem raised by the forbiddenness of the $\mu^+ \rightarrow e^+ + \gamma$ decay.

It remains to understand the nature of the 6 "shower" events. All of these events were obtained in the first part of the run during conditions in which there was certainly some neutron background. It is not unlikely that some of the events are small neutron produced stars. One or two could, in fact, be μ mesons. It should also be remarked that of the order of one or two electron events are expected from the neutrinos produced in the decays $K^+ \rightarrow e^+ + \nu_e + \pi^0$ and

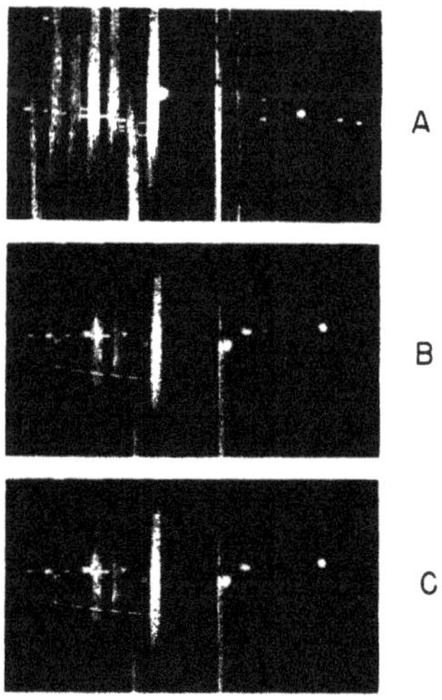

FIG. 8. 400-MeV electrons from the Cosmotron.

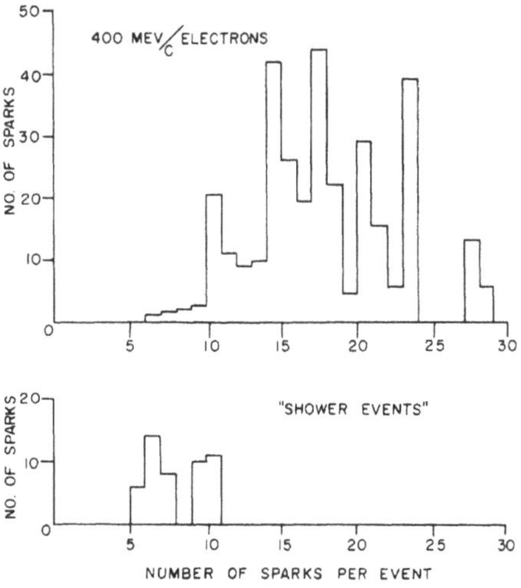

FIG. 9. Spark distribution for 400-MeV/c electrons normalized to expected number of showers. Also shown are the "shower" events.

VOLUME 9, NUMBER 1 PHYSICAL REVIEW LETTERS JULY 1, 1962

$K_2^0 \to e^{\pm} + \nu_e + \pi^{\mp}$.

The intermediate boson. It has been pointed out[1] that high-energy neutrinos should serve as a reasonable method of investigating the existence of an intermediate boson in the weak interactions. In recent years many of the objections to such a particle have been removed by the advent of V-A theory[14] and the remeasurement of the ρ value in μ decay.[15] The remaining difficulty pointed out by Feinberg,[4] namely the absence of the decay $\mu \to e + \gamma$, is removed by the results of this experiment. Consequently it is of interest to explore the extent to which our experiment has been sensitive to the production of these bosons.

Our neutrino intensity, in particular that part contributed by the K-meson decays, is sufficient to have produced intermediate bosons if the boson had a mass m_w less than that of the mass of the proton (m_p). In particular, if the boson had a mass equal to $0.6\,m_p$, we should have produced ~20 bosons by the process $\nu + p \to w^+ + \mu^- + p$. If $m_w = m_p$, then we should have observed 2 such events.[16]

Indeed, of our vertex events, 5 are consistent with the production of a boson. Two events, with two outgoing prongs, one of which is shown in Fig. 6(B), are consistent with both prongs being muons. This could correspond to the decay mode $w^+ \to \mu^+ + \nu$. One event shows four outgoing tracks, each of which leaves the chamber after traveling through 9 in. of aluminum. This might in principle be an example of $w^+ \to \pi^+ + \pi^- + \pi^+$. Another event, by far our most spectacular one, can be interpreted as having a muon, a charged pion, and two gamma rays presumably from a neutral pion. Over 2 BeV of energy release is seen in the chamber. This could in principle be an example of $w^+ \to \pi^+ + \pi^0$. Finally, we have one event, Fig. 6(A), in which both a muon and an electron appear to leave the same vertex. If this were a boson production, it would correspond to the boson decay mode $w^+ \to e^+ + \nu$. The alternative explanation for this event would require (i) that a neutral pion be produced with the muon; and (ii) that one of its gamma rays convert in the plate of the interaction while the other not convert visibly in the chamber.

The difficulty of demonstrating the existence of a boson is inherent in the poor resolution of the chamber. Future experiments should shed more light on this interesting question.

Neutrino cross sections. We have attempted to compare our observations with the predicted cross sections for reactions (2) using the theory.[1-3] To include the fact that the nucleons in (2) are, in fact, part of an aluminum nucleus, a Monte Carlo calculation was performed using a simple Fermi model for the nucleus in order to evaluate the effect of the Pauli principle and nucleon motion. This was then used to predict the number of "elastic" neutrino events to be expected under our conditions. The results agree with simpler calculations based on Fig. 2 to give, in terms of number of circulating protons,

from $\pi \to \mu + \nu$,	0.60 events/10^{16} protons,
from $K \to \mu + \nu$,	0.15 events/10^{16} protons,
Total	0.75 events/10^{16} ± ~30%.

The observed rates, assuming all single muons are "elastic" and all vertex events "inelastic" (i.e., produced with pions) are

"Elastic": 0.84 ± 0.16 events/10^{16} (29 events),

"Inelastic": 0.63 ± 0.14 events/10^{16} (22 events).

The agreement of our elastic yield with theory indicates that no large modification to the Fermi interaction is required at our mean momentum transfer of 350 MeV/c. The inelastic cross section in this region is of the same order as the elastic cross section.

Neutrino flip hypothesis. Feinberg, Gursey, and Pais[17] have pointed out that if there were two different types of neutrinos, their assignment to muon and electron, respectively, could in principle be interchanged for strangeness-violating weak interactions. Thus it might be possible that

$$\begin{matrix} \pi^+ \to \mu^+ + \nu_1 & & K^+ \to \mu^+ + \nu_2 \\ & \text{while} & \\ \pi^+ \to e^+ + \nu_2 & & K^+ \to e^+ + \nu_1. \end{matrix}$$

This hypothesis is subject to experimental check by observing whether neutrinos from $K_{\mu 2}$ decay produce muons or electrons in our chamber. Our calculation of the neutrino flux from $K_{\mu 2}$ decay indicates that we should have observed 5 events from these neutrinos. They would have an average energy of 1.5 BeV. An electron of this energy would have been clearly recognizable. None have been seen. It seems unlikely therefore that the neutrino flip hypothesis is correct.

The authors are indebted to Professor G. Feinberg, Professor T. D. Lee, and Professor C. N. Yang for many fruitful discussions. In particular, we note here that the emphasis by Lee and Yang on the importance of the high-energy behavior of weak interactions and the likelihood of the existence of two neutrinos played an important part in stimulating this research.

We would like to thank Mr. Warner Hayes for technical assistance throughout the experiment. In the construction of the spark chamber, R. Hodor and R. Lundgren of BNL, and Joseph Shill and Yin Au of Nevis did the engineering. The construction of the electronics was largely the work of the Instrumentation Division of BNL under W. Higinbotham. Other technical assistance was rendered by M. Katz and D. Balzarini. Robert Erlich was responsible for the machine calculations of neutrino rates, M. Tannenbaum assisted in the Cosmotron runs.

The experiment could not have succeeded without the tremendous efforts of the Brookhaven Accelerator Division. We owe much to the cooperation of Dr. K. Green, Dr. E. Courant, Dr. J. Blewett, Dr. M. H. Blewett, and the AGS staff including J. Spiro, W. Walker, D. Sisson, and L. Chimienti. The Cosmotron Department is acknowledged for its help in the initial assembly and later calibration runs.

The work was generously supported by the U. S. Atomic Energy Commission. The work at Nevis was considerably facilitated by Dr. W. F. Goodell, Jr., and the Nevis Cyclotron staff under Office of Naval Research support.

---

*This research was supported by the U. S. Atomic Energy Commission.

†Alfred P. Sloan Research Fellow.

[1]T. D. Lee and C. N. Yang, Phys. Rev. Letters 4, 307 (1960).

[2]Y. Yamaguchi, Progr. Theoret. Phys. (Kyoto) 6, 1117 (1960).

[3]N. Cabbibo and R. Gatto, Nuovo cimento 15, 304 (1960).

[4]G. Feinberg, Phys. Rev. 110, 1482 (1958).

[5]Several authors have discussed this possibility. Some of the earlier viewpoints are given by: E. Konopinski and H. Mahmoud, Phys. Rev. 92, 1045 (1953); J. Schwinger, Ann. Phys. (New York) 2, 407 (1957); I. Kawakami, Progr. Theoret. Phys. (Kyoto) 19, 459 (1957); M. Konuma, Nuclear Phys. 5, 504 (1958); S. A. Bludman, Bull. Am. Phys. Soc. 4, 80 (1959); S. Oneda and J. C. Pati, Phys. Rev. Letters 2, 125 (1959); K. Nishijima, Phys. Rev. 108, 907 (1957).

[6]T. D. Lee and C. N. Yang (private communications). See also Proceedings of the 1960 Annual International Conference on High-Energy Physics at Rochester (Interscience Publishers, Inc., New York, 1960), p. 567.

[7]D. Bartlett, S. Devons, and A. Sachs, Phys. Rev. Letters 8, 120 (1962); S. Frankel, J. Halpern, L. Holloway, W. Wales, M. Yearian, O. Chamberlain, A. Lemonick, and F. M. Pipkin, Phys. Rev. Letters 8, 123 (1962).

[8]B. Pontecorvo, J. Exptl. Theoret. Phys. (U.S.S.R.) 37, 1751 (1959) [translation: Soviet Phys.-JETP 10, 1236 (1960)].

[9]M. Schwartz, Phys. Rev. Letters 4, 306 (1960).

[10]W. F. Baker et al., Phys. Rev. Letters 7, 101 (1961).

[11]R. L. Cool, L. Lederman, L. Marshall, A. C. Melissinos, M. Tannenbaum, J. H. Tinlot, and T. Yamanouchi, Brookhaven National Laboratory Internal Report UP-18 (unpublished).

[12]These will be published in a more complete report.

[13]H. L. Anderson, T. Fujii, R. H. Miller, and L. Tau, Phys. Rev. 119, 2050 (1960); G. Culligan, J. F. Lathrop, V. L. Telegdi, R. Winston, and R. A. Lundy, Phys. Rev. Letters 7, 458 (1961); R. Hildebrand, Phys. Rev. Letters 8, 34 (1962); E. Bleser, L. Lederman, J. Rosen, J. Rothberg, and E. Zavattini, Phys. Rev. Letters 8, 288 (1962).

[14]R. Feynman and M. Gell-Mann, Phys. Rev. 109, 193 (1958); R. Marshak and E. Sudershan, Phys. Rev. 109, 1860 (1958).

[15]R. Plano, Phys. Rev. 119, 1400 (1960).

[16]T. D. Lee, P. Markstein, and C. N. Yang, Phys. Rev. Letters 7, 429 (1961).

[17]G. Feinberg, F. Gursey, and A. Pais, Phys. Rev. Letters 7, 208 (1961).

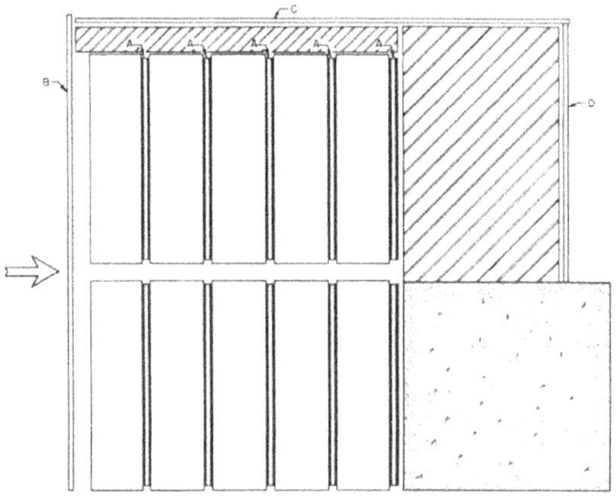

FIG. 3. Spark chamber and counter arrangement. *A* are the triggering slabs; *B*, *C*, and *D* are anticoincidence slabs. This is the front view seen by the four-camera stereo system.

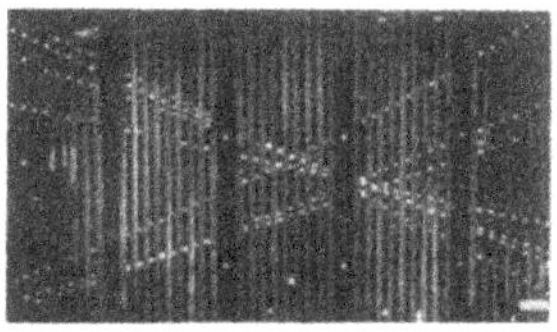

FIG. 4. Land print of Cosmic-ray muons integrated over many incoming tracks.

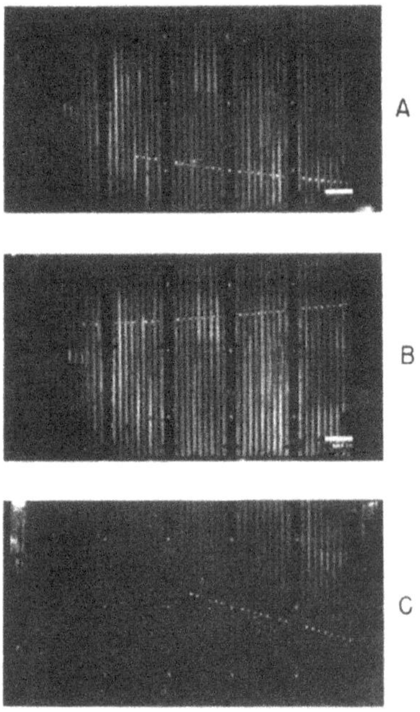

FIG. 5. Single muon events. (A) $p_\mu > 540$ MeV and δ ray indicating direction of motion (neutrino beam incident from left); (B) $p_\mu > 700$ MeV/c; (C) $p_\mu > 440$ with δ ray.

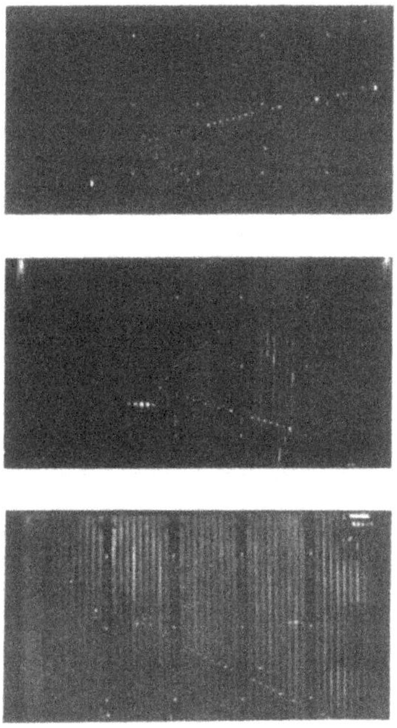

A

B

C

FIG. 6. Vertex events. (A) Single muon of $p_\mu > 500$ MeV and electron-type track; (B) possible example of two muons, both leave chamber; (C) four prong star with one long track of $p_\mu > 600$ MeV/c.

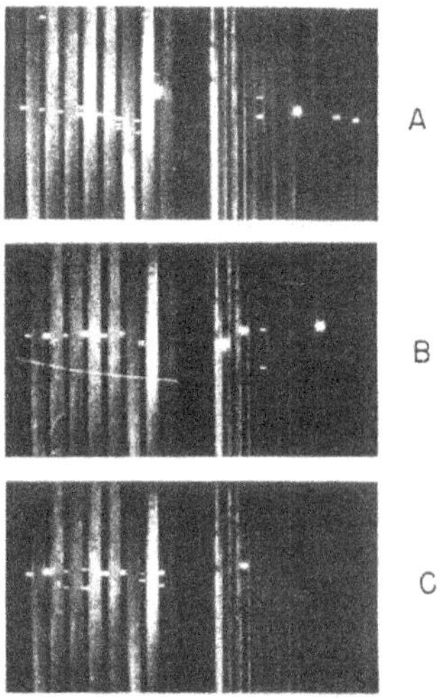

FIG. 8. 400-MeV electrons from the Cosmotron.

Volume 48B, number 5 PHYSICS LETTERS 4 March 1974

A NEW DETERMINATION OF THE $K^0 \to \pi^+\pi^-$ DECAY PARAMETERS

C. GEWENIGER*[1], S. GJESDAL*[2], G. PRESSER*[1], P. STEFFEN,
J. STEINBERGER, F. VANNUCCI*[3] and H. WAHL,
CERN, Geneva, Switzerland

F. EISELE*[1], H. FILTHUTH, K. KLEINKNECHT*[1], V. LÜTH and G. ZECH*[4]
Institut für Hochenergiephysik, Heidelberg, Germany

Received 4 February 1974

In a short neutral beam we have measured the proper time-dependence of the decay $K^0 \to \pi^+\pi^-$. This time structure exhibits the interference between the short- and long-lived states and is in agreement with the general expectations of the *CP* violation phenomenology.

This experiment gives new and more precise measurements of the following three parameters:
i) the decay width of the short-lived K_S component: $\Gamma_S = (1.119 \pm 0.006) \times 10^{10}\ \mathrm{sec}^{-1}$;
ii) the modulus of the *CP* violating parameter η_{+-} : $|\eta_{+-}| = (2.30 \pm 0.035) \times 10^{-3}$;
iii) the phase of η_{+-} as a function of the $K_S - K_L$ mass difference Δm: $\phi_{+-} = (49.4 \pm 1.0)^o + [(\Delta m - 0.540)/0.540] \times 305^o$.

The result of $|\eta_{+-}|$ may be compared with the result of the foregoing letter on Re ϵ in the frame of the superweak model. Good agreement is observed.

The experiment presented here studies the time dependence of the $\pi^+\pi^-$ decay mode of the K^0 meson in a short neutral beam. This time dependence has been previously studied [1, 2] and the main purpose of this experiment is to improve the precision, statistically as well as systematically, in order to provide a more accurate measure of the amplitude ratio

$$\eta_{+-} = \frac{\langle \pi^+\pi^- |T| K_L \rangle}{\langle \pi^+\pi^- |T| K_S \rangle} = |\eta_{+-}| \exp(i\phi_{+-}).$$

There is a substantial interest in experimental precision, because of the predictions of the superweak [3] and other models. If *CPT* is assumed to be conserved and if ϵ is the admixture of the *CP* odd (*CP* even) state in the dominantly *CP* even (*CP* odd) decay state, then the superweak model predicts

$$\eta_{+-} = \epsilon. \qquad (1)$$

Together with unitarity the model further specifies the phase of both parameters:

$$\phi = \tan^{-1} \frac{2\Delta m}{\Gamma_S - \Gamma_L}.$$

A whole class of other models is less specific in its predictions, but deviations from eq. (1) are expected to be of the order of the admixture of the $I = 2$ isospin state in the dominantly $I = 0$, $K_S \to \pi^+\pi^-$ decay. The smallness of this admixture (~ 4%) explains the interest in precision in the experimental verification of eq. (1).

This experiment presents a very substantial effort over a number of years, and all the relevant details of apparatus and analyses unfortunately cannot be included in this letter. The interested reader must be referred to a future, more detailed publication elsewhere.

The apparatus has been presented in the preceding letter [4]. Without going into details we point out its important properties.

The decay region which extends from 2.2 m to 11.6 m after the target permits detection in the proper time interval

$$3.5 \times 10^{-10}\ \mathrm{sec} < \tau < 30 \times 10^{-10}\ \mathrm{sec}.$$

The use of multiwire proportional chambers allows

*[1] Now at Universität Dortmund, Abteilung Physik, Germany.
*[2] Now at Districtshøgskolen, Stavanger, Norway.
*[3] Now at Institut de Physique Nucléaire, Orsay, France.
*[4] Now at Gesamthochschule, Siegen, Germany.

a high data-taking rate and limits the amount of matter in the beam.

The spectrometer section is followed by a trigger plane of twelve thin (1.6 mm) counters. A right-left coincidence in this plane is required to initiate an event, and the final read-out system requires two – and only two – hits in each plane of the multiwire proportional chambers.

The following requirements are imposed on the selection of events:

i) each chamber must have exactly two vertical and two horizontal wires hit;

ii) a χ^2 deviation, formed of the vertical kink angle of each track (after correction for vertical focusing) and of the skewness of the vertex, must be less than 12;

iii) there must be no signal in the Čerenkov counter and no coincidence between the two muon counter planes;

iv) the longitudinal distance target-decay vertex must be greater than 2.2 m;

v) the momenta of both charged secondaries must lie in the interval 1.50 GeV/c to 8.50 GeV/c (the minimum range for traversal of the muon detector is 1.45 GeV/c and the threshold for pion detection in the Čerenkov is 8.40 GeV/c);

vi) only events with inwards bending in the magnet are retained;

vii) events for which simultaneously $m_{p\pi}$ is within 10 MeV of the Λ^0 mass, and $p_+/(p_+ + p_-)$ is greater than 0.74, are withdrawn from the sample to reduce the Λ contamination.

These criteria are designed to select $K^0 \to \pi^+\pi^-$ decays and reject other decays as cleanly as necessary.

The remaining sample is plotted in the histograms of figs. 1a and 1b as a function of the invariant mass $m_{\pi\pi}$ and of ρ_T^2, the squared distance of the reprojected momenta in the target plane, from the target centre.

A two-dimensional linear background subtraction in $m_{\pi\pi}$ and ρ_T^2 was performed in each momentum bin ($\Delta p = 0.5$ GeV/c) and each proper time bin ($\Delta\tau = 0.5 \times 10^{-10}$ sec). The amount of subtracted events varies from 2.2% for K_S to 7.3% for K_L of the accepted events.

The final data sample is accumulated in a two-dimensional histogram in the kaon proper time and momentum. There are 6 million events in total and perhaps more significantly ~ 5000 events per 10^{-10} sec time bin at long times. The momentum distribution of the observed events is shown in fig. 2.

Extraction of the information from the experi-

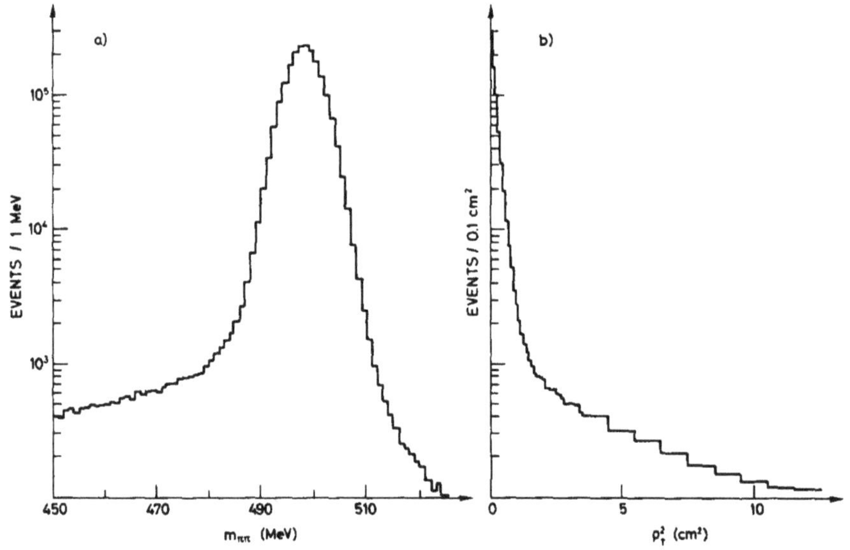

Fig. 1. (a) $m_{\pi\pi}$ distribution. (b) ρ_T^2 is the squared distance of the reprojected momenta in the target plane, from the target centre.

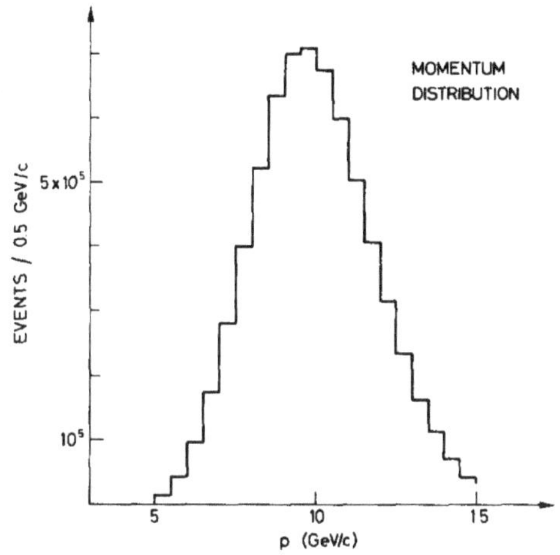

Fig. 2. Kaon momentum spectrum of the accepted events.

mental curve demands a good knowledge of the acceptance of the apparatus. This was achieved by simulating 6.6 million decays with a Monte Carlo program. These events were treated in the same way as the data. Fig. 3 shows the acceptance curves for different momenta as well as a weighted acceptance for events with momentum between 5 and 12.5 GeV/*c*.

The time distribution of events summed over the momentum interval 5–12.5 GeV/*c* with limits im-

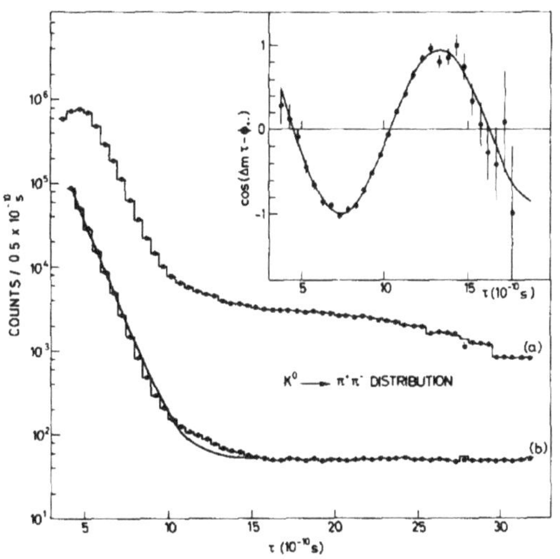

Fig. 4. Time distribution of $K \to \pi^+\pi^-$ events. a) Events (histogram) and fitted distribution (dots). b) Events corrected for detection efficiency (histogram), fitted distribution with interference term (dots) and fitted distribution without interference term (solid line). Insert: Interference term as extracted from data (dots) and fitted term (line).

posed by the decay volume is shown as the histogram a in fig. 4. The curve b of fig. 4 shows the data corrected for the detection efficiency.

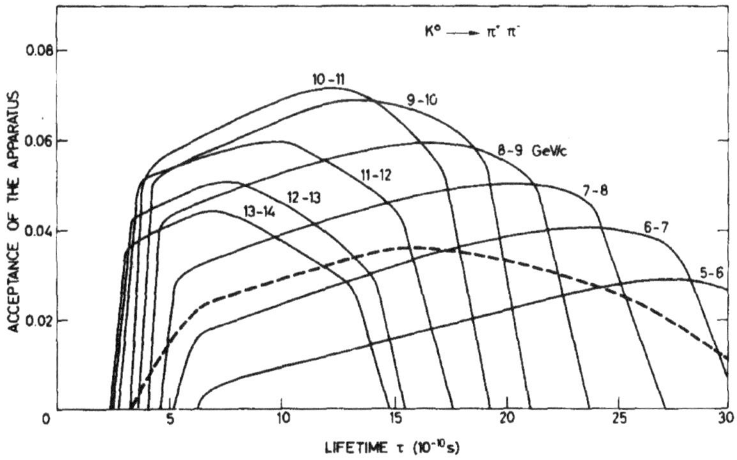

Fig. 3. Acceptance curves for different kaon momenta. The dashed line represents the average acceptance of the apparatus.

The theoretically expected distribution in proper time is:

$$I_{2\pi}(\tau) = [S(p) + \bar{S}(p)]\{\exp(-\Gamma_S\tau)$$

$$+ 2A(p)|\eta_{+-}| \exp[-(\Gamma_L+\Gamma_S)\tau/2] \cos(\Delta m\tau - \phi_{+-})$$

$$+ |\eta_{+-}|^2 \exp(-\Gamma_L\tau)\},$$

where $S(p)$ and $\bar{S}(p)$ are the production intensities of K^o and $\bar{K}^o$ and $A(p)$ measures the initial admixture of K^o and $\bar{K}^o$:

$$A(p) = \frac{S(p) - \bar{S}(p)}{S(p) + \bar{S}(p)}.$$

This expression is fitted to the data in 0.5 GeV/c momentum bins to find Γ_S, $|\eta_{+-}|$, ϕ_{+-} and unparametrized $S(p)$ and $\bar{S}(p)$ assuming Δm and Γ_L to be known. The experimentally-determined phase ϕ_{+-} is a linear function of Δm.

The χ^2 of the fit is 421 for 444 degrees of freedom. The result of the fit is shown as the dots in figs. 4a and 4b. The cosine part of the interference term is extracted from the full curve as presented in the insert of fig. 4.

The final results are the following:

$$\Gamma_S = (1.119 \pm 0.006) \times 10^{10}\ \mathrm{sec}^{-1}$$

$$|\eta_{+-}| = (2.30 \pm 0.035) \times 10^{-3}$$

$$\phi_{+-} = (49.4 \pm 1.0) + \left(\frac{\Delta m - 0.540}{0.540}\right) \times 305°.$$

The stability of the results was checked by varying the momentum range and the time interval used in the fit, as well as by changing the positions of the cuts mentioned previously. Also different background subtractions gave consistent results. The stated errors include estimated systematic uncertainties. The correction on the phase ϕ_{+-} amounts to $(0.4 \pm 0.3)°$ for the effects of a γ-ray absorber following the target, the scattering on the collimator walls, the K^o's produced by the nucleons absorbed in the collimator and the regeneration in the helium. A 0.5% uncertainty in the magnetic field is also taken into account.

The value of Γ_S, although in disagreement with earlier results [5], agrees with the most recently reported measurement [6].

The value of $|\eta_{+-}|$, in disagreement with earlier results [7] has been confirmed [8] since it was first reported by this group [9]. A check measurement was made in order to confirm the result on $|\eta_{+-}|$ by comparing the rates of the processes $K_L \to \pi^+\pi^-$ and $K_L \to \pi e \nu$. This result, which is systematically less reliable, is $|\eta_{+-}| = (2.30 \pm 0.06) \times 10^{-3}$.

The measurement of Δm, undertaken with the same apparatus, is close to completion and we prefer to wait for this result before drawing a conclusion on the compatibility of the phase measurement with the superweak model.

In any case, the new value of $|\eta_{+-}|$, together with the more precise charge asymmetry measurements of the previous letter, can be compared with the prediction of the superweak model supplemented with unitarity:

$$|\eta_{+-}| \frac{\Gamma_S - \Gamma_L}{\sqrt{(\Gamma_S - \Gamma_L)^2 + (2\Delta m)^2}} = \mathrm{Re}\,\epsilon.$$

In the foregoing letter it is found that

$$\mathrm{Re}\,\epsilon = (1.67 \pm 0.08) \times 10^{-3}.$$

This has to be compared with the result of this letter:

$$|\eta_{+-}| \frac{\Gamma_S - \Gamma_L}{\sqrt{(\Gamma_S - \Gamma_L)^2 + (2\Delta m)^2}}$$

$$= (2.30 \pm 0.035) \times (0.721 \pm 0.005) \times 10^{-3}$$

$$= (1.66 \pm 0.03) \times 10^{-3}.$$

The agreement is very good, the precision being approximately 5%.

We wish to express our thanks to Dr. E.M. Rimmer for excellent programming assistance, to Messrs. H. Dieperink, J. Olsfors, P. Schilly and M. Vysočansky for their remarkable technical assistance throughout the experiment. We also wish to thank Dr. G. Petrucci for the design of the proton beam splitting technique which was essential to the success of this experiment, and the CERN PS staff, particularly Dr. L. Hoffmann for the design and setting-up of the beam.

References

[1] A. Böhm et al., Nucl. Phys. B9 (1969) 605.
[2] D.A. Jensen et al., Phys. Rev. Lett. 23 (1969) 615.

[3] L. Wolfenstein, Phys. Rev. Lett. 13 (1964) 562.
[4] C. Geweniger et al., Phys. Lett. 48B (1974) 000.
[5] D.G. Hill et al., Phys. Rev. 171 (1968) 1418;
R.A. Donald et al., Phys. Lett. 27B (1968) 58;
L. Kirsch and P. Schmidt, Phys. Rev. 147 (1966) 939.
[6] O. Skjeggestad et al., Nucl. Phys. B48 (1972) 343.
[7] V.L. Fitch, R.F. Roth, J. Russ and W. Vernon, Phys. Rev. 164 (1967) 1711;
M. Bott-Bodenhausen et al., Phys. Lett. 23 (1966) 277.
[8] R. Messner et al., Phys. Rev. Lett. 30 (1973) 876.
[9] C. Rubbia, Rapporteurs' talk, Proc. 16th Intern. Conf. on High-energy physics, Batavia, 1972 (NAL, Batavia, 1972), vol. 4, p. 157.

Volume 52B, number 1 PHYSICS LETTERS 16 September 1974

A MEASUREMENT OF THE K_L-K_S MASS DIFFERENCE FROM THE CHARGE ASYMMETRY IN SEMI-LEPTONIC KAON DECAYS

S. GJESDAL [*1], G. PRESSER [*2], T. KAMAE [*3], P. STEFFEN,
J. STEINBERGER, F. VANNUCCI [*4] and H. WAHL
CERN, Geneva, Switzerland

F. EISELE [*2], H. FILTHUTH, V. LÜTH [*5] and G. ZECH [*6]
Institut für Hochenergiephysik, Universität Heidelberg, Germany

K. KLEINKNECHT
Universität Dortmund, Dortmund, Germany

Received 31 July 1974

The charge asymmetry in semi-leptonic kaon decays has been measured as a function of the kaon lifetime. High statistics data of K^0_{e3} and $K^0_{\mu 3}$ decay modes agree with each other and with the general expectation of the *CP* violation phenomenology together with the ΔS–ΔQ rule. The K_L-K_S mass difference obtained is $\Delta m = (0.533 \pm 0.004) \times 10^{10}\ s^{-1}$.

Introduction. The experiment reported in this paper is part of an extensive study aiming principally at a more precise determination of ϕ_{+-}, the phase of η_{+-}, the parameter describing *CP*-violation in $K^0 \to \pi^+\pi^-$ decays. New values of $K^0 \to \pi^+\pi^-$ decay parameters have been reported previously [1]. The data on semi-leptonic kaon decays recorded simultaneously and in the same apparatus as the pionic decays, are presented here to the extent that they improve the accuracy of the phase ϕ_{+-}.

A measurement of the K_L-K_S mass difference from leptonic kaon decays is of interest for the following reason: the determination of ϕ_{+-} by an observation of the interference term $\sim|\eta_{+-}|\cos(\Delta m\tau - \phi_{+-})$ in the time distribution of $\pi^+\pi^-$ decays suffers from the strong correlation between the phase and the mass difference $\Delta m = m_L - m_S$. On the other hand, the charge asymmetry of semi-leptonic decays exhibits an interference term $\sim \cos(\Delta m\tau)$. In principle it is, therefore, the difference in phase in the interference terms of pionic and leptonic decays which gives a measure of ϕ_{+-} without explicit, accurate knowledge of the mass difference Δm. In other words, a determination of ϕ_{+-} using the mass difference obtained from leptonic decays has the advantage of reduced sensitivity to systematic errors in common to the two decay modes.

The charge asymmetry in semi-leptonic decay $K^0 \to \pi^\mp \ell^\pm \nu$ is defined as:

$$\delta = (N^+ - N^-)/(N^+ + N^-)\,, \tag{1}$$

where N^+ and N^- denote the number of decays observed with a positive and negative lepton ℓ (electron or muon) respectively. The transition amplitudes:

$$f = \langle \pi^- \ell^+ \nu | T | K^0 \rangle\,, \quad \Delta S = \Delta Q\,; \qquad g = \langle \pi^- \ell^+ \nu | T | \bar{K}^0 \rangle\,, \quad \Delta S = -\Delta Q\,,$$

[*1] Now at Districtshogskolen Stavanger, Norway.
[*2] Now at Institut für Physik, Universität Dortmund, Germany.
[*3] Now at Institut for Nuclear Study, Tokyo University, Japan.
[*4] Visitor from Institut de Physique Nucleaire, Orsay, France.
[*5] Now at SLAC, Stanford University, USA.
[*6] Now at Gesamthochschule Siegen, Germany.

are related by *CPT* invariance, neglecting final state interactions, to the complex conjugate amplitudes:

$$f^* = \langle \pi^+ \ell^- \nu | T | \bar{K}^0 \rangle , \quad \Delta S = \Delta Q , \qquad g^* = \langle \pi^+ \ell^- \nu | T | K^0 \rangle , \quad \Delta S = -\Delta Q .$$

Since $\Delta S = -\Delta Q$ transitions are strongly suppressed, it is convenient to introduce the parameter $x = g/f$ such that the $\Delta S = \Delta Q$ selection rule implies $x = 0$.

The charge asymmetry for an initial pure K^0 or $\bar{K}^0$ state is a function of the lifetime τ in the kaon rest system:

$$\delta(\tau) = \frac{2(1-|x|^2)\,[\mathrm{Re}\,\epsilon\,(\exp(-\Gamma_L\tau)+\exp(-\Gamma_S\tau)) + A(p)\exp(-\bar{\Gamma}\tau)\cos(\Delta m\tau)]}{|1-x|^2\exp(-\Gamma_L\tau)+|1+x|^2\exp(-\Gamma_S\tau)+4A(p)\,\mathrm{Im}\,x\,\exp(-\bar{\Gamma}\tau)\sin(\Delta m\tau)} \qquad (2)$$

with the average decay width $\bar{\Gamma} = \frac{1}{2}(\Gamma_S + \Gamma_L)$, and ϵ being the usual *CP* mixing parameter in the K^0 mass matrix.

The amplitude of the interference term $A(p)$ is $+1$ for an initial K^0 state and -1 for an initial $\bar{K}^0$ state. In general, the "dilution factor" for an incoherent mixture of K^0 and $\bar{K}^0$ is:

$$A(p) = \frac{S(p) - \bar{S}(p)}{S(p) + \bar{S}(p)} .$$

The production yields $S(p)$ and $\bar{S}(p)$ of K^0 and $\bar{K}^0$ in the beam are determined as a function of the kaon momentum p from the analysis of $K^0 \to \pi^+\pi^-$ decays [1]. In the limit $\mathrm{Re}\,\epsilon \ll 1$, $\mathrm{Im}\,x \ll 1$ and $\tau \gtrsim 3/\Gamma_S$ eq. (2) can be simplified as follows†1:

$$\delta(\tau) = 2\,\frac{(1-|x|^2)}{|1-x|^2}\,[A(p)\exp(-\bar{\Gamma}\tau)\cos(\Delta m\tau) + \mathrm{Re}\,\epsilon] . \qquad (3)$$

According to this equation, Δm, $\mathrm{Re}\,\epsilon$ and $y = (1-|x|^2)/|1-x|^2 \approx (1+2\,\mathrm{Re}\,x)$ can be determined from a measurement of the time distribution of the semi-leptonic charge asymmetry. The results on the long time asymmetry $\delta(\tau > 14/\Gamma_S)$ have been reported previously [2]. The following sections describe the determination of Δm from the data.

Apparatus. The experiment was performed in a short neutral beam at the CERN Proton Synchrotron, providing neutral kaons over the momentum range 3–15 GeV/*c*. A 2 m long uranium collimator defines the beam, followed by a 9 m long helium-filled decay region and a magnetic spectrometer to measure the two charged tracks. Further details of the set-up can be found in ref. [2]. The decay path permits one to record kaon decays in the proper time interval between 3×10^{-10} s and 40×10^{-10} s.

A 6 m long threshold Cerenkov counter filled with hydrogen at atmospheric pressure is employed to identify electrons. The $270 \times 140\ \mathrm{cm}^2$ cross section is subdivided by focussing mirrors into 12 optically independent cells, 6 above and 6 below the beam line. The detection efficiency was investigated as a function of the particle momentum for identified muons and kinematically constrained pions from $K^0 \to \pi^+\pi^-$ decays. By extrapolation to $\beta = 1$ the efficiency for electrons was found to be bigger than 99% in all but cells, corresponding to 4.8 ± 0.5 photoelectrons per track. The remaining 3 cells are 4–6% inefficient. The threshold momenta for pions (8.4 GeV/*c*) and muons (6.3 GeV/*c*) check well with the expected values. The pulseheight of the phototubes was recorded for every event and checked to be insensitive to field reversals in the analysing magnet.

Muons are identified by a coincidence signal from a set of two scintillator hodoscopes positioned behind an absorber of 800 g/cm^2 of light concrete. The efficiency for muons from $K_{\mu 3}$ decays, with momenta above the threshold at 1.45 GeV/*c*, was found to be ~75% compatible with the calculated geometrical acceptance. The misidentification of pions or protons due to penetration varies as a function of momentum. An average probability of 0.8% was found in the momentum range between 1.6 GeV/*c* and 6 GeV/*c*.

The accidental rates in both lepton detectors were monitored continuously using delayed data strobes. The probability for a non-leptonic decay to pick up an accidental electron or muon signature was found to be

†1 The correct form of eq. (2) has been used in the analysis, the simplified formula is given only for illustration.

$(2.3 \pm 0.1) \times 10^{-3}$ and $(0.5 \pm 0.1) \times 10^{-3}$ respectively, with a charge asymmetry compatible with zero.

In total, more than 10^9 events with exactly two charged tracks in the spectrometer were recorded on magnetic tape. 140 million of these have at least one signal in a lepton counter. The magnet polarity has been reversed approximately every half hour.

Data reduction and analysis. Leptonic decays are selected from the data by an unambiguous lepton signature on one of the two tracks. A series of cuts is applied as follows:

– The geometrical reconstruction of the vertex is constrained by cuts on the skweness and the kinks of the tracks in the vertical projection. The decay vertex in confined to a fiducial volume.

– A minimum opening angle between the two tracks of $(0.1\ \mathrm{GeV}/c)/p_K$ is required to remove electron pairs and events with poor resolution of the vertex position along the beam line.

– The reconstructed momenta of the secondaries are restricted to p_e and $p_\pi < 8\ \mathrm{GeV}/c$ to eliminate pions above the Cerenkov threshold from the K_{e3} sample. In the $K_{\mu 3}$ sample, only $p_\mu > 1.6\ \mathrm{GeV}/c$ is required, well above the threshold of the muon counter.

– Events compatible with $K_{\pi 3}$ decays are removed from the data.

– Limits are set on the longitudinal neutrino momentum in the K^0 rest system in order to reject events that are not compatible with the kinematics of leptonic decays.

After these cuts there is still some background left due to the accidental rate of the Cerenkov counter and due to the misidentification of particles in the muon counter because of π-μ decay in flight or penetration of pions or protons through the concrete absorber. The bulk of this background was found to be two-body decays ($\Lambda \to p\pi$, $K_S \to 2\pi$) which have been excluded kinematically from the data.

For the K_{e3} sample the only significant remaining background are $\Lambda \to pe\nu$ decays, which, though small in fraction, are dangerous because of their charge asymmetry $\delta = -1$. These events are completely removed by a combined, charge symmetric cut in the $p\pi$-mass and in the ratio of the secondary momenta.

For the $K_{\mu 3}$ sample there are two important remaining sources of background:

1) $K_{\pi 3}$ events with one of the pions giving a muon signal. The distribution of these events is obtained by a Monte Carlo calculation feeding in the π-μ decay probability, the measured penetration probability of pions, and the relative branching ratio of $K_{\mu 3}$ versus $K_{\pi 3}$ events. The calculated background is subtracted statistically from the $K_{\mu 3}$ data. The effect of the correction is small since it affects only the sum distribution (the measured raw charge asymmetry of $K_{\pi 3}$ events is $\delta_{\pi 3} = (6.8 \pm 1.7) \times 10^{-3}$).

2) The neutral beam produces events in the helium of the decay volumes that give a muon signal. The asymmetry of these events if high (around -0.3) thus the correction is substantial. The effect inside the $K_{\mu 3}$ sample cannot be studied directly. A clean sample of those events can however be recorded from the data in the region outside the kinematical limits of $K_{\mu 3}$ decays. Additional data from absorbers put into the decay volume at different positions allows one to determine the charge asymmetry for events inside the $K_{\mu 3}$ region relative to that of events outside. In this way the amount of background events that have to be subtracted from the data has been determined crudely as a function of momentum and time. The resultant correction turned out to be especially large for low momenta and vertices close to the end of the collimator. In order to reduce the systematic uncertainties arising from that correction, more stringent cuts have been applied for the final selection: the vertex position was required to be more than 270 cm distant from the target. The total effect of this correction on Δm is then reduced to one half standard deviation of the statistical error.

A considerable effort was devoted to the study of other sources of background and systematic biases. Effects due to pion interaction and decay, delta-ray production and positron annihilation were found to be of the order of 100 ppm in the charge asymmetry [2] and largely independent of the decay time. Consequently they have been included as a constant overall correction to the long lived asymmetry [2].

Results. The kaon laboratory momentum is determined with a twofold ambiguity inherent to the kinematical reconstruction of three-body decays. The lower momentum solution p' is favoured by the acceptance of the detector. Therefore, the lifetime τ' derived from p' is chosen as a measure of the true proper lifetime τ of the kaon.

In total 6 million K_{e3} decays and 2 million $K_{\mu 3}$ decays with lifetimes $\tau' < 12.75 \times 10^{-10}$ s remain after cuts.

Their charge asymmetry is evaluated as a function of τ' and p' in bins of width $\Delta\tau' = 0.5 \times 10^{-10}$ s and $p' =$ 2 GeV/c starting at $\tau'_{min} = 2.25 \times 10^{-10}$ s and $p_{min} = 7$ GeV/c.

The mass difference Δm is determined by a comparison of the time dependence of the measured charge asymmetry with the theoretical expectation $\delta(\tau', \Delta m, y)$ for the set of parameters to be determined. The theoretical function δ and its derivatives are calculated by Monte Carlo techniques from eq. (2). This treatment accounts for the following:

– The $K_{\ell 3}$ matrix elements according to V–A theory with linear formfactors for the hadronic current [3] and radiative corrections [4].

– The observed beam profile, and the experimental resolution and acceptance.

– Transformation from the true kaon momentum p and lifetime τ to the measured quantities p' and τ'.

– The shape of the kaon momentum spectrum and the dilution factor $A(p)$ as obtained in the $K_{\pi 2}$ analysis [1]

The influence of the actual form of the matrix element on the charge asymmetry is weak. The shape of the momentum spectrum enters only indirectly in the transformation from τ to τ'. The K_S lifetime and the K_L charge asymmetries are taken from previous results of the same experiment [1, 2]. The results of the best fits to the measured charge asymmetries are shown in figs. 1 and 2. The ΔS–ΔQ factor y is left free in the fits. The uncorrected values for the K_L-K_S mass difference are:

$$\Delta m(K_{e3}) = (0.5287 \pm 0.0040) \times 10^{10}\,s^{-1}\,, \quad \Delta m(K_{\mu 3}) = (0.526 \pm 0.0085) \times 10^{10}\,s^{-1}\,.$$

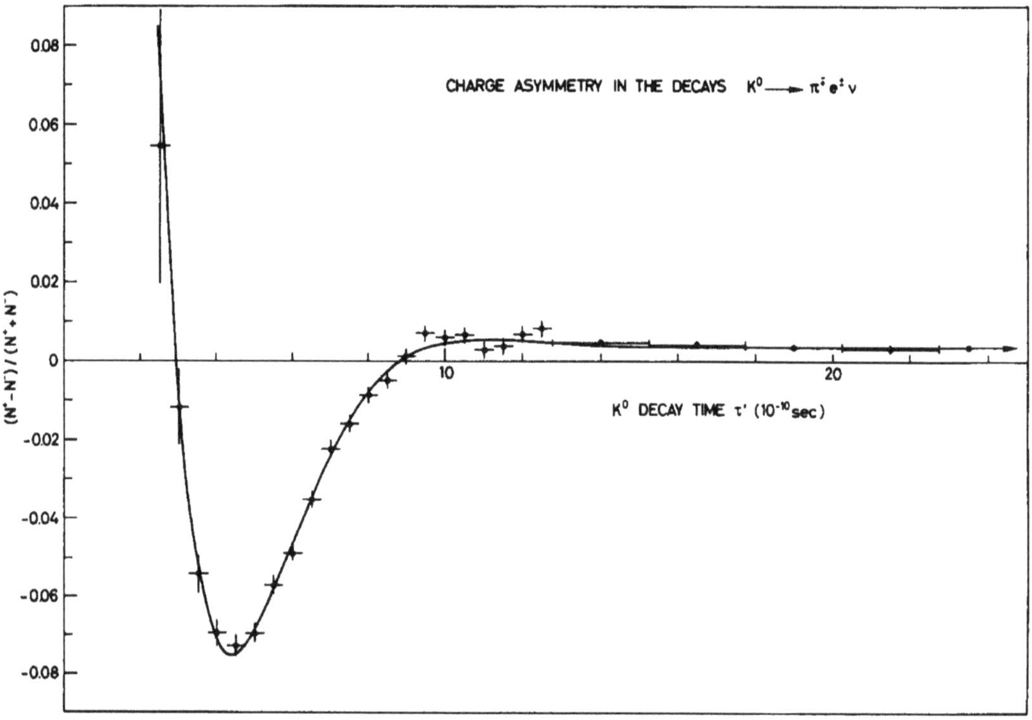

Fig. 1. The charge asymmetry as a function of the reconstructed decay time τ' for the K_{e3} decays. The experimental data are compared to the best fit as indicated by the solid line.

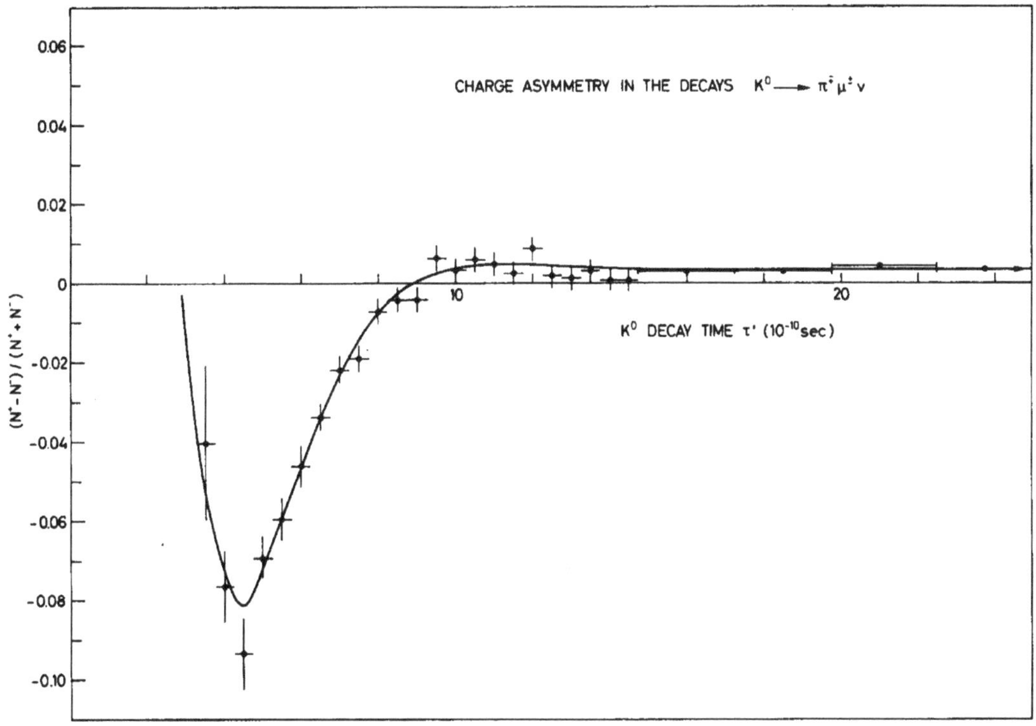

Fig. 2. The charge asymmetry as a function of the reconstructed decay time τ' for the $K_{\mu3}$ decays. The experimental data are compared to the best fit as indicated by the solid line.

The quoted error includes the statistical error as well as the uncertainties in the dilution factor. χ^2 values per degree of freedom of 17/23 and 16/24 are obtained.

The above result is obtained assuming an incoherent mixture of K^o and $\bar{K}^o$ produced at the centre of the primary target. The following corrections account for the accumulated effects due to secondary interactions of the kaons in the beam line. These effects can be described as a common initial phase change of 0.4° ± 0.3° [1] and results in a correction of $(+0.0018 \pm 0.0013) \times 10^{10}\,s^{-1}$ in Δm. Kaons produced in the beam dump lead to an independent correction of $(+0.0012 \pm 0.005) \times 10^{10}\,s^{-1}$. Furthermore, K_{e3} radiative decays cause a $(-0.45 \pm 0.1)\%$ shift in the reconstructed kaon momentum implying a correction of $(+0.0024 \pm 0.0005) \times 10^{10}\,s^{-1}$. The final corrected values of Δm and the average from K_{e3} and $K_{\mu3}$ decays are:

$$\Delta m(K_{e3}) = (0.5341 \pm 0.0043) \times 10^{10}\,s^{-1}\,, \quad \Delta m(K_{\mu3}) = (0.529 \pm 0.010) \times 10^{10}\,s^{-1}\,,$$

$$\Delta m(\mathrm{av}) = (0.5334 \pm 0.0040) \times 10^{10}\,s^{-1}\,.$$

The quoted error includes the estimated uncertainties of the corrections including the uncertainty in the background subtraction of the $K_{\mu3}$ data. In addition, a 0.3% systematic error has to be allotted to the uncertainty in the momentum calibration and the associated uncertainty in the K_S lifetime [1].

The results compare well with an independent determination of Δm by the two-regenerator method [5]

We are grateful for the programming assistance of E.M. Rimmer and the technical help provided by H. Dieperink, J. Olsfors, P. Schilly and M. Vysocansky. The Heidelberg members of the collaboration acknowledge the financial support of the Bundesministerium für Forschung und Technologie.

References

[1] C. Geweniger et al., Phys. Lett. 48B (1974) 487.
[2] C. Geweniger et al., Phys. Lett. 48B (1974) 483.
[3] S. Gjesdal, G. Presser, T. Kamae, P. Steffen, J. Steingerger, F. Vannucci, H. Wahl, F. Eisele, K. Kleinknecht, V. Lüth and G. Zech, to be published.
[4] Virtual radiative corrections are treated according to E.S. Ginsberg, Phys. Rev. 171 (1968) 1675 and Phys. Rev. 187 (1969) 2280. Internal Bremsstrahlung with the emission of real photons is evaluated separately for the conditions of this experiment.
[5] C. Geweniger et al., Phys. Lett. 51B (1974) 000.

Volume 206, number 1 PHYSICS LETTERS B 12 May 1988

FIRST EVIDENCE FOR DIRECT *CP* VIOLATION

CERN–Dortmund–Edinburgh–Mainz–Orsay–Pisa–Siegen Collaboration

H. BURKHARDT [1], P. CLARKE, D. COWARD [2,3], D. CUNDY, N. DOBLE, L. GATIGNON, V. GIBSON, R. HAGELBERG, G. KESSELER, J. VAN DER LANS, I. MANNELLI [4], T. MICZAIKA [5], A.C. SCHAFFER [6], J. STEINBERGER, H. TAUREG, H. WAHL, C. YOUNGMAN [7]
CERN, CH-1211 Geneva 23, Switzerland

G. DIETRICH, W. HEINEN [8]
Institut für Physik, Universität Dortmund, D-4600 Dortmund 50, Fed. Rep. Germany [9]

R. BLACK, D.J. CANDLIN, J. MUIR, K.J. PEACH, B. PIJLGROMS [10], I.P. SHIPSEY [11], W. STEPHENSON
Physics Department, University of Edinburgh, Edinburgh EH9 3JZ, UK

H. BLÜMER, M. KASEMANN, K. KLEINKNECHT, B. PANZER, B. RENK
Institut für Physik, Universität Mainz, D-6500 Mainz, Fed. Rep. Germany [9]

E. AUGÉ, R.L. CHASE, M. CORTI, D. FOURNIER, P. HEUSSE, L. ICONOMIDOU-FAYARD, A.M. LUTZ, H.G. SANDER [1]
Laboratoire de l'Accélérateur Linéaire, Université de Paris-Sud, F-91405 Orsay, France [12]

A. BIGI, M. CALVETTI [13], R. CAROSI, R. CASALI, C. CERRI, G. GARGANI, E. MASSA, A. NAPPI, G.M. PIERAZZINI
Dipartimento di Fisica e Sezione INFN, I-56100 Pisa, Italy

C. BECKER [5], D. HEYLAND [5], M. HOLDER, G. QUAST, M. ROST, W. WEIHS and G. ZECH
Fachbereich Physik, Universität Siegen, D-5900 Siegen 21, Fed. Rep. Germany [14]

Received 31 March 1988

[1] Present address: Fachbereich Physik, Universität Siegen, D-5900 Siegen 21, Fed. Rep. Germany.
[2] On leave from SLAC, Stanford, CA 94305, USA.
[3] Work supported in part by the US Department of Energy contract DE-AC03-765F00515.
[4] Present address: Scuola Normale Superiore e Sezione INFN, I-56100 Pisa, Italy.
[5] Present address: DVFLR, D-5000 Cologne, Fed. Rep. Germany.
[6] Present address: LAL, Université Paris-Sud, F-91405 Orsay, France.
[7] Present address: II. Institut für experimentelle Physik, Universität Hamburg, D-2000 Hamburg, Fed. Rep. Germany.
[8] Present address: IP-Systems, D-7500 Karlsruhe, Fed. Rep. Germany.
[9] Funded by the German Federal Minister for Research and Technology (BMFT) under contract 05 4MZ18.
[10] Present address: FWI, University of Amsterdam, 1018 XE Amsterdam, The Netherlands.
[11] Present address: Syracuse University, Syracuse, NY 13244-1130, USA.
[12] Funded by Institut National de Physique des Particules et de Physique Nucléaire, France.
[13] Present address: Dipartimento di Fisica e Sezione INFN, I-06100 Perugia, Italy.
[14] Funded by the German Federal Minister for Research and Technology (BMFT) under contract 054 Si74.

The double ratio R of the relative decay rates of the short- and long-lived neutral kaons into two charged and two neutral pions was measured to be $0.980 \pm 0.004 \pm 0.005$. The deviation of R from unity implies CP violation in the transition of the CP-odd K_2 into two pions with $\epsilon'/\epsilon = (3.3 \pm 1.1) \times 10^{-3}$.

Since its first observation in the decay of the long-lived neutral kaon into two pions [1], CP violation remains one of the enigmas in particle physics. While CP violation is manifest in neutral kaon decays, the search for CP-violating effects elsewhere has been unsuccessful. In the phenomenology of CP violation in the neutral kaon system [2] it is convenient to define the CP eigenstates $K_1 = (K^0 + \bar{K}^0)/\sqrt{2}$ and $K_2 = (K^0 - \bar{K}^0)/\sqrt{2}$ with $K_1 = +CP\,K_1$ and $K_2 = -CP\,K_2$. The short- and long-lived K^0 are the mass eigenstates which can be written as $K_S \approx K_1 + \epsilon K_2$ and $K_L \approx K_2 + \epsilon K_1$. The parameter ϵ describes CP violation induced by kaon state mixing. Direct CP violation may also occur in the decay of the K_2 into two pions with a relative amplitude of [2]

$$\epsilon' = i/\sqrt{2}\,\mathrm{Im}(A_2/A_0)\,\exp[i(\delta_2 - \delta_0)]\ ,$$

where A_0 and A_2 are the amplitudes for the decay into isospin 0 and 2 two-pion states; δ_0 and δ_2 are the corresponding $\pi\pi$ scattering phase shifts at the mass of the K^0. With these definitions the ratios of K_L and K_S decay amplitudes into $2\pi^0$ and $\pi^+\pi^-$ respectively are

$$\eta_{00} \equiv \langle 2\pi^0 | T | K_L \rangle / \langle 2\pi^0 | T | K_S \rangle$$

$$\equiv |\eta_{00}|\,\exp(i\Phi_{00}) = \epsilon - 2\epsilon'$$

and

$$\eta_{+-} \equiv \langle \pi^+\pi^- | T | K_L \rangle / \langle \pi^+\pi^- | T | K_S \rangle$$

$$\equiv |\eta_{+-}|\,\exp(i\Phi_{+-}) = \epsilon + \epsilon'\ .$$

The relevant experimental measurements are the magnitudes and phases of these two parameters, and the real part of ϵ determined from the charge asymmetry in semileptonic K_L decays: $|\eta_{+-}| = (2.27 \pm 0.02) \times 10^{-3}$ [3]; $\Phi_{+-} = 44.6° \pm 1.2°$ [4]; $|\eta_{00}/\eta_{+-}| = 1.00 \pm 0.01$ [5]; $\Phi_{00} = 55° \pm 5°$ [6]; and $\mathrm{Re}\,\epsilon = (1.62 \pm 0.09) \times 10^{-3}$ [7]. All experimental results are compatible with $\epsilon = 2.27 \times 10^{-3} \exp(i\,43.7°)$ and the superweak model [8] in which state mixing is the only source of CP violation and $\epsilon' = 0$ [#1].

[#1] We disregard here the two-standard-deviation discrepancy in Φ_{00}. A considerably more sensitive measurement of $\Phi_{00} - \Phi_{+-}$ is being carried out by this group.

In the theory of six weakly interacting quarks [9], direct CP violation as well as state mixing are introduced by transitions via heavy-quark intermediate states. Based on this, a small but non-zero value of ϵ'' is predicted [10]. From $\pi\pi$ scattering, its phase is determined to be $61° \pm 3°$ [11]. This angle is close to the phase of ϵ, so that to a good approximation $\mathrm{Re}\,\epsilon'/\epsilon = 1/6(1 - |\eta_{00}/\eta_{+-}|^2)$. This relation is used to determine ϵ'/ϵ from the double ratio of K_S and K_L decay rates into charged and neutral pions:

$$R = \left|\frac{\eta_{00}}{\eta_{+-}}\right|^2 = \frac{\Gamma(K_L \to 2\pi^0)/\Gamma(K_L \to \pi^+\pi^-)}{\Gamma(K_S \to 2\pi^0)/\Gamma(K_S \to \pi^+\pi^-)}\ .$$

This experiment has been performed at the CERN Super Proton Synchrotron. It is based on the concurrent detection of $2\pi^0$ and $\pi^+\pi^-$ decays. Collinear K_S and K_L beams are employed alternately, changing frequently from one to the other to reduce time-dependent effects. Details of the apparatus and beams have been given elsewhere [12]. Kaons with energies around 100 GeV are produced by 450 GeV protons incident upon one of two targets at an angle of 3.6 mrad with respect to the kaon beam line. The K_L are derived from $\sim 10^{11}$ protons per pulse and are selected by two-stage collimation at distances of 48 m and 120 m, respectively, from the first production target. Alternatively, $\sim 10^7$ protons per pulse are brought onto the second target, from which the K_S are selected by collimation after 7 m. The retractable K_S target station, sweeping magnet and collimator system are mounted on a train which can be moved through 48 m of the K_L decay region. The K_S data are taken with the beam train displaced in 1.2 m steps so that both the K_S and K_L decay distributions become effectively uniform in the fiducial region (the average K_S decay length is 6 m).

The detector is based on calorimetry and is designed for good stability and high efficiency, large acceptance and fast data-acquisition. A schematic layout of the apparatus is shown in fig. 1. The principal features are summarized as follows:

- both K_S and K_L beams are transported in vacuum;
- an anticounter in the K_S beam, preceded by a 7 mm

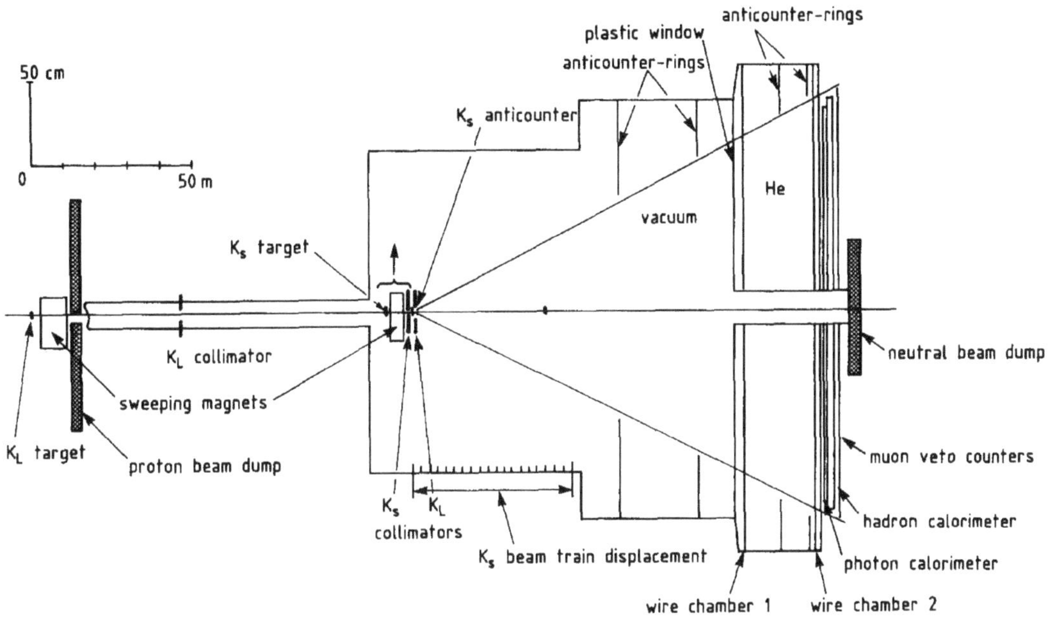

Fig. 1. Schematic layout of apparatus and beams.

lead converter, vetos decays in the collimator, defines the upstream edge of the decay region, and permits the relative calibration of the $2\pi^0$ and $\pi^+\pi^-$ energy scales to a precision better than $\pm 10^{-3}$;

- four ring-shaped anticounters surrounding the decay region detect large-angle photons and thus reduce unwanted three-body decays;
- two wire chambers spaced 25 m apart, with ± 0.5 mm resolution in each projection, track charged pions;
- a hodoscope of scintillation counters triggers on $\pi^+\pi^-$ decays by a coincidence of hits in opposite quadrants;
- a liquid-argon/lead sandwich calorimeter with strip readout measures photons with ± 0.5 mm position and $\pm 7.5\%/\sqrt{E}$ (GeV) energy resolution;
- a plane of scintillation counters, installed in the liquid argon after 13 radiation lengths of material, triggers on $2\pi^0$ decays;
- an iron/scintillator sandwich calorimeter measures, in conjuction with the liquid-argon calorimeter, the energy of charged pions with $\pm 65\%/\sqrt{E}$ (GeV) resolution;
- two planes of scintillators, after a total of 3 m of iron equivalent, reject $K^0 \to \pi\mu\nu$ decays.

The decay region is evacuated and the volume between the chambers is filled with helium. A thin composite Kevlar window of 3×10^{-3} radiation length separates the decay region from the wire chamber section. A tube of 20 cm diameter, through the centre of the window and the detectors, allows the neutral beam to continue in vacuum as far as the final beam dump.

Single counting rates are typically 10^5 Hz, originating predominantly from K^0 decays and beam-associated muons. The trigger on two-body K^0 decays is done in three steps. A pretrigger signal is generated from a coincidence of hits in opposite quadrants of the scintillator hodoscope, or from a left–right coincidence of the liquid-argon scintillators, with a veto from the ring and muon anticounters. A trigger signal is accepted, subject to further conditions on calorimeter energies, the number of hits in the first chamber, and the number of peaks in the liquid-argon calorimeter. After digitization of pulse heights and chamber information, three-body decays are rejected using on-line processors. The pretrigger rate is about 10 kHz,

Volume 206, number 1 PHYSICS LETTERS B 12 May 1988

and typically 1000 events are recorded per burst in the K_L beam.

The $K^0 \to 2\pi^0 \to 4\gamma$ decays are reconstructed from the measured positions and energies of the photons. Details of the reconstruction method may be found in ref. [13]. Events with extra photons of more than 2.5 GeV are rejected. For accepted events, the photon energies have to be above 5 GeV with at least 5 cm separation between shower centres, and the centre of gravity of the energies of all photons has to lie within the beam region. The K^0 energy is measured with typically 1% accuracy. The distance of the decay vertex from the calorimeter is calculated, using the K^0 mass as a constraint, with similar precision. Constraints on the masses of two-photon pairs are used to reduce the background, which is primarily due to $K_L \to 3\pi^0 \to 6\gamma$ decays with undetected photons. This background is uniformly distributed in a two-dimensional scatter plot of photon-pair masses. The π^0 mass resolution is ~2 MeV. Signal and background events are counted in equal-area χ^2 contours around the region defined for accepted events (see fig. 2). The signal region is taken as $\chi^2 < 9$. Background is subtracted by linear extrapolation into the signal region. It is about 4% in the K_L beam but depends strongly on the longitudinal vertex position because of the apparent vertex shift due to the missing energy in $3\pi^0$ decays with undetected photons. It is negligible in the K_S beam.

The $K^0 \to \pi^+\pi^-$ decays are reconstructed from space points defined by at least three hits out of the four planes in each of the two wire chambers. Events with extra space points in the first chamber are rejected for both charged and neutral decays. The longitudinal vertex resolution is better than 1 m. The K^0 energy is calculated with 1% precision from the kaon mass, the opening angle between the two tracks, and the ratio of track energies as measured in the calorimeter. This ratio is limited to a maximum of 2.5, in order to achieve this resolution and also to reduce the contribution of $\Lambda \to p\pi$ decays to a negligible level. Events with isolated photons, such as $K^0 \to \pi^+\pi^-\pi^0$ decays and events with accidental photons, are rejected. The $K^0 \to \pi e\nu$ events are identified and re-

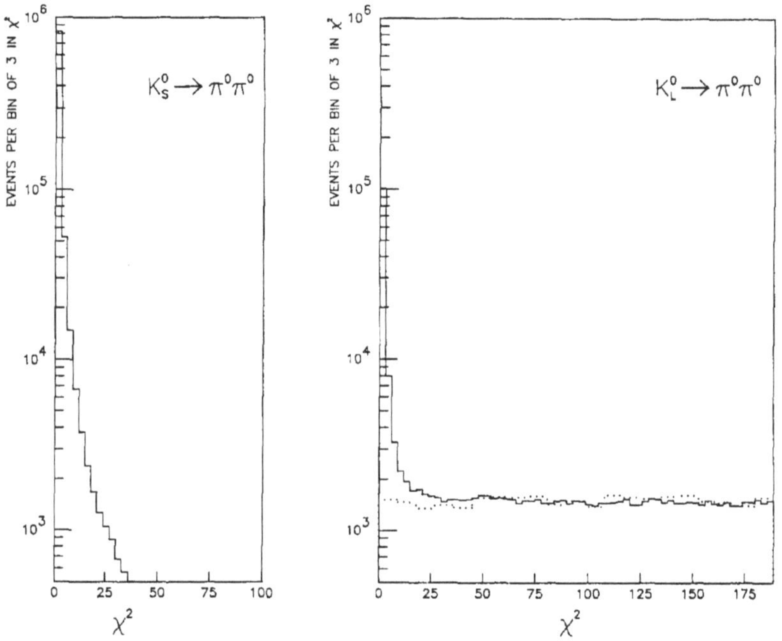

Fig. 2. Number of accepted 4γ events as a function of χ^2, for $K_S \to 2\pi^0$ and $K_L \to 2\pi^0$ data, and a Monte Carlo calculation for background originating from $K_L \to 3\pi^0$ decays. The signal region is taken as $\chi^2 < 9$.

jected by comparing, for each track, the energy deposited in the front half of the photon calorimeter with the energy deposited in the hadron calorimeter. About half of the detected $\pi^+\pi^-$ events are lost equally from K_S and K_L because of all these requirements. Possible variations of the rejection rate due to systematic changes in detector response are monitored by the observed $\pi^+\pi^-$ mass. The response of the hadron calorimeter is evaluated to be constant within $\pm 0.5\%$, leading to an uncertainty ($<0.1\%$) in the measured ratio of $K_S \to \pi^+\pi^-$ and $K_L \to \pi^+\pi^-$ event rates. After cuts on the $\pi^+\pi^-$ mass and on the reconstructed centre of gravity with respect to the beam axis, a residual background of the three-body decays must be subtracted.

In a two-body decay the decay plane should contain the production target, but because of measurement errors and multiple scattering, a certain distribution of the perpendicular distance d_T of this plane to the target is expected, and can be measured with $K_S \to \pi^+\pi^-$ decays. In the three-body decays of K_L, because of the non-coplanarity of the decay, a much broader d_T distribution is expected. This is illustrated in fig. 3, where the d_T distributions of accepted two-track events in the K_S and K_L beams are shown separately. The d_T distribution in K_S decays is scaled geometrically in order to compare directly with K_L.

The signal region is taken to be $d_T < 5$ cm, and the control region for the background extrapolation is taken as 7 cm $< d_T <$ 12 cm. The fraction of events in the background region is $(3.6 \pm 0.1) \times 10^{-3}$ of the signal. This background consists mainly of $K_L \to \pi e \nu$ decays and has contributions from $K \to \pi \mu \nu$ where the muon loses its energy by bremsstrahlung in the hadron calorimeter, from $K \to \pi^+\pi^+\pi^0$ where one photon overlaps the shower of one of the charged pions, and a small amount of K_S production in the final K_L collimator. The $K^0 \to \pi^+\pi^-\pi^0$ background is subtracted directly by counting events with identified photons as a function of the distance between the photon and the nearest track. The remaining $K^0 \to \pi e \nu$ candidates are identified by the well-defined electron shower width in the photon calorimeter and longitudinal energy deposition, and the $K^0 \to \pi \mu \nu$ candidates by the shower width in the hadron calorimeter for events in

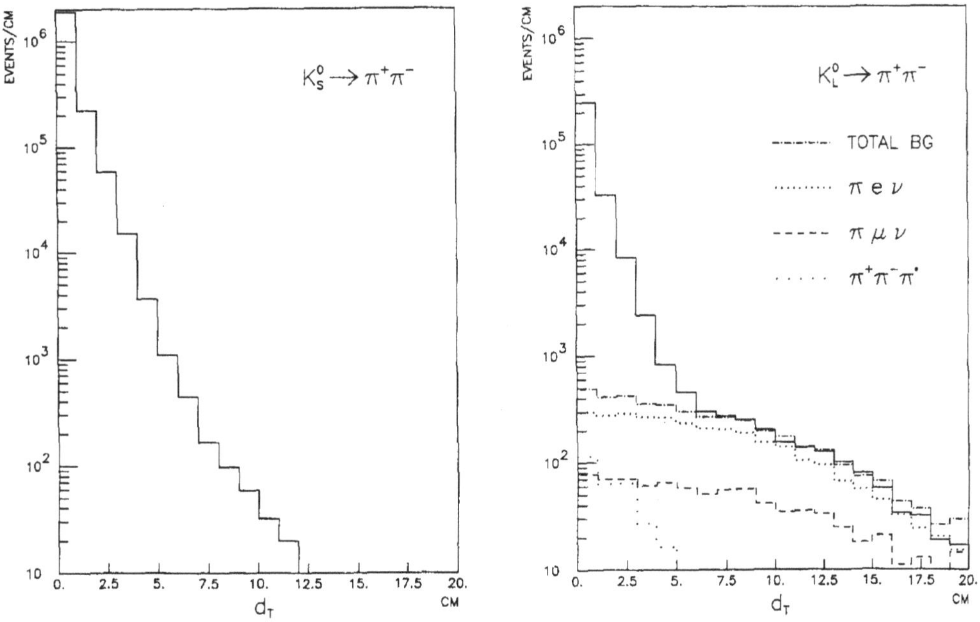

Fig. 3. Event distribution for charged decays as a function of distance d_T between the decay plane and the production target, for K_S and K_L decays and for various background components.

Table 1
Background composition for $K_L \to \pi^+\pi^-$ decays.

Background	Control region	Signal region
$K^0 \to \pi^+\pi^-\pi^0$	$(0.1 \pm 0.1) \times 10^{-3}$	$(1.0 \pm 1.0) \times 10^{-3}$
$K^0 \to \pi e \nu$	$(2.8 \pm 0.2) \times 10^{-3}$	$(4.4 \pm 0.3) \times 10^{-3}$
$K^0 \to \pi \mu \nu$	$(0.5 \pm 0.2) \times 10^{-3}$	$(0.7 \pm 0.3) \times 10^{-3}$
regenerated K_S	$(0.2 \pm 0.1) \times 10^{-3}$	$(0.4 \pm 0.2) \times 10^{-3}$
total	$(3.6 \pm 0.1) \times 10^{-3}$	$(6.5 \pm 2.0) \times 10^{-3}$

which both charged particles deposit less than 5 GeV in the photon calorimeter. The shape of the d_T distributions for the background events is determined from data samples for $K^0 \to \pi e \nu$ and residual $K^0 \to \pi^+\pi^-\pi^0$ decays, and by Monte Carlo for $K^0 \to \pi\mu\nu$ decays. The inelastic K_S regeneration on the K_L beam collimator has been determined from events with the vertex close to the collimator. The average background subtracted by extrapolation in d_T is $(6.5 \pm 2.0) \times 10^{-3}$, including systematic uncertainties (see table 1). In the case of K_S the background is negligible.

The total available statistics is $\sim 10^6$ K_L and $\sim 10^7$ K_S two-pion decays. The energy spectra of accepted $2\pi^0$ and $\pi^+\pi^-$ events are shown in fig. 4. After recon-

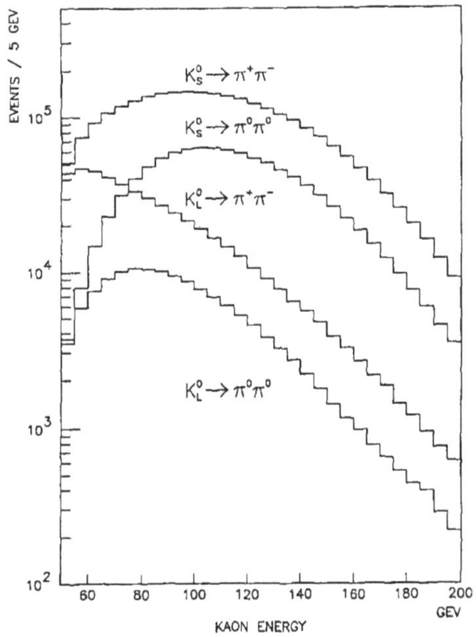

Fig. 4. Energy spectra for K_S and K_L decays into two pions.

struction, the relative energy scales of neutral and charged decays are adjusted to be the same within $\pm 10^{-3}$ by fits of the vertex distributions to the position of the anticounter in the K_S beam (see fig. 5). In this analysis, the data were selected in the energy range 70–170 GeV and with vertices between 10.5 and 48.9 m from the position of the final K_L beam collimator. A breakdown of event statistics is given in table 2 second column. The double ratio is evaluated in 10×32 bins in energy and vertex position, for each of 16 self-contained data sets of K_S and K_L. The weighted average, corrected for acceptance and resolution, is $R = 0.977 \pm 0.004$ (statistical error). In principle, the detection efficiencies for the two decay modes cancel. A Monte Carlo calculation has been used to determine the acceptance ratio. It includes the effect of the known difference in K_S and K_L beam divergences (0.7%) and the scattering of the K_S beam in the anticounter and collimator (0.3%, as measured from events without centre-of-gravity and d_T cuts, see table 2, fourth column), and the effects due to finite bin size and to energy and vertex resolution. The net total Monte Carlo correction amounts to 0.3% on R for the weighted average of all bins.

The trigger system and analysis procedures are designed such that no significant bias should result in the events retained for analysis. Event losses due to inefficiencies of the pretrigger hodoscope counters and of the trigger system itself are measured using a sample of events with relaxed trigger conditions. Since those results are consistent with the expectation of no bias (see table 2, fifth and sixth columns), no correction has been applied. Gains and losses of good events due to accidentals are measured by overlaying a sample of events with events taken with a random trigger, at a rate proportional to the neutral beam intensity. These are primarily due to the cuts on extra space points in the first wire chamber on the number of photons. An asymmetry between charged and neutral decays is observed (table 2, last column), and a correction is applied $(0.34 \pm 0.1\%)$. This asymmetry is mainly due to a loss of charged decays in the K_S beam caused by additional background in the first chamber. Another correction of 0.06% accounts for the difference in efficiency, for charged and neutral decays, of the anticounter in the K_S beam.

The various systematic uncertainties are listed in table 3. Of these, the dominant ones are due to a pos-

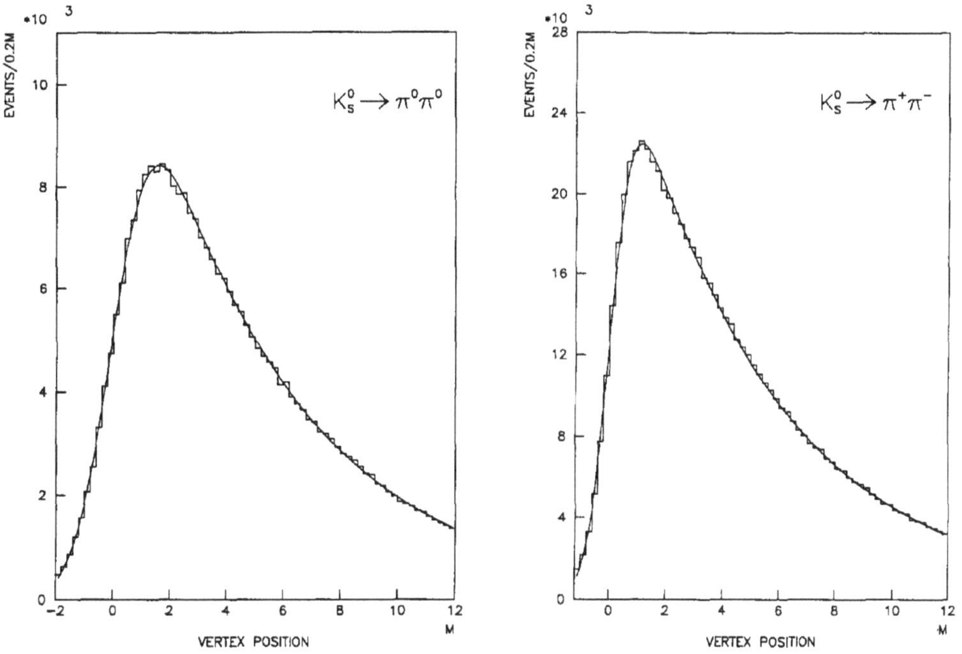

Fig. 5. $K_S \to 2\pi^0$ and $K_S \to 2\pi^+\pi^-$ event distributions as a function of distance from the anticounter in the K_S beam. The continuous lines show the best fits to the data.

sible energy scale difference between charged and neutral decays, the subtraction of backgrounds in the K_L beam, and the net losses of good events by accidentals. After all corrections we obtain the final result, $R=0.980\pm0.004\pm0.005$, with statistical and systematic uncertainties given separately. With combined errors this corresponds to $\mathrm{Re}\,\epsilon'/\epsilon=(3.3\pm1.1)\times10^{-3}$. This is the first time that evidence of *CP*-violating effects is seen in the decay of the *CP*-odd K_2 into two pions, as implied by a non-zero value of ϵ'. It is at the level predicted recently by several evaluations of the standard model for a t-quark mass in the range 50–100 GeV [14] and does not agree with the superweak model [8].

We are indebted to our colleagues L. Bertanza, F. Eisele and P. Steffen for their participation in the early stages of the experiment. We wish to thank the tech-

Table 2
Event statistics and corrections.

	Signal events (×1000)	Background (%)	Scattering (%)	Inefficiencies pretrigger (%)	Inefficiencies trigger (%)	Accidental losses (%)
$K_L \to 2\pi^0$	109	4.0	<0.1	0.06 ± 0.06	0.20 ± 0.10	$2.6\ \pm0.07$
$K_L \to \pi^+\pi^-$	295	0.6		0.37 ± 0.07	0.05 ± 0.06	$2.6\ \pm0.05$
$K_S \to 2\pi^0$	932	<0.1	0.3	0.04 ± 0.02	0.12 ± 0.03	$2.5\ \pm0.05$
$K_S \to \pi^+\pi^-$	2300	<0.1		0.48 ± 0.03	0.01 ± 0.01	$2.8\ \pm0.05$
effect on *R*			0.3	-0.12 ± 0.10	-0.03 ± 0.12	-0.34 ± 0.10

Table 3
Systematic uncertainties on the double ratio R (in %).

background subtraction for $K_L \to 2\pi^0$	0.2
background subtraction for $K_L \to \pi^+\pi^-$	0.2
$2\pi^0/\pi^+\pi^-$ difference in energy scale	0.3
regeneration in the K_L beam	<0.1
scattering in the K_S beam	0.1
K_S anticounter inefficiency	<0.1
difference in K_S/K_L beam divergence	0.1
calorimeter instability	<0.1
Monte Carlo acceptance	0.1
gains and losses by accidentals	0.2
pretrigger and trigger inefficiency	0.1
total systematic uncertainty	±0.5%

nical staff of the participating laboratories and universities for their dedicated effort in establishing the beam, the detectors, and the data-acquisition system for the experiment. We are indebted, in particular, to H.W. Atherton, C. Bizeau, F. Blin, M. Clément, G. Di Tore, G. Dubail, G. Dubois, G.P. Ferri, G. Fersuella, D. Guyon, R. Harfield, D. Jacobs, G. Juban, Ch. Lasseur, G. Laverrière, P. Le Cossec, K.D. Lohmann, A. Lovell, R. Maleyran, M. Marin, M.L. Mathieu, R. McLaren, L. Pregernig, P. Ponting, E.M. Rimmer, P. Schilly and B. Tomat (CERN); H.J. Büttner, U. Dretzler, K. Noffke and K. Wydinski (Institute of Physics, University of Dortmund); P. McInnes and A. Main (Physics Department, University of Edinburgh); the UK Science and Engineering Research Council, and the Rutherford Appleton Laboratory; C. Arnault, A. Bellemain, R. Bernier, A. Bozzone, J.P. Coulon, J.C. Drulot, J.P. Marolleau, E. Plaige, J.P. Richer and A. Roudier (Linear Accelerator Laboratory, Orsay); the IN2P3 Computing Centre at Lyons; C. Avanzini, R. Fantechi, S. Galeotti, G. Gennaro, F. Morsani, G. Pagani, D. Passuello, R. Ruberti, P. Salvadori and L. Zaccarelli (Department of Physics and INFN, University of Pisa); G. Iksal, M. Roschangar and R. Seibert (Physics Department, University of Siegen).

References

[1] J.H. Christenson et al., Phys. Rev. Lett. 13 (1964) 138.
[2] T.T. Wu and C.N. Yang, Phys. Rev. Lett. 13 (1964) 380.
[3] C. Geweniger et al., Phys. Lett. B 48 (1974) 487;
R. Messner et al., Phys. Rev. Lett. 30 (1973) 876;
D.P. Coupal et al., Phys. Rev. Lett. 55 (1985) 566.
[4] C. Geweniger et al., Phys. Lett. B 52 (1974) 119.
[5] J.K. Black et al., Phys. Rev. Lett. 54 (1985) 1628;
R.H. Bernstein et al., Phys. Rev. Lett. 54 (1985) 1631;
M. Woods et al., Chicago University preprint EFI 88-03, submitted to Phys. Rev. Lett. (1988).
[6] J.H. Christenson et al., Phys. Rev. Lett. 43 (1979) 1209.
[7] C. Geweniger et al., Phys. Lett. B 48 (1974) 483.
[8] L. Wolfenstein, Phys. Rev. Lett. 13 (1964) 562.
[9] M. Kobayashi and K. Maskawa, Prog. Theor. Phys. 49 (1973) 652.
[10] J. Ellis, M.K. Gaillard and D.V. Nanopoulos, Nucl. Phys. B 109 (1976) 213;
F.J. Gilman and M.B. Wise, Phys. Lett. B 83 (1979) 83;
Phys. Rev. D 20 (1979) 2392;
B. Guberina and R. Peccei, Nucl. Phys. B 163 (1980) 289;
L. Wolfenstein, Annu. Rev. Nucl. Sci. 36 (1986) 137, and references therein.
[11] N.N. Biswas et al., Phys. Rev. Lett. 47 (1981) 1378.
[12] H. Burkhardt et al., The beam and detector for a high precision measurement of *CP* violation in neutral kaon decays, preprint CERN-EP/87-166 (1987), Nucl. Instrum. Methods, to be published.
[13] H. Burkhardt et al., Phys. Lett. B 199 (1987) 139.
[14] A.J. Buras and J.-M. Gerard, ϵ'/ϵ in the standard model, Munich preprint MPI-PAE/PTh 84-87 (1987), and references therein.

Inclusive Interactions of High-Energy Neutrinos and Antineutrinos in Iron

J. G. H. de Groot, T. Hansl, M. Holder, J. Knobloch, J. May, H. P. Paar,
P. Palazzi, A. Para, F. Ranjard, D. Schlatter, J. Steinberger, H. Suter,
W. von Rüden, H. Wahl, S. Whitaker, E. G. H. Williams, F. Eisele,
K. Kleinknecht, H. Lierl, G. Spahn, H. J. Willutzki, W. Dorth, F. Dydak,
C. Geweniger, V. Hepp, K. Tittel, J. Wotschack, P. Bloch, B. Devaux,
S. Loucatos, J. Maillard, J. P. Merlo, B. Peyaud, J. Rander,
A. Savoy-Navarro, R. Turlay and F. L. Navarria

Zeitschrift für Physik C - Particles and Fields
Volume 1, 143–162 (1979)
https://doi.org/10.1007/BF01445406
Received: 13 December 1978
Issue Date: June 1979

Abstract

We present results on charged current inclusive neutrino and antineutrino scattering in the neutrino energy range 30–200 GeV. The results include a) total cross-sections; b) y distributions; c) structure functions; and d) scaling violations observed in the structure functions. The results, as well as their comparison with the results of electron and muon inclusive scattering, are in agreement with the expectations of the quark parton model and QCD.

[Note: The full text manuscript is not included in this memorial volume.]

Volume 82B, number 3,4 PHYSICS LETTERS 9 April 1979

QCD ANALYSIS OF CHARGED-CURRENT STRUCTURE FUNCTIONS

J.G.H. de GROOT, T. HANSL, M. HOLDER, J. KNOBLOCH, J. MAY, H.P. PAAR, P. PALAZZI, A. PARA, F. RANJARD, D. SCHLATTER, J. STEINBERGER, H. SUTER, W. von RÜDEN, H. WAHL, S. WHITAKER and E.G.H. WILLIAMS
CERN, Geneva, Switzerland

F. EISELE, K. KLEINKNECHT, H. LIERL, G. SPAHN and H.J. WILLUTZKI
Institut für Physik [1] der Universität, Dortmund, Germany

W. DORTH, F. DYDAK, C. GEWENIGER, V. HEPP, K. TITTEL and J. WOTSCHACK
Institut für Hochenergiephysik [1] der Universität Heidelberg, Germany

P. BLOCH, B. DEVAUX, S. LOUCATOS, J. MAILLARD, J.P. MERLO, B. PEYAUD, J. RANDER, A. SAVOY-NAVARRO and R. TURLAY
D.Ph.P.E., CEN-Saclay, France

and

F.L. NAVARRIA
Istituto di Fisica dell'Università, Bologna, Italy

Received 19 December 1978

The structure functions $F_2(x, Q^2)$ and $xF_3(x, Q^2)$ measured in high-energy neutrino charged-current interactions on nuclei are compared with QCD predictions. Solutions to the moment equations of QCD are found which are in good agreement with the data and yield simple parametrisations of the structure functions. For the scale parameter Λ we find Λ = 0.5 ± 0.2 GeV. The analysis also results in values for the width of the gluon distribution as a function of Q^2. We find $\langle x \rangle_{\text{gluons}} = 0.16 \pm 0.03$ for $Q^2 = 10\ \text{GeV}^2$.

In this letter we give the results of an analysis of the structure functions $F_2(x, Q^2)$ and $xF_3(x, Q^2)$ measured in high-energy neutrino interactions based on the predictions of quantum chromodynamics (QCD) [1]. A description of the experimental procedure and the experimental results has been submitted for publication elsewhere [2]. The data on $F_2(x, Q^2)$ and $xF_3(x, Q^2)$ relevant for this analysis, corrected for resolution, acceptance, radiative effects and non-isoscalarity of the target are reproduced in figs. 1a and 1b. The errors given are statistical errors and do not include an overall scale error of ±6% for F_2 and of ±8% for xF_3 which is due mainly to the uncertainty in the neutrino flux. Additional systematic point to point errors are estimated to be smaller or at most equal to the statistical errors.

The predictions of QCD which are tested and exploited in this analysis concern the evolution of the structure functions with Q^2 once these are measured for a fixed value Q_0^2. These predictions are unambiguous for high Q^2 where the strong interaction constant $\alpha_s(Q^2)$ is small compared to unity and where quark and target mass effects can be neglected. The QCD

[1] Supported by the Bundesministerium für Forschung and Technology.

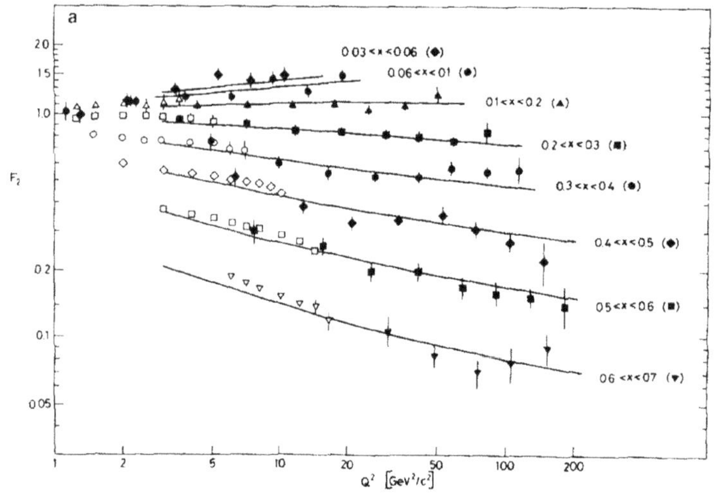

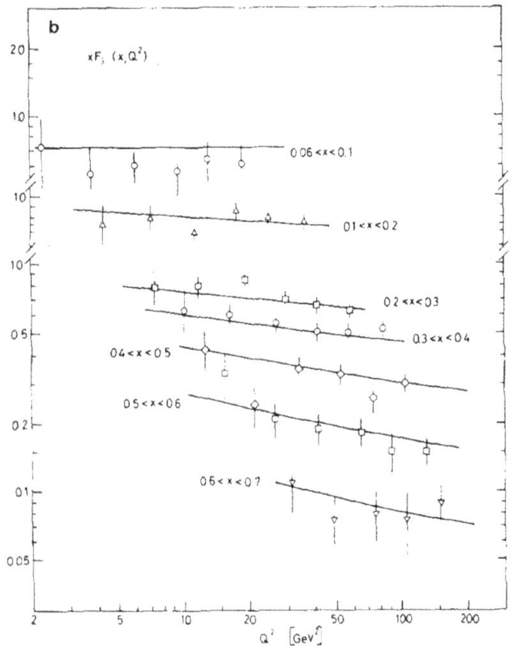

Fig. 1. (a) F_2 in different x bins as a function of ln Q^2 (solid symbols). The values of F_2 measured in electron–deuteron scattering are also shown as open symbols. The solid lines represent the QCD fit to the neutrino data. (b) Same as (a) for the structure function $xF_3(x, Q^2)$.

predictions are expressed in terms of the Q^2 dependences of the moments of the structure functions which are given in eqs. (1), (2) and (3):

$$M_3(n, Q^2) = M_3(n, Q_0^2)\exp(-\gamma^n_{\psi\psi} s)\,, \qquad (1)$$

$$\begin{aligned} M_2(n, Q^2) &= [(1-\alpha_n)M_2(n, Q_0^2) - \beta_n M_G(n, Q_0^2)]\exp(-\gamma^n_+ s) \\ &+ [\alpha_n M_2(n, Q_0^2) + \beta_n M_G(n, Q_0^2)]\exp(-\gamma^n_- s)\,, \qquad (2) \end{aligned}$$

$$\begin{aligned} M_G(n, Q^2) &= [\alpha_n M_G(n, Q_0^2) - \epsilon_n M_2(n, Q_0^2)]\exp(-\gamma^n_+ s) \\ &+ [(1-\alpha_n)M_G(n, Q_0^2) + \epsilon_n M_2(n, Q_0^2)]\exp(-\gamma^n_- s)\,. \qquad (3) \end{aligned}$$

Here

$$M_3(n, Q^2) = \int_0^1 x^{n-2} xF_3(x, Q^2)\,\mathrm{d}x\,,$$

$$M_2(n, Q^2) = \int_0^1 x^{n-2} F_2(x, Q^2)\,\mathrm{d}x\,,$$

$$M_G(n, Q^2) = \int_0^1 x^{n-2} G(x, Q^2)\,\mathrm{d}x\,,$$

$$s = \ln[\ln(Q^2/\Lambda^2)/\ln(Q_0^2/\Lambda^2)]\,.$$

$G(x, Q^2)$ is the momentum distribution of the gluons inside the nucleon and $\alpha_n, \beta_n, \epsilon_n, \gamma^n_{\psi\psi}, \gamma^n_\pm$ are numbers predicted by QCD. We refer to ref. [3] for further details. Λ is the characteristic length of the strong interaction which is related to the effective strong-interaction coupling constant

$$\alpha_s(Q^2) = 12\,\pi/(33 - 2m_f)\ln(Q^2/\Lambda^2)\,,$$

where m_f is the number of excited flavours which is taken to be four.

Eqs. (1)–(3) for the moments of the structure functions are not easily translated into predictions for the Q^2-dependence of the structure functions themselves. This is especially true for $F_2(x, Q^2)$ whose development with Q^2 depends not only on quark distributions but also on the gluon distribution which cannot be measured directly in lepton–nucleon interactions.

In our analysis we follow closely a method developed by Buras and Gaemers [3]. These authors have demonstrated that the parametrisation

$$xF_3(x,Q^2) = [3/B(\eta_1,\eta_2+1)]\,x^{\eta_1(s)}(1-x)^{\eta_2(s)}\,,$$

with linear s dependence of the exponents,

$$\eta_1(s) = \eta_{10} + \eta_{11}s\cdot\tfrac{4}{25}\,,$$

$$\eta_2(s) = \eta_{20} + \eta_{21}s\cdot\tfrac{4}{25}\,,$$

can satisfy the QCD moment relations in the present range of s for the first ten moment equations with sufficient accuracy. The normalization factor $3/B$, where B is Euler's beta function is imposed by the requirement of having three valence quarks inside the nucleon, $\int_0^1 F_3(x)\,dx = 3$.

QCD-fits to xF_3. The fits to the valence structure function have been performed for different values of Q_0^2 giving consistent results for $\eta_1(Q^2)$, $\eta_2(Q^2)$ and Λ. The data used consist of all measurements of xF_3 as well as the measurements of F_2 for x greater than 0.4, always for $Q^2 > 3$ GeV$^2/c^2$. It is possible to justify the inclusion of the F_2-data which are more precise than xF_3-data, by the experimental fact that the sea-quark distribution is negligible in this region of large x.

All points used in the fit correspond to values of $M^2x^2/Q^2 < 0.03$ such that corrections due to target mass effects are expected to be small [4].

For $Q_0^2 = 20$ GeV$^2/c^2$ which corresponds to the average Q^2 of our data we obtain the following results with $\chi^2/\mathrm{DF} = 56/63$:

$$\eta_1 = (0.51 \pm 0.02) - 0.83s\cdot\tfrac{4}{25}\,,$$

$$\eta_2 = (3.03 \pm 0.09) + 5.0s\cdot\tfrac{4}{25}\,,$$

$$\Lambda = 0.55 \pm 0.15\ (\pm 0.1\ \text{systematic})\ \text{GeV}\,.$$

This simple parametrisation gives a good fit to our data as shown in figs. 1a and 1b. The value of Λ, which is the most sensitive parameter, is affected by the uncertainty in the $\bar{\nu}/\nu$ flux ratio of $\pm 5\%$, the uncertainty of $\pm 2\%$ in the K/π ratio [2] and most probably by higher-order effects which have been neglected in this analysis. The overall uncertainty for Λ due to systematic errors in the data is estimated to be ± 0.1.

Common QCD fits to F_2 and xF_3. The QCD prediction for $F_2(x,Q^2)$ depends on the gluon distribution which is not measured directly except for its second moment $M_G(2,Q^2)$ which is obtained from four-momentum conservation: $M_G(2,Q^2) = 1 - M_2(2,Q^2)$. Given the limited accuracy of our data, we have chosen to follow the approximate method of ref. [3] by using the parametrisation

$$F_2(x,Q^2) = xF_3(x,Q^2) + A(s)(1-x)^{P(s)}\,.$$

This simple parametrisation cannot satisfy all QCD moment equations for F_2 but is reasonably good for a narrow sea distribution [5]. It has the advantage that $A(s)$ and $P(s)$ can be calculated from the second and third moment equation for F_2 with only three additional parameters: $A(0)$, $P(0)$ and $M_G(3,Q_0^2)$.

Common fits for $F_2(x,Q^2)$ and $xF_3(x,Q^2)$ have been obtained. The charm content of the sea has been fixed to be zero at $Q_0^2 = 1.8$ GeV$^2/c^2$ following ref. [3].

For $Q_0^2 = 5$ GeV$^2/c^2$ wj obtain with $\chi^2/\mathrm{DF} = 101/89$:

$$\eta_1 = (0.56 \pm 0.2) - 0.92s\cdot\tfrac{4}{25}\,,$$

$$\eta_2 = (2.71 \pm 0.11) + 5.08s\cdot\tfrac{4}{25}\,,$$

$$A(0) = 0.99 \pm 0.07\,,\qquad P(0) = 8.1 \pm 0.7\,,$$

$$M_G(2,Q_0^2=5) = 0.515\,,$$

$$M_G(3,Q_0^2=5) = 0.105 \pm 0.02\,,$$

$$\Lambda = 0.47 \pm 0.11\ (\pm 0.1\ \text{systematic})\ \text{GeV}\,.$$

Volume 82B, number 3,4 PHYSICS LETTERS 9 April 1979

Again the agreement with the data is good, as shown in fig. 1a, and the parameters for the valence part and Λ are in good agreement with the values found in the fit to xF_3 only. The measurements of F_2^{eD} obtained in eD scattering experiments [6] (not used in the fit) are also shown in fig. 1a multiplied by 9/5. The QCD fits to the neutrino data describe these eD data equally well within the relative normalisation errors of the two experiments.

Q^2-dependence of gluon moments. The Q^2-dependence of the second and third moments for all nucleon constituents can be calculated from their fitted values for $Q_0^2 = 5\ \mathrm{GeV}^2/c^2$ given above using the moment equations (1), (2) and (3). It should be noted, however, that the QCD fit to F_2 and xF_3 makes no use of eq. (3) for the moments of the gluon distribution except for $M_G(2, Q^2)$ which is required to satisfy four-momentum conservation. $M_G(3, Q_0^2)$ is obtained in the fit as a free parameter which is relevant for the Q^2-dependence of the sea quarks. Eq. (3) for the gluon moments is trivially satisfied, if eq. (2) is satisfied [‡1]. This allows a consistency check of the fitting method for the low-x region, where the method is approximative only. This is done by repeating the fit to $F_2(x, Q^2)$ for different values of Q_0^2, keeping the parameters for the valence distribution and Λ fixed to the values obtained at $Q_0^2 = 5\ \mathrm{GeV}^2/c^2$. The only free parameters left are then the second and third moment of the sea distribution and the third moment of the gluon distribution at $Q^2 = Q_0^2$. The results of these fits are summarized in fig. 2, where the total fractional momentum of all constituents and their average value $\langle x \rangle$ are shown for the Q^2 range $3 \leqslant Q^2 \leqslant 20\ \mathrm{GeV}^2/c^2$ for which this experiment has relevant data in the sea region.

These QCD fits give a definite increase of the total momentum carried by the sea quarks, a decrease of the total momentum of quarks plus antiquarks and a corresponding increase of the momentum carried by the gluons. The average x of the gluon distribution $\langle x \rangle_G$ (points with error bars in fig. 2) is found comparable to that of the total quark distribution at $Q^2 = 5\ \mathrm{GeV}^2/c^2$ and shows a substantial shrinking with increasing Q^2. The dotted line gives the QCD prediction

[‡1] We wish to thank C. Sachrajda for calling our attention to this fact.

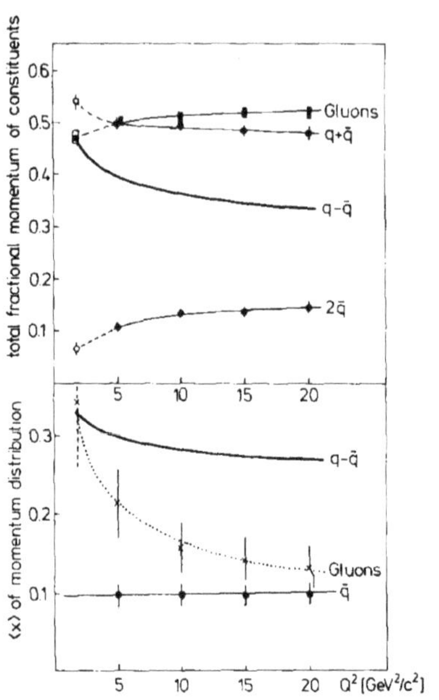

Fig. 2. The total fractional momentum of all nucleon constituents and the average value of their momentum distributions as a function of Q^2 as obtained from QCD fits to $xF_3 = q - \bar{q}$ and $F_2 = q + \bar{q}$. The points at $Q^2 = 1.8\ \mathrm{GeV}^2/c^2$ shown as open symbols suffer from a larger systematic uncertainty since they require an extrapolation to a Q^2-region where higher-order effects are important. The dotted line gives the QCD prediction for the third gluon moment which was not used in the fit.

for $\langle x \rangle_G$ which is calculated from the moment equation (3) (which was not used in the fits) starting with the measured values $M_2(3, Q_0^2 = 5\ \mathrm{GeV}^2/c^2) = 0.127$ and $M_G(3, Q_0^2 = 5\ \mathrm{GeV}^2/c^2) = 0.105$. It should be noted that in the present fitting method, every point shown in fig. 2 represents the same data and carries therefore the full satistical weight of all data, such that there are no statistical point to point errors. The good agreement between the fitted values of $\langle x \rangle$ and the calculated QCD curve gives confidence that the approximate method used is able to describe the Q^2-dependence of the sea distribution quite well in our Q^2-range. Since the gluon distribution cannot be directly measured, the results of this analysis for $\langle x \rangle_G$ are of interest. At, e.g., $Q^2 = 10$ GeV we find $\langle x \rangle_G = 0.16 \pm 0.03$, which can be compared with $\langle x \rangle_{q-\bar{q}} = 0.28 \pm 0.01$ and $\langle x \rangle_{\bar{q}} = 0.10 \pm 0.01$. The width of the gluon distribution has been studied earlier in an analysis of muon and electron production data with similar results [7].

The points for the total momentum of the constituents in fig. 2 have a common scale error of ±6% and all errors given are relative errors with respect to the valence quarks $q - \bar{q}$. The measurements of the average x of the gluon distribution show systematic variations with different cuts in Q^2 to the data. This is probably due to the approximate nature of the fitting method. They are also slightly dependent on the assumptions about the charmed sea and variations of Λ and the valence structure function within errors. These effects can lead to an overall shift of all points by about one standard deviation but do not change the Q^2-behaviour.

In summary, parametrized solutions of the Buras–Gaemers type to the QCD moment equations can be found which are in good agreement with the neutrino and electron deep inelastic scattering data. This constitutes an important quantitative confirmation of the validity of QCD. For the gluon constituents of the nucleon this fit gives the total fractional momentum and the average value of the gluon momentum distribution.

We thank A.J. Buras, K.J.F. Gaemers and E. Reya for fruitful discussions.

References

[1] For a general review of QCD see: H.D. Politzer, Phys. Rep. 14C (1974) 129;
W. Marciano and H. Pagels, Phys. Rep. 36C (1977) 137.
[2] J.G.H. de Groot et al., Inclusive charged current interactions of high-energy neutrinos and antineutrinos in iron, submitted to Z. Phys.
[3] A.J. Buras and K.J.F. Gaemers, Nucl. Phys. B132 (1978) 249.
[4] O. Nachtmann, Nucl. Phys. B63 (1973) 237; B78 (1974) 455.
[5] J.F. Owens and E. Reya, Phys. Rev. D17 (1978) 3003.
[6] E.M. Riordan et al., SLAC-PUB-1634 (1975).
[7] H.L. Anderson et al., Phys. Rev. Lett. 40 (1978) 1061.

Z. Phys. C – Particles and Fields 12, 289–295 (1982)

Zeitschrift für Physik C Particles and Fields

Determination of the Gluon Distribution in the Nucleon from Deep Inelastic Neutrino Scattering

H. Abramowicz[2], F. Dydak, J. G. H. de Groot, J. Knobloch, J. May, P. Palazzi, A. Para[2], F. Ranjard, D. Schlatter, J. Steinberger, H. Taureg, W. von Rüden, H. Wahl, J. Wotschack
CERN, CH-1211 Geneva, Switzerland

J. Duda, F. Eisele, H. P. Klasen, K. Kleinknecht, B. Pszola, B. Renk, H. J. Willutzki
Institut für Physik[1] der Universität, D-4600 Dortmund, Federal Republic of Germany

T. Flottmann, C. Geweniger, J. Królikowski[2], J. Rothberg[3], K. Tittel
Institut für Hochenergiephysik[1] der Universität, D-6900 Heidelberg, Federal Republic of Germany

C. Guyot, J. P. Merlo, B. Peyaud, J. Rander, J. P. Schuller, R. Turlay
D.Ph. P. E., CEN-Saclay, F-91190 Gif-sur-Yvette, France

J. T. He, T. Z. Ruan, W. M. Wu
Institute of High-Energy Physics, Beijing, People's Republic of China

Received 21 December 1981

Abstract. The observed scaling violations of the nucleon structure functions F_2 and $\bar{q}$ have been analysed in the framework of perturbative QCD to determine the shape and magnitude of the gluon distribution. The data are in good agreement with leading order QCD, and the simultaneous use of F_2 and $\bar{q}$ structure functions permits, for the first time, a reliable determination of the gluon structure function.

1. Introduction

It is well known that quarks and antiquarks carry only about 50% of the nucleon momentum, the other half being attributed to gluons. Apart from their task of binding the hadrons, gluons are expected to show up directly as elementary hadron constituents in hard hadron-hadron scattering processes such as Drell-Yan μ-pair production, direct photon production, and high-p_T scattering, and in heavy quark production by photons, muons, and neutrinos. It is therefore important to know the gluon distribution in order to be able to make predictions for these reactions. In deep inelastic lepton scattering, gluons do not take part directly in the elementary scattering processes. However, QCD predicts that their interaction with quarks inside the nucleon leads to characteristic scaling violations in the observable structure functions via gluon bremsstrahlung and quark pair production. If the observed scaling violations are predominantly due to QCD effects, the measured Q^2-slopes of the structure functions are directly related to the gluon distribution and can be used to determine its shape and magnitude.

2. Determination of Structure Functions F_2 and $\bar{q}^{\bar{\nu}}$

The present analysis is based on the measurement of the structure functions $F_2(x, Q^2)$ and $\bar{q}^{\bar{\nu}}(x, Q^2) = x(\bar{u}+\bar{d}+2\bar{s})$ by high-energy neutrino and antineutrino interactions in iron, using the CERN-Dortmund-

1 Supported by the Bundesministerium für Forschung und Technologie, Bonn, FRG
2 On leave from the Institute for Nuclear Research, Warsaw, Poland
3 On leave from the Dept. of Physics, University of Washington, Seattle, USA

0170-9739/82/0012/0289/$01.40

Heidelberg-Saclay (CDHS) detector [1]. The structure function F_2 is measured using narrow-band beam data only, whereas the measurement of $\bar{q}$, which needs high statistics, uses wide-band beam data in addition. In total we use 94,000 neutrino and 25,000 antineutrino charged-current events from narrow-band beams and 35,000 neutrino and 155,000 antineutrino events from wide-band beams. The data selection, extraction of differential cross-sections, and the determination of structure functions will be described in detail in a forthcoming paper [2].

The structure function F_2 is obtained from the sum of neutrino and antineutrino cross-sections, whereas $\bar{q}^{\bar{v}}$ is measured by antineutrino scattering at high y:

$$F_2^{\nu N} \equiv x(u+d+s+c+\bar{u}+\bar{d}+\bar{s}+\bar{c})(1+R)$$
$$= \frac{1+R}{1+(1-y)^2}\left\{\frac{\pi}{\mathrm{MEG}^2}\left[\frac{d^2\sigma^\nu}{dxdy}+\frac{d^2\sigma^{\bar{\nu}}}{dxdy}\right]\right.$$
$$\left.-2F_L(1-y)-4x\mathcal{S}-4xc(1-y)^2\right\}$$
$$+x(s+\bar{s}+c+\bar{c})(1+R), \tag{1}$$

$$\bar{q}^{\bar{\nu}} \equiv x(\bar{u}+\bar{d}+2\mathcal{S})$$
$$= \frac{1}{1-(1-y)^4}\left\{\frac{\pi}{\mathrm{MEG}^2}\left[\frac{d^2\sigma^{\bar{\nu}}}{dxdy}-(1-y)^2\frac{d^2\sigma^\nu}{dxdy}\right]\right.$$
$$-F_L[(1-y)-(1-y)^3]$$
$$\left.+(2x\mathcal{S}-2xc)[(1-y)^2-(1-y)^4]\right\}. \tag{2}$$

Here $F_L = F_2 - 2xF_1$ is the longitudinal structure function related to $R=\sigma_L/\sigma_T = F_L/2xF_1$; u, d, s, c are the up, down, strange, and charmed quark densities in the proton; and $x\mathcal{S}(x,Q^2)$ is the strange-quark structure function in the nucleon as seen by the weak charge current. Note that in our Q^2-range $x\mathcal{S}(x,Q^2)$ exhibits a threshold behaviour in contrast to $xs(x,Q^2)$. This is due to the fact that the transition $s \to c$ is Cabibbo favoured and the charmed quark in the final state has a high mass. For the sake of transparency, mass correction terms of the order of Q^2/ν^2, which were included in the analysis, have been omitted in (1) and (2).

To extract the structure functions $F_2^{\nu N}$ and $\bar{q}^{\bar{\nu}}$ from the measured differential cross-sections we have to know $R=\sigma_L/\sigma_T$, and we need information about the shape and magnitude of charmed and strange sea and the effect of charm threshold. We take $R(x,Q^2)$ from a previous analysis of the narrow-band beam data [3] and $x\mathcal{S}(x,Q^2)$ from an analysis [4] of opposite sign dimuon data. Since the uncertainties in both quantities are large, the present analysis uses different sets of structure functions $F_2^{\nu N}$ and $\bar{q}^{\bar{\nu}}$ which are extracted under various assumptions about the strange and charmed sea and the magnitude of $R(x,Q^2)$.

Radiative corrections have been applied according to the prescription of De Rújula et al. [5].

3. Method of Extracting the Gluon Structure Function

The Q^2-evolution of the structure functions F_2, $\bar{q}^{\bar{\nu}}$, and the gluon distribution $G(x)$ due to QCD effects are given by the Altarelli-Parisi equations [6]:

$$\frac{dF_2(x,Q^2)}{d\ln Q^2} = \frac{\alpha_s(Q^2)}{2\pi}\int_x^1\left[P_{qq}\left(\frac{x}{z}\right)F_2(z,Q^2)\right.$$
$$\left.+2N_{F_2}P_{qg}\left(\frac{x}{z}\right)G(z,Q^2)\right]\frac{xdz}{z^2}, \tag{3a}$$

$$\frac{d\bar{q}(x,Q^2)}{d\ln Q^2} = \frac{\alpha_s(Q^2)}{2\pi}\int_x^1\left[P_{qq}\left(\frac{x}{z}\right)\bar{q}(z,Q^2)\right.$$
$$\left.+N_{\bar{q}}P_{qg}\left(\frac{x}{z}\right)G(z,Q^2)\right]\frac{xdz}{z^2}, \tag{3b}$$

$$\frac{dG(x,Q^2)}{d\ln Q^2} = \frac{\alpha_s(Q^2)}{2\pi}\int_x^1\left[F_2(z,Q^2)P_{gq}\left(\frac{x}{z}\right)\right.$$
$$\left.+P_{gg}\left(\frac{x}{z}\right)G(z,Q^2)\right]\frac{xdz}{z^2}. \tag{3c}$$

Here N_{F_2} and $N_{\bar{q}}$ are the number of flavours which contribute to the evolution of the F_2 and $\bar{q}$ structure functions, respectively. Our standard set of structure functions has $N_{F_2}=N_{\bar{q}}=4$. The splitting functions P_{ij} are known (in leading order) and the structure functions F_2 and $\bar{q}$ are measured. Given the structure functions $F_2(x,Q_0^2)$ and $\bar{q}(x,Q_0^2)$ at some Q_0^2, and their Q^2-evolution at this Q_0^2, we can determine

$$\alpha_s(Q_0^2) = \frac{12\pi}{25\ln(Q_0^2/\Lambda_{\mathrm{LO}}^2)}$$

(*or the scale parameter* Λ_{LO}) and the unknown gluon distribution $G(x,Q_0^2)$ on the basis of (3a) and (3b). The predictions of the Altarelli-Parisi equations for the Q^2-evolution of F_2 and $\bar{q}$ are compared directly to the measured slopes of our structure function set *I* (Table 1) in Fig. 1a, b for $Q_0^2 = 4.5\,\mathrm{GeV}^2/c^2$, where the contributions due to gluon bremsstrahlung and gluon pair production are given separately. The curves shown correspond to the best QCD fit to the data as described below. The data are not precise enough to measure the slopes locally (in a small Q^2 range). The data points in Fig. 1 are therefore obtained by linear fits to F_2 in $\ln\ln Q^2$ for fixed x using all available Q^2 values. This procedure is only a rough approximation to the expected QCD behaviour, such that the data points do not have to agree completely with the correct QCD solution. However, they show very well that significant scaling violations are observed both for F_2 and $\bar{q}$, and their pattern is in good agreement with the QCD prediction. A measurement of F_2 alone does not allow a good simultaneous determination of $\alpha_s(\Lambda)$ and the gluon distribution since both are very strongly correlated. If α_s were increased in Fig. 1a, the negative bremsstrahlung contribution to $dF_2/d\ln Q^2$ would become larger than the observed scaling violations at large x. This could be compensated by the positive contribution of a broader gluon distribution. However, the additional measurement of $\bar{q}^{\bar{\nu}}$, whose scaling violations are also very strongly linked to the gluon distribution, allows us to decouple Λ and the width of the gluon distribution. In particular, the convolution $P_{gq}\otimes G(x)$ has to be zero where no antiquarks are observed, i.e. for $x \gtrsim 0.45$.

The quantitative confrontation of the data with QCD is based on the simultaneous numerical integration of (3a)–(3c) using leading-order expressions for α_s and the splitting functions*. Equations (3a)–(3c) constitute a system of coupled ordinary differential equations (one equation for each measured value of x and each structure function). As boundary conditions the structure functions are parametrized for $Q^2 = Q_0^2$ in the following way:

$$\begin{aligned} F_2(x, Q_0^2) &= a_2(1+b_2x)(1-x)^{c_2} \\ G(x, Q_0^2) &= a_g(1+b_gx)(1-x)^{c_g} \\ \bar{q}(x, Q_0^2) &= a_q(1-x)^{c_q}. \end{aligned} \tag{4}$$

The shape parameters a_i, b_i, c_i and the parameter Λ_{LO} are then determined by a least squares fit to the data. The distribution $G(x, Q_0^2)$ cannot be compared directly with the data, in contrast to F_2 and $\bar{q}$. It is therefore important to choose a reasonable parametrization. Our choice (4) is motivated by counting rule predictions at large x. In addition, it allows for substantial variations in shape at small x.

The numerical solution of the Altarelli-Parisi equations has the advantage that it can make full use of all available data and that it depends only weakly on unmeasured kinematical regions at large x. This is because the convolution $\int_x^1 P_{qq}F_i(z)\cdot x/z^2 dz$ gets contributions mainly from z values which are near to x where the structure function is measured.

4. Fit Results

Our standard set of structure functions (set I) [2] is obtained from (1) and (2) assuming that the strange sea and $x(\bar{u}+\bar{d})$ have the same x distributions but that $x\mathscr{S}$

* The method has been introduced by L. F. Abbott and M. Barnett. We use a modified version of their program

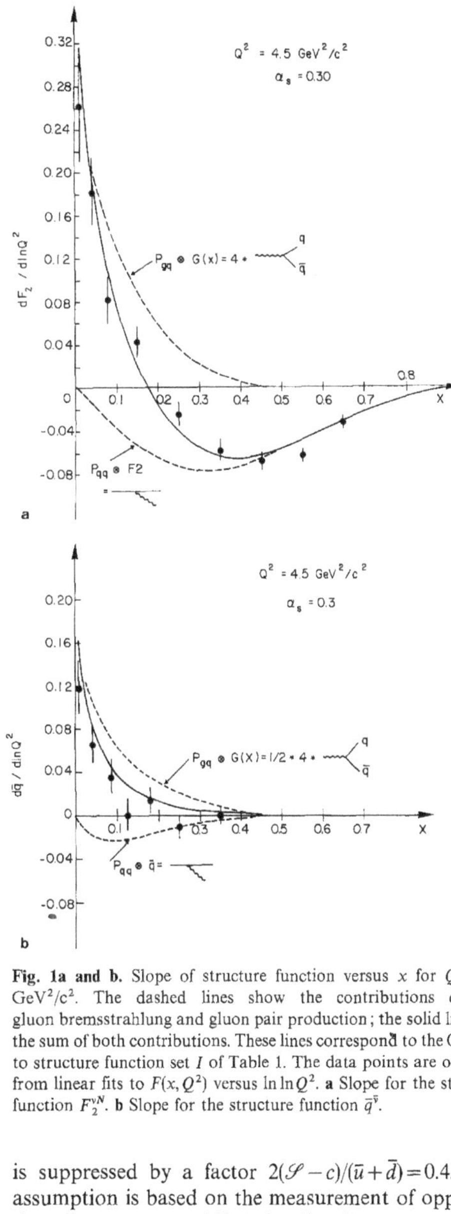

Fig. 1a and b. Slope of structure function versus x for $Q^2 = 4.5$ GeV^2/c^2. The dashed lines show the contributions due to gluon bremsstrahlung and gluon pair production; the solid lines are the sum of both contributions. These lines correspond to the QCD fit to structure function set I of Table 1. The data points are obtained from linear fits to $F(x, Q^2)$ versus $\ln\ln Q^2$. **a** Slope for the structure function $F_2^{\nu N}$. **b** Slope for the structure function $\bar{q}^{\bar{\nu}}$.

is suppressed by a factor $2(\mathscr{S}-c)/(\bar{u}+\bar{d}) = 0.4$. This assumption is based on the measurement of opposite-sign dimuon events [4]. It implies that no correction for charm threshold is done and the contribution of the charmed sea is small. The value of $R = \sigma_L/\sigma_T$ has been fixed to $R = 0.1$ in agreement with the published result [3]. In addition, a small correction has been applied to correct for the non-isoscalarity of the iron nucleus.

In the QCD fits we have used all data points for F_2 and $\bar{q}$ with $Q^2 > 2\,\mathrm{GeV/c}^2$ and the invariant hadron mass squared $W^2 > 11\,\mathrm{GeV}^2$. The latter cut is imposed

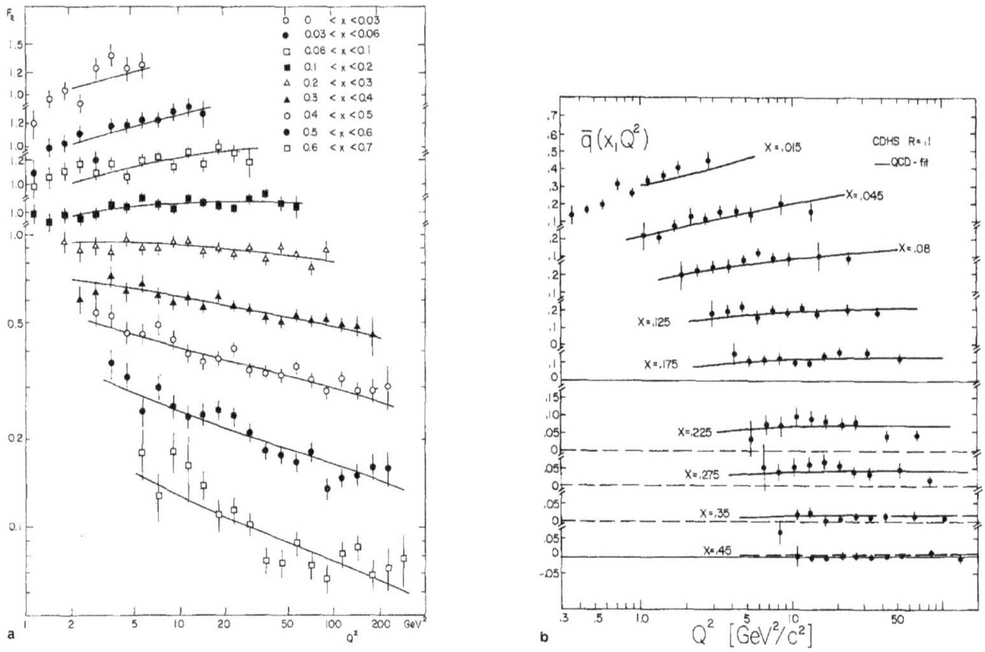

Fig. 2 a and b. The measured structure functions (set *I*) versus Q^2 for different bins in *x*. The solid linees are the result of the leading-orderQCD fit. **a** Structure function $F_2^{\nu N}(x, Q^2)$. **b** Structure function $\bar{q}(x, Q^2)$

to avoid the kinematical region where higher-twist contributions might be important and where second-order and non-leading corrections are substantial. Target mass corrections are applied according to the prescription of Barbieri et al. [7]. We have also included a propagator term with a mass $m_W = 80$ GeV.

In a first step we enforce the energy momentum sum rule, i.e. the total gluon and quark momenta have to add up to the nucleon momentum. We obtain the following fit results for structure function set *I* at $Q_0^2 = 5\,\text{GeV}^2/\text{c}^2$:

Structure function set *I*: $N_{F_2} = 4$, $N_{\bar{q}} = 4$

$$\Lambda_{LO} = 0.18 \pm 0.02$$
$$\langle F_2 \rangle_2 = 0.45 \pm 0.02$$
$$\langle \bar{q} \rangle_2 = 0.055 \pm 0.002$$
$$\langle G \rangle_2 = 1 - \langle F_2 \rangle_2$$
$$\left.\begin{array}{l} \langle x \rangle_{F_2} = 0.25 \pm 0.01 \\ \langle x \rangle_q = 0.095 \pm 0.004 \\ \langle x \rangle_{\text{glue}} = 0.16 \pm 0.012 \end{array}\right\} \begin{array}{l} Q_0^2 = 5\,\text{GeV}^2/\text{c}^2 \\ \chi^2/\text{DF} = 209/196\,. \end{array}$$

Here $\langle F_i \rangle_2$ and $\langle x \rangle_{F_i}$ are the integral and the average value of *x* for the structure function F_i, respectively, and the errors are purely statistical. This QCD fit describes the observed scaling violations very well, as can be seen in Fig. 1a, b and in Fig. 2a, b, which show the structure functions F_2 and $\bar{q}$ versus Q^2 for all *x*-bins compared with the result of this fit. The correlation between Λ and the width of the gluon distribution is small. The obtained shape of the gluon distribution is displayed in Fig. 3a, b for two values of Q^2 and compared with F_2 and $\bar{q}$. It should be noted that the gluon distribution is determined independently at only one value of Q_0^2. It can then be calculated for every other value of Q^2 using the evolution equations. A simple parametrization including the Q^2-dependence is given in the Appendix.

The validity of the energy momentum sum rule can be tested by leaving also the normalization factor for $G(x, Q_0^2)$, a_g, as a free parameter in the fit. At $Q_0^2 = 5\,\text{GeV}^2/\text{c}^2$ we obtain

$$\langle G \rangle_2 = 0.55 \pm 0.11\,, \tag{5}$$

in good agreement with the expectation: $1 - \langle F_2 \rangle = 0.55$. Alternatively, assuming again the energy momentum sum rule, we can check the number of effective flavours by leaving N_{F_2} and $N_{\bar{q}}$ in (3a) and (3b) as free parameters. We obtain $N_{F_2} = 4.4 \pm 0.4$ and $N_{\bar{q}} = 3.8 \pm 0.5$, which is well compatible with the expectation.

The structure function set *I* has also been used to study some systematic dependences. Target mass corrections give a negligible effect in this range of W^2. We note that the choice of Q_0^2 is not completely free if we stick to the gluon parametrization of (5). This specific form is not able to describe the very steep rise of the gluon distribution at large Q_0^2 (as shown in Fig. 3b). If it were enforced for a $Q_0^2 \gtrsim 15\,\mathrm{GeV}^2/c^2$, the gluon distribution would become negative at low values of Q^2. For the same reason the simple assumption $G(x, Q_0^2) \sim (1-x)^p$, which is often found in the literature, can only hold in an even more limited Q_0^2-range.

The above determination of the gluon distribution and of Λ_{LO} is valid if the observed scaling violations are only due to QCD effects. There are at least two other known effects which give rise to scaling violations: *i)* the charm threshold for the transitions $s \to c$ and $d \to c$, and *ii)* the uncertainty in the magnitude and Q^2-dependence of $R = \sigma_L/\sigma_T$. Both effects have been studied systematically. With the present knowledge they contribute a substantial uncertainty to the measured slopes of $\bar{q}$. In the following fits we have therefore no longer used the slopes of $\bar{q}$, but only require that $\bar{q}$ is close to zero at large x within the well-established experimental limits. This requirement still gives a very good constraint on the width of the gluon distribution. The effect of R on F_2 is rather small, within the experimental uncertainties. The threshold for the charm production from the strange sea, however, contributes substantially to the slopes of F_2 at small x. We have evaluated two different sets of structure functions F_2' in order to estimate the effect on the gluon distribution and Λ_{LO}. The effect of charm threshold has been corrected using the slow rescaling model [8] with two values of the charm mass and the amount of strange sea:

Structure function set II:

$$m_c = 1.5\,\mathrm{GeV}\,; \quad 2s/(\bar{u}+\bar{d}) = 0.5$$

Structure function set III:

$$m_c = 1.8\,\mathrm{GeV}\,; \quad 2s/(\bar{u}+\bar{d}) = 1$$

$$\left.\right\} \quad N_{F_2} = 3, \quad R = 0.1\,.$$

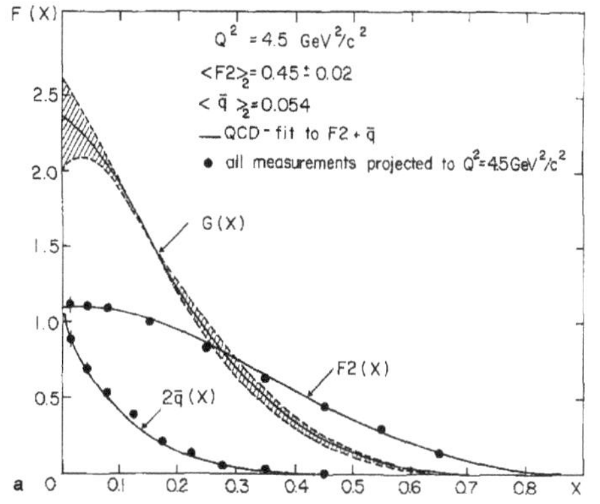

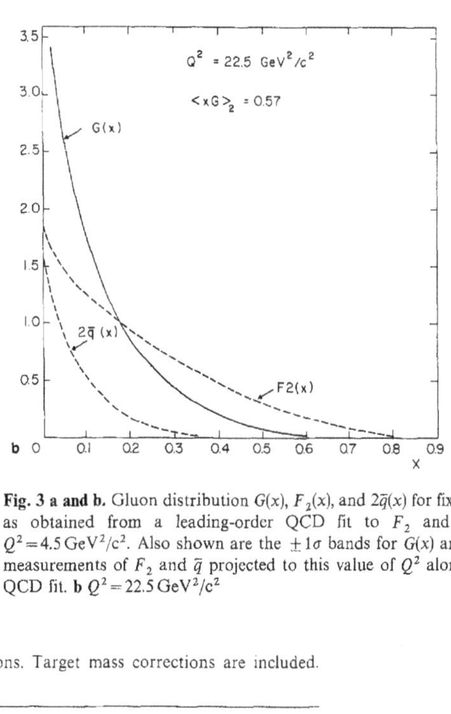

Fig. 3 a and b. Gluon distribution $G(x)$, $F_2(x)$, and $2\bar{q}(x)$ for fixed Q^2 as obtained from a leading-order QCD fit to F_2 and $\bar{q}$. **a** $Q^2 = 4.5\,\mathrm{GeV}^2/c^2$. Also shown are the $\pm 1\sigma$ bands for $G(x)$ and the measurements of F_2 and $\bar{q}$ projected to this value of Q^2 along the QCD fit. **b** $Q^2 = 22.5\,\mathrm{GeV}^2/c^2$

Table 1. Results of LO QCD-fits to different sets of structure functions. Target mass corrections are included. $Q^2 > 2\,\mathrm{GeV}^2/c^2$, $W^2 > 11\,\mathrm{GeV}^2$

Structure function	Assumptions about $s(x)$, $c(x)$	Fit results	χ^2/DF
I) $F_2, \bar{q}$ all x	No threshold effect $2(s-c)/(\bar{u}+\bar{d}) = 0.4$	$\Lambda_{\mathrm{LO}} = (0.18 \pm 0.02)$ GeV $G(x, Q_0^2 = 5) = 2.62\,(1+3.5x)\,(1-x)^{5.9 \pm 0.5}$	209/196
II) F_2'	$c(x) = 0$ for $Q^2 = 1\,\mathrm{GeV}^2/c^2$ $2s/(\bar{u}+\bar{d}) = 0.5$	$\Lambda_{\mathrm{LO}} = (0.20 \pm 0.02)$ GeV	166/137
$\bar{q}$ for $x > 0.3$	Slow rescaling with $m_c = 1.5$ GeV	$G(x, Q_0^2 = 5) = 2.86\,(1+2.8x)\,(1-x)^{6.3 \pm 0.8}$	
III) F_2'	$c(x) = 0$ for $Q^2 = 1\,\mathrm{GeV}^2/c^2$ $2s/(\bar{u}+\bar{d}) = 1$	$\Lambda_{\mathrm{LO}} = (0.206 \pm 0.02)$ GeV	173/137
$\bar{q}$ for $x > 0.3$	Slow rescaling with $m_c = 1.8$ GeV	$G(x, Q_0^2 = 5) = 2.98\,(1+2.84x)\,(1-x)^{6.65 \pm 0.9}$	

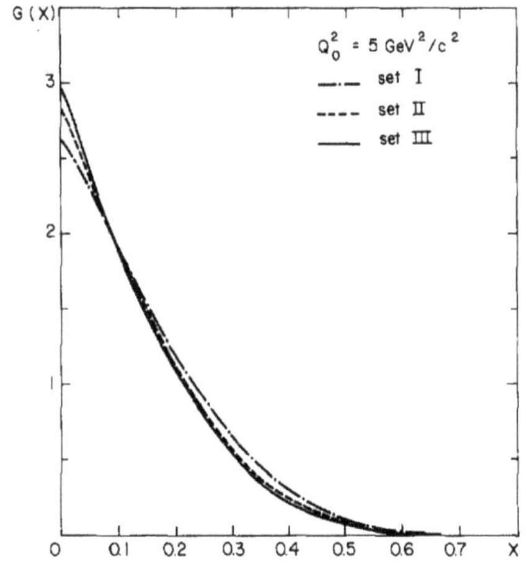

Fig. 4. Gluon distributions at $Q_0^2=5\,\mathrm{GeV}^2/\mathrm{c}^2$ for the fits to the three different structure function sets of Table 1

Both assumptions are compatible with the data on opposite-sign dimuons [4]. The charmed sea has been set to zero for $Q_0^2=1\,\mathrm{GeV}^2/\mathrm{c}^2$, and its Q^2-evolution has been calculated using the QCD equations for the evolution of $xc(x, Q^2)$, F_2, and G, where $F_2(x, Q_0^2)$ and $G(x, Q_0^2)$ have been fixed to the result of fit I. This charmed sea was then subtracted from the expression in (1), leading to a very small correction. The remaining structure function F_2' has only three flavours. We believe that the three data sets give a conservative estimate of the effects due to charm threshold and the amount of strange and charmed sea. The results of the fit to all three data sets are summarized in Table 1.

5. Conclusions

The uncertainties related to the strange and charmed sea and to $R=\sigma_L/\sigma_T$ have only a moderate effect on the results of the present analysis. The value of Λ_{LO} is remarkably stable and in good agreement with the results of a QCD analysis of the non-singlet structure functions xF_3 [9] with $\Lambda_{\mathrm{LO}}=0.2^{+0.15}_{-0.10}\,\mathrm{GeV}$. The gluon distributions from the fits to all three structure function sets are compared in Fig. 4 for $Q_0^2=5\,\mathrm{GeV}^2/\mathrm{c}^2$. It can be seen that the shape and magnitude of $G(x)$ are only moderately affected by the uncertainties due to charm threshold and the magnitude of strange and charmed sea. Actually this is not too surprising since the integral of $G(x, Q^2)$ is well constrained by the energy momentum sum rule, and the width by the absence of antiquarks at large x. Higher-twist contributions are not expected to give large effects at small x, i.e. in the gluon region. They may, however, be present at large x and affect the result through a change of Λ_{LO}. An analysis [9] of the present data at all W^2 in combination with SLAC e–d data indicates that higher-twist contributions for $W^2>11\,\mathrm{GeV}^2$ are most likely small. We conclude that the combined analysis of F_2 and $\bar{q}$ has provided a reliable measurement of the gluon distribution.

Acknowledgements. We thank M. Barnett for making his fitting program available to us and for useful discussions.

Appendix

Parametrization of $G(x, Q^2)$

The structure function for the gluons $G(x, Q_0^2)$ and for the quarks $F_2(x, Q_0^2)$ have been determined in the QCD fits at a reference value $Q_0^2=5\,\mathrm{GeV}^2/\mathrm{c}^2$. For the structure function set I we found:

$$\left.\begin{array}{l} G(x)=2.62\times(1+3.5x)(1-x)^{5.9} \\ F_2(x)=1.10\times(1+3.7x)(1-x)^{3.19} \end{array}\right\} \text{ for } Q_0^2=5\,\mathrm{GeV}^2/\mathrm{c}^2$$

The Q^2-evolution of the gluon distribution can then be obtained by the numerical integration of the Altarelli-Parisi equations (3a) and (3c).

In order to facilitate applications, we have parametrized $G(x, Q^2)$ as given by this numerical integration, using the functional dependence

$$G(x, Q^2)=a(1+bx^c)(1-x)^d(Q^2)^e+f e^{-(gx+hx^3)},$$

where the parameters a, b, c, d, e, f, g, and h are Q^2-dependent. This parametrization is entirely ad hoc, i.e. it has no theoretical basis. We find:

$$a=2.616+3.99s+4.46s^2$$
$$b=3.5-6.83s+80s^2$$
$$c=1+0.806s$$
$$d=5.9+40s+84.2s^2-64s^3$$
$$e=(-0.033-0.28x+59.1x^2)(s-s^2)$$
$$f=5.16s-0.955s^2$$
$$g=-0.48+12.2s+0.38s^2$$
$$h=29.72-32.4s,$$

where

$$s=\ln\left[\frac{\ln(Q^2/0.04)}{\ln(5/0.04)}\right].$$

This parametrization approximates the QCD fit to the structure function set I and is reasonably good for $Q^2>2\,\mathrm{GeV}^2/\mathrm{c}^2$ and energy transfer $\nu<2\times10^5\,\mathrm{GeV}$. It should be noted that apart from the uncertainties in shape which can be estimated from Table 1 and Fig. 4, the structure function $G(x, Q^2)$ has also a scale error of $\pm 6\%$ due to the uncertainty in the absolute neutrino cross-sections.

References

1. M. Holder et al. CDHS Collaboration: Nucl. Instrum. Methods **148**, 235 (1978)
2. H. Abramowicz et al. CDHS Collaboration: Neutrino and antineutrino charged-current inclusive scattering in iron in the energy range $20 < E_{\nu,\bar{\nu}} < 300$ GeV (in preparation)
3. H. Abramowicz et al.: Phys. Lett. **B107**, 141 (1981)
4. J. Knobloch et al. CDHS Collaboration: New results on dimuons from CDHS. In: Proc. Neutrino '81 Conf., Hawaii, 1981
   B. Peyaud et al. CDHS Collaboration: New results on opposite-sign and same-sign dimuons produced by neutrinos and antineutrinos at high energies. In: Proc. EPS Int. Conf. on High-Energy Physics, Lisbon, 1981
5. A. De Rújula et al.: Nucl. Phys. B**154**, 394 (1979)
6. G. Altarelli, G. Parisi: Nucl. Phys. **B126**, 298 (1979)
7. R. Barbieri et al.: Nucl. Phys. B**117**, 50 (1976)
8. R. M. Barnett: Phys. Rev. D**14**, 70 (1976)
9. F. Eisele et al. CDHS Collaboration: New results from the CDHS experiment. In: Proc. Neutrino '81 Conf., Hawaii, 1981

Z. Phys. C - Particles and Fields 15, 19-31 (1982)

Zeitschrift für Physik C Particles and Fields

Experimental Study of Opposite-Sign Dimuons Produced in Neutrino and Antineutrino Interactions

H. Abramowicz[1], J.G.H. de Groot, J. Knobloch, J. May, P. Palazzi, A. Para[1], F. Ranjard, J. Rothberg[2], W. von Rüden, W.D. Schlatter, J. Steinberger, H. Taureg, H. Wahl, J. Wotschack

CERN, CH-1211 Geneva, Switzerland

F. Eisele, H.P. Klasen, K. Kleinknecht, H. Lierl, B. Pszola, B. Renk, H.J. Willutzki

Institut für Physik der Universität, D-4600 Dortmund, Federal Republic of Germany

F. Dydak, T. Flottmann, C. Geweniger, J. Królikowski[1], K. Tittel

Institut für Hochenergiephysik der Universität, D-6900 Heidelberg, Federal Republic of Germany

C. Guyot, J.P. Merlo, P. Perez, B. Peyaud, J. Rander, J.P. Schuller, R. Turlay

D.Ph.P.E., CEN-Saclay, F-91190 Bures-sur-Yvette, France

J.T. He, T.Z. Ruan and W.M. Wu

Institute of High Energy Physics, Beijing, China

Received 21 June 1982

Abstract. A large sample of opposite-sign dimuons, produced by the interaction of neutrinos and antineutrinos in iron, is analysed. The data agree very well with the hypothesis that the extra muon is the product of charm decay. They yield information on the strength and space-time structure of the charm-producing weak current. The strange-sea structure function $xs(x)$ is determined. The difference between neutrino and antineutrino dimuon production is analysed to provide a value of the Kobayashi-Maskawa weak mixing angle θ_2.

1. Introduction

After the first observation of opposite-sign dimuon events in muon neutrino interactions [1], several experiments using electronic detectors [2, 3], as well as bubble chambers [4-6] have confirmed the hypothesis that single charm production and decay is the origin of opposite-sign dimuon events. In the quark parton model the charm-producing processes are expected to be the following:

$$\nu + d \to \mu^- + c$$
$$\nu + s \to \mu^- + c$$
$$\bar{\nu} + \bar{d} \to \mu^+ + \bar{c}$$
$$\bar{\nu} + \bar{s} \to \mu^+ + \bar{c}.$$

The second muon, of opposite charge, is the result of the semileptonic decay of the charmed particle:

$$c \to s + \mu^+ + \nu_\mu$$
$$\bar{c} \to s + \mu^- + \bar{\nu}_\mu.$$

These earlier experiments have also shown that new heavy particles, quarks, or leptons, can be excluded as a major source of dimuon events at present neutrino energies [2-7]. The new experiment that we describe here involves the reconstruction of a large sample of dimuon events from extended exposures in wide-band and narrow-band neutrino and antineutrino beams at the CERN SPS with the CDHS (CERN-Dortmund-Heidelberg-Saclay) detector [8]. The results are based on 10,381 neutrino and 3,513 antineutrino opposite-sign dimuon events from wide-band beams (WBBs) and 660 neutrino and 171 antineutrino events from narrow-band beams (NBBs). This is to be compared with 257 neutrino events and 58 antineutrino events in the previous publication [3]. The large increase in statistics permits a more thorough check of the charm hypothesis

1 On leave from the Institute of Experimental Physics, Warsaw University, Warsaw, Poland
2 On leave from the University of Washington, Seattle, USA

0170-9739/82/0015/0019/$02.60

and allows the study of the $V-A$ structure of the charm-producing current as well as the determination of the Kobayashi-Maskawa angle θ_2 [9], the measurement of the structure function of the strange sea, and the charm fragmentation function.

In the next section we describe the data reduction. In Sect. 3 we present the raw data sample. Section 4 outlines the theoretical formalism and Sect. 5 describes the Monte Carlo simulation. The results of the analysis of the data are presented in Sects. 6-11.

2. Exposures and Data Reduction

2.1. Beams

The neutrino and antineutrino WBBs were produced by protons of 350 and 400 GeV energy on a *Be* target followed by 300 m of decay space. The total intensities on target were respectively 5.2×10^{18} and 4.0×10^{18} protons. The horn and reflector focusing used for the WBB leads to neutrino spectra peaked around 20 GeV. The contamination of antineutrinos in the neutrino beam is small (less than 1%); however the contamination of neutrinos in the antineutrino beam is substantial and becomes dominant above 150 GeV. These characteristics are shown in Fig. 1. The neutrino and antineutrino beam spectra as well as background in the WBBs were obtained from the analysis of single-muon events of appropriate muon charge. Although lower in statistics, data

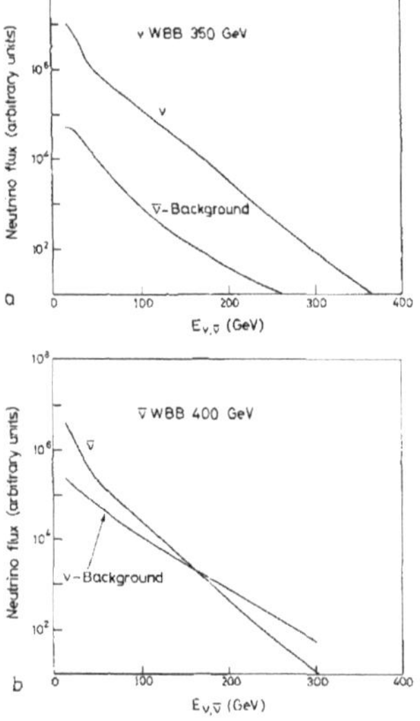

Fig. 1a and b. Neutrino spectra in **a** neutrino and **b** antineutrino WBBs

from NBB exposures are interesting because of different beam characteristics: the wrong-sign background is negligible, there is a correlation between neutrino energy and radial position of the interaction vertex in the detector, and the shape of the energy spectrum is accurately known and it is flat [10].

2.2. Apparatus and Trigger

The events were observed with the detector described previously [8]. In the WBB the multimuon trigger requires two or more hits in at least two of the three planes of four or more adjacent drift chambers. This requirement implies that each muon must have a minimum energy of 3.5 GeV. The capability of the readout system is such that up to four hits on the same drift wire can be detected if separated by at least 7 mm in space (140 ns in drift-time). In addition to the drift chamber requirement, a minimum of 7 GeV muonic or hadronic energy deposition in the calorimeter was demanded. In the NBB exposures no special trigger is required for dimuon events; they are included in the normal single-muon event trigger (deposition of hadron energy of at least 3 GeV, or minimum ionizing particle in three modules). The dimuons are selected later by software, using the same criteria as those for the WBB dimuons.

2.3. Event Selection

Initially all NBB events, and a subsample of 2,000 neutrino WBB events were scanned and measured by hand using an interactive pattern recognition and track-fitting program. These interactively-measured events were used to determine the efficiency and to evaluate the biases of the off-line reconstruction program with which subsequently the entire sample was processed. The off-line program accepts events only if both muons traverse at least five chambers and fulfil the following requirements:

- One μ^- and one μ^+ are found and successfully fitted.
- The fit of each track has to start at least in the third drift chamber downstream from the vertex.
- The separation ΔR of the two tracks at the vertex is less than 15 cm. (The mean value of the ΔR distribution is 7.5 cm.)

After these cuts $80 \pm 5\%$ of those $\mu^+\mu^-$ events survived where both muons traverse five chambers. The comparison with the sample of hand-measured events shows that no significant shifts or biases on the measured quantities are introduced by the off-line reconstruction program.

Dimuon events are accepted for the final analysis if they satisfy the following selection criteria:

- The event vertex is required to be within the first 14 of the 19 modules.
- The vertex has to be within a circle of 1.6 m radius around the centre of the toroids.
- Events near the centre of the apparatus are excluded (cut 51 cm wide, 26 cm high).
- Both muon momenta have to be larger than 5 GeV/c.
- The total visible energy E_{vis} (as defined in Sect. 3) is required to be at least 20 GeV.

A total of 10,381 events in the neutrino WBB and 3,513 events in the antineutrino WBB, and 660 and 171 respectively in the NBB survive these cuts.

2.4. Separation of Neutrino- and Antineutrino-Induced Events in Antineutrino Exposure

In the neutrino WBB the contamination of antineutrinos is sufficiently low to be neglected in the analysis. In the antineutrino WBB, however, one third of the observed dimuon events is due to the neutrino background in the beam. In this case the antineutrino-induced dimuon events can be extracted with acceptable purity on the basis of the transverse muon momenta p_T^+ and p_T^- with respect to the hadron shower direction. The latter is defined as the direction which balances the negative muon momentum in the neutrino beam and the positive muon momentum in the antineutrino beam. The scatter plot p_T^+ versus p_T^- for the events from the neutrino WBB is shown in Fig. 2a. The opposite-sign transverse momentum p_T^+ is in general limited. Figure 2b shows the same scatter plot for the antineutrino WBB events. The bulk of the large p_T^- events in this plot must be attributed to the neutrino contamination. Antineutrino-induced events have been selected by the requirement $p_T^+ > p_T^-$. The correction for migration of neutrino events into the "$\bar{\nu}$ region" and for the loss of $\bar{\nu}$ events due to this selection has been evaluated using a Monte Carlo program. The net correction of $6 \pm 1\%$ is almost independent of energy. The procedure is verified with data taken in the relatively pure neutrino beam. The systematic uncertainties are small compared with the statistical errors of the sample of antineutrino-induced events. Table 1 gives the decomposition of the dimuon events from the antineutrino WBB as a result of this separation procedure.

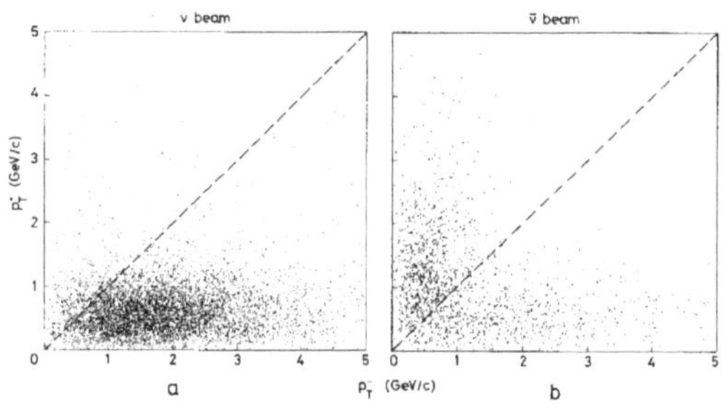

Fig. 2a and b. Scatter plot of p_T^+ versus p_T^- for dimuons from **a** neutrino and **b** antineutrino beams. p_T is the transverse momentum of the muon relative to the hadron shower direction

Table 1. Antineutrino- and neutrino-induced dimuon events, $N_{\bar{\nu}}$ and N_ν, as obtained from the antineutrino WBB after separation and correction for migration

Visible energy (GeV)	$N_{\bar{\nu}}$	N_ν
30– 40	261	69
40– 60	534	194
60– 80	523	212
80–100	353	185
100–120	200	163
120–140	113	130
140–160	70	89
160–200	55	85

2.5. Background from π or K Decay

There is a background due to non-charmed muonic decays, chiefly π and K decay. Details of the Monte Carlo calculations of this background are given elsewhere [11]. The average contamination due to π and K decay is 13% in neutrino, and 6% in antineutrino interactions. The variation with respect to the neutrino energy is small as seen in Fig. 3.

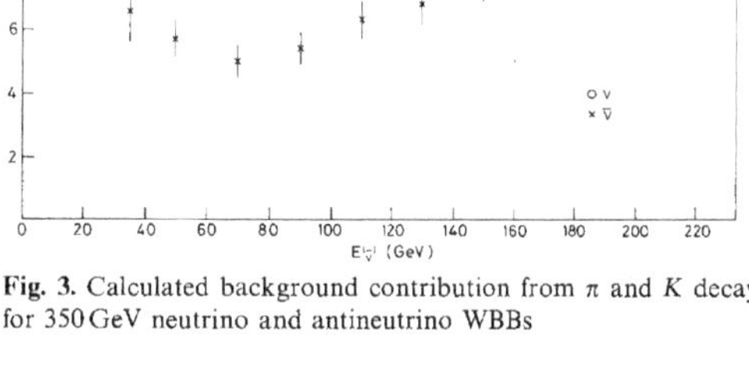

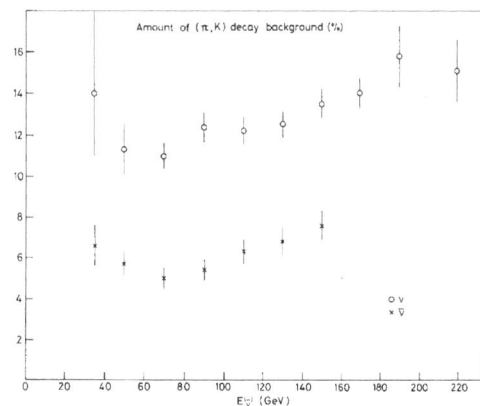

Fig. 3. Calculated background contribution from π and K decays for 350 GeV neutrino and antineutrino WBBs

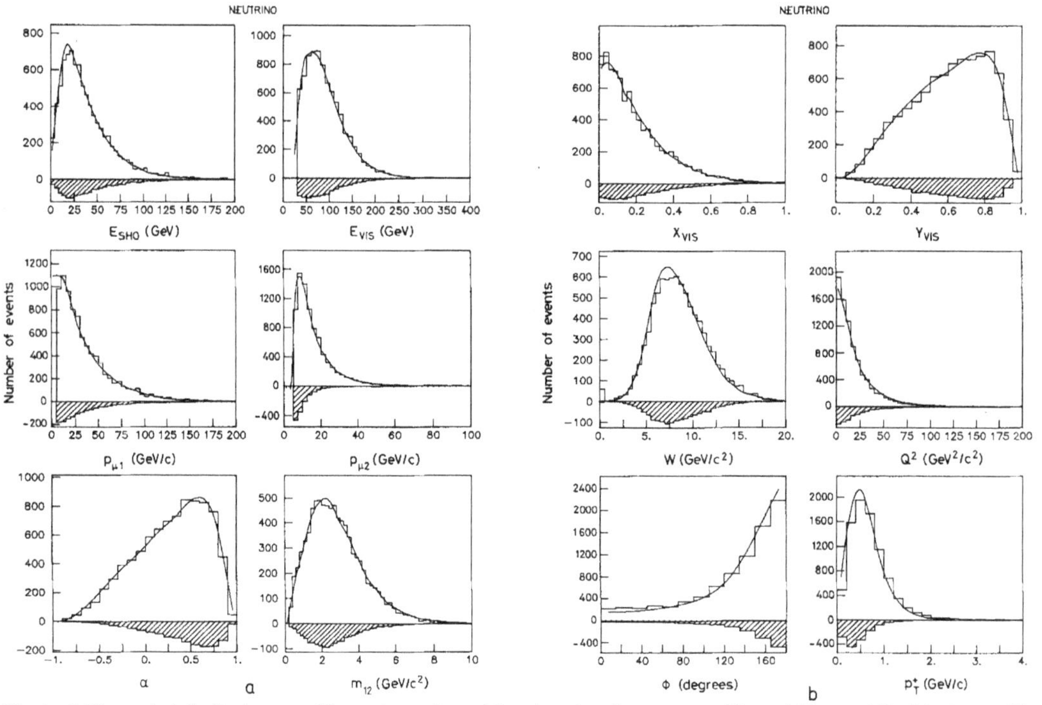

Fig. 4a–d. Kinematical distributions **a** and **b** neutrino and **c** and **d** antineutrino dimuon events. The variables are defined in the text. The dashed area represents the π and K background which has been subtracted from the raw data. The curves are results of the Monte Carlo simulation

Throughout this analysis these backgrounds have been subtracted from the experimental samples.

3. Raw Data

We define here the variables used in the following analysis. The neutrino direction is assumed to be parallel to the axis of the detector.

- p_{μ_1} = momentum of the muon carrying the same lepton number as the incident neutrino. This is the "leading muon", i.e. μ^- for neutrino-induced events and μ^+ for antineutrino-induced events.
- p_{μ_2} = momentum of the "second" muon
- E_{sho} = hadron shower energy
- $E_{\text{had}} = E_{\text{sho}} + p_{\mu_2}$
- $E_{\text{vis}} = E_{\text{sho}} + p_{\mu_2} + p_{\mu_1}$
- E_ν = energy of incident neutrino
- $\nu = E_\nu - p_{\mu_1}$ = energy transfer
- θ = angle between $\vec{p}_{\mu_1}$ and the neutrino direction
- $Q^2 = 4E_{\text{vis}}\, p_{\mu_1} \sin^2\theta/2$
- $x = 4E_\nu\, p_{\mu_1} \sin^2(\theta/2)/(2M_N \cdot \nu)$
- $x_{\text{vis}} = Q^2/(2M_N \cdot E_{\text{had}})$
- M_N = mass of the nucleon
- $y = \nu/E_\nu$
- $y_{\text{vis}} = E_{\text{had}}/E_{\text{vis}}$
- Φ = angle between the projections of $\vec{p}_{\mu_1}$ and $\vec{p}_{\mu_2}$ on a plane perpendicular to the neutrino direction
- $\vec{p}_{\text{had}} = \vec{p}_\nu - \vec{p}_{\mu_1}$ = hadron shower momentum
- p_T = transverse momentum of muon (either leading muon or second muon) relative to $\vec{p}_{\text{had}}$
- $W_{\text{vis}} = (2M_N \cdot E_{\text{had}} + M_N^2 - Q^2)^{1/2}$
- m_{12} = invariant mass of the muon pair $= [(E_1 + E_2)^2 - (\vec{p}_{\mu_1} + \vec{p}_{\mu_2})^2]^{1/2}$
- α = momentum asymmetry $= (p_{\mu_1} - p_{\mu_2})/(p_{\mu_1} + p_{\mu_2})$
- E_D = energy of D meson
- E_c = energy of primary charmed quark
- $z = E_D/E_c$
- $z_L = p_{\mu_2}/(p_{\mu_2} + E_{\text{sho}})$.

Distributions of the data are presented in Figs. 4a–d for the neutrino and antineutrino samples. Some general comments can be made:

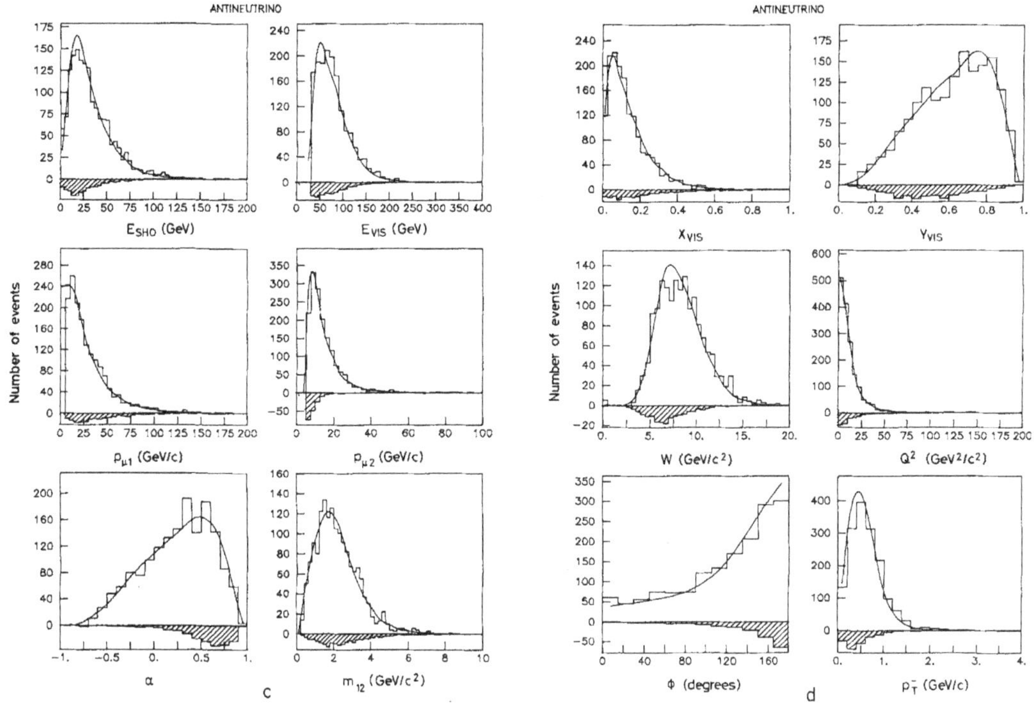

Fig. 4c and d

i) The momentum of the second muon p_{μ_2} is smaller than that of the leading muon p_{μ_1}.

ii) The p_T distribution, as well as the Φ distribution which is peaked at 180°, shows that the second muon is associated with the hadronic shower.

iii) The shape of the y_{vis} distribution is biased by the minimum momentum requirement for the muons. The cut-off on p_{μ_2} and consequently on E_{had} depopulates the small y_{vis} region, whereas the cut-off on p_{μ_1} depopulates the large y_{vis} region. The Monte Carlo curves shown in Fig. 4 correspond to a flat generated y distribution.

iv) The x_{vis} distributions are narrower in the antineutrino data than in the neutrino data. These in turn are narrower than the structure functions $F_2(x)$ and $xF_3(x)$ of single-muon reactions.

v) The W_{vis} and m_{12} mass plots do not show significant structure.

The dimuon event numbers are presented in Table 2. The single-muon event numbers also shown in the table are based on a normalized subsample of the data. In this table, the visible energy for single-muon events equals the neutrino energy, while the visible energy for dimuon events is smaller because of the energy lost by an escaping neutrino from the semi-leptonic decay producing the second muon. In order to compare dimuon and single-muon event rates, in

Table 2. Number of observed neutrino-induced (N_{-+}) and antineutrino-induced (N_{+-}) dimuon events, together with normalized single-muon event rates, N_- and N_+

E_{vis} (GeV)	ν		$\bar{\nu}$	
	N_{-+}	$N_- \times 10^{-6}$	N_{+-}	$N_+ \times 10^{-6}$
30– 40	749	1.9	261	0.54
40– 60	1,850	1.49	534	0.51
60– 80	2,037	1.00	523	0.25
80–100	1,697	0.71	353	0.17
100–120	1,306	0.47	200	0.08
120–140	876	0.27	113	0.04
140–160	694	0.17	70	0.02
160–180	338	0.09	38	
180–200	195	0.04	17	0.01
200–240	189	0.04	9	
>240	91	0.01	5	
30–500	9,922	6.19	2,123	1.62

bins of true neutrino energy, a correction must be applied to the visible energy (see Sect. 6).

4. Theoretical Frame for the Analysis

A model of charm-producing weak currents was proposed by Glashow, Iliopoulos and Maiani (GIM) [12], even before charm was experimentally discovered. The GIM current has the following structure:

$$J_\mu^{\mathrm{GIM}} = (\bar{u}, \bar{c})\, \gamma_\mu (1+\gamma_5) \begin{pmatrix} \cos\theta_c & \sin\theta_c \\ -\sin\theta_c & \cos\theta_c \end{pmatrix} \begin{pmatrix} d \\ s \end{pmatrix} + \text{h.c.},$$

where θ_c is the Cabibbo angle.

Charm is produced with amplitude $\sin\theta_c$ from d quarks, and with amplitude $\cos\theta_c$ from s quarks. We now know that there are not four, but at least five quarks: the extension to three flavour pairs is due to Kobayashi and Maskawa (KM) [9] and involves three mixing angles and one phase in place of the single Cabibbo angle:

$$J_\mu^{\mathrm{KM}} = (\bar{u}, \bar{c}, \bar{t})\, \gamma_\mu (1+\gamma_5)\, \mathscr{U} \begin{pmatrix} d \\ s \\ b \end{pmatrix} + \text{h.c.}$$

where $\mathscr{U}$ is the 3×3 KM matrix.

The important contributions to charm production are the processes:

$$\nu + d \to \mu^- + c$$

$$\nu + s \to \mu^- + c,$$

and their charge conjugates. The conversion of a bottom quark to a charm quark can be expected to be negligible at our energies because of the high threshold and the smallness of the bottom structure function. The conversion of a charm quark to either d or s, with production of its associated (anti) charm quark can also be ignored because of the smallness of the charm structure function and our experimental biases, which make the detection of charm produced essentially at rest very inefficient.

If the charm-producing currents are left-handed, as the other known charged weak currents, then the cross-sections on isoscalar nuclei are

$$\frac{d\sigma^\nu}{dx\,dy} = \frac{G^2 M E_\nu x}{\pi} \left[U_{cd}^2 [u(x) + d(x)] + |U_{cs}^2|\, 2s(x) \right] \quad (1)$$

and

$$\frac{d^2\sigma^{\bar{\nu}}}{dx\,dy} = \frac{G^2 M E_\nu x}{\pi} \left[U_{cd}^2 [\bar{u}(x) + \bar{d}(x)] + |U_{cs}^2|\, 2\bar{s}(x) \right], \quad (2)$$

where $u(x)$, $d(x)$, and $s(x)$ are the quark density distributions in the proton. We denote the integrals of the quark structure functions by $U = \int x u(x)\, dx$, $D = \int x d(x)\, dx$, $S = \int x s(x)\, dx$, and analogously for antiquarks. In the GIM [12] model $U_{cd} = \sin\theta_c$ and $U_{cs} = \cos\theta_c$. In the KM [9] notation $U_{cd} = \sin\theta_1 \cos\theta_2$ and $U_{cs} = \cos\theta_1 \cos\theta_2 \cos\theta_3 - \sin\theta_2 \sin\theta_3 e^{i\delta}$.

At present only two of these angles are known with any confidence: $\sin\theta_1 = 0.228 \pm 0.011$ and $\cos\theta_3 = 0.96^{+0.04}_{-0.09}$ [13]. As we will see, neutrino charm-production experiments can contribute to the understanding of θ_2.

We note that in distinction to the usual (i.e. not charm-producing) cross-sections, the charm-producing cross-sections are uniform in y; the $(1-y)^2$ terms are absent because the interaction proceeds solely on quarks for neutrino collisions, and solely on antiquarks for antineutrino collisions. Right-handed weak charm-producing currents would be characterized by $(1-y)^2$ dependences, and this provides the possibility of an experimental test of the chirality of the charm-producing current.

The data must be corrected for the threshold effect due to the mass m_c for the charm quark, the so-called slow rescaling. To correct for this kinematical effect [14] the structure functions in (1) and (2) should be replaced by functions of the variable $\xi = x + m_c^2/(2M_N \nu)$, $F(x) \to F(\xi)$ for $Q^2 \to \infty$, and the expressions multiplied by the factor $(1 - y + xy/\xi)$:

$$\frac{d^2\sigma^\nu}{dx\,dy} = \frac{G^2 M E \xi}{\pi} \left(1 - y + \frac{xy}{\xi}\right) \{ U_{cd}^2 [d(\xi) + u(\xi)] + U_{cs}^2\, 2s(\xi) \} \quad (3)$$

and

$$\frac{d^2\sigma^{\bar{\nu}}}{dx\,dy} = \frac{G^2 M E \xi}{\pi} \left(1 - y + \frac{xy}{\xi}\right) \{ U_{cd}^2 [\bar{d}(\xi) + \bar{u}(\xi)] + U_{cs}^2\, 2\bar{s}(\xi) \}. \quad (4)$$

We have set $m_c = 1.5$ GeV in the following.

5. Monte Carlo Simulation of Charm Production

The second muon, the one which is the product of charm decay, is the result of a complex process, including charm-quark fragmentation and charmed-particle decay. Comparison of our data with the hypothesis of charm production has been carried out with the help of a Monte Carlo simulation in which:

i) The y distribution is taken to be constant.

ii) The valence x distribution is $\sqrt{x}(1-x)^{3.5}$ and the sea x-distribution $(1-x)^7$.

iii) The production of charmed baryons is not considered. This is justified by the observation [19] that, for $E_\nu > 30$ GeV, out of twenty-six decays of charmed particles at most two are interpreted as Λ_c.

iv) The charm-quark fragmentation function is chosen to fit the experimental results, as discussed in Sect. 8.

v) The transverse momentum distribution of the charmed mesons with respect to the direction of the hadron shower is

$$\frac{dN}{dp_t^2} \propto \exp[-6(p_t^2 + M_D^2)^{1/2}],$$

to fit the experimental p_t distribution.

vi) The branching ratios for the decay modes $K^*\mu\nu$, $K\mu\nu$, and $\pi\mu\nu$ are taken to be 0.37/0.56/0.06 [15].

In every case the analysis has been performed with and without the assumption of slow rescaling (see Sect. 4).

6. Missing Energy

If the interpretation of charm decay is correct, some of the energy of the incident neutrino is missing in the observed event because the neutrino emitted in the charm decay escapes detection. This can be studied in the NBB, because here the incident neutrino energy is related to the impact radius of the neutrino event in the detector, as can be seen in Fig. 5b

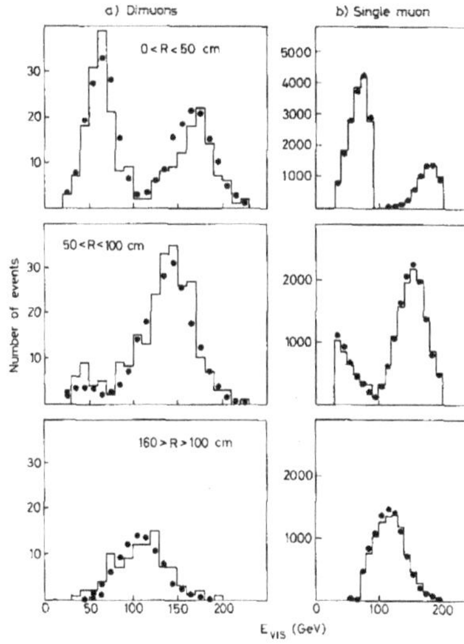

Fig. 5a and b. Energy spectrum of **a** dimuon events and **b** single-muon events for different radial bins of the detector; 200 GeV NBB data (histogram). The points are the result of a Monte Carlo simulation

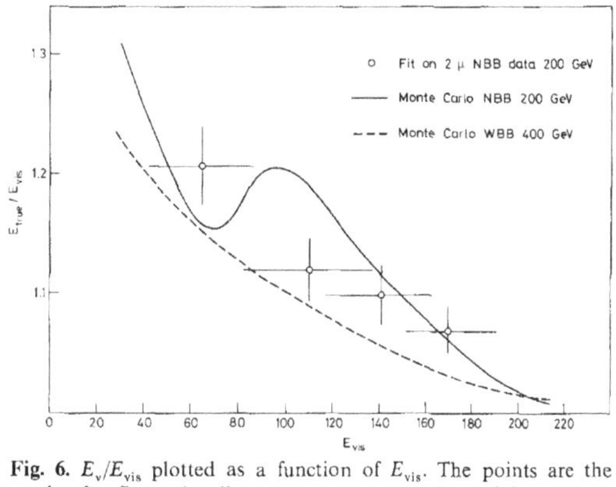

Fig. 6. E_ν/E_{vis} plotted as a function of E_{vis}. The points are the result of a fit to the dimuon energy spectra in radial bins (Fig. 5a). The curves are Monte Carlo results for 200 GeV NBB and 400 GeV WBB

for the single-muon events. The observed energy distributions agree well with the kinematic expectations. For dimuon events the observed energy distribution is shifted to lower energies (Fig. 5a) on the average by $12 \pm 1\%$. This is in good agreement with the expectation of charm origin on the basis of a Monte Carlo calculation (Fig. 6). The average predicted energy loss is $12.3 \pm 2\%$. The missing energy has already been demonstrated in an earlier publication [3], but the present result is more precise.

7. y Distributions and $V-A$ Structure of the Charm Producing Current

As discussed in Sect. 4, the neutrino and antineutrino y distributions for single-charm production are expected to be flat for left-handed, and of the form $(1-y)^2$ for right-handed currents.

As explained before the observed $d\sigma/dy$ distributions are seriously distorted by the minimum energy requirement for the two muons. The simulation of this bias, which is less severe at high than at low neutrino energy, depends on the choice of the charm-fragmentation function. In the analysis the experimentally observed distribution (see next section) was taken.

The solid curves in Fig. 7 are the expectations for purely left-handed currents. The dashed curves in Fig. 7 show the expected y_{vis} distributions for purely right-handed currents with the same fragmentation function. The good agreement with $V-A$ shows that only a small admixture of $V+A$ coupling can be accommodated by the data.

A parametrization of the form $d\sigma/dy \propto \beta(1-y)^2 + (1-\beta)$ gives the following upper limit on the pa-

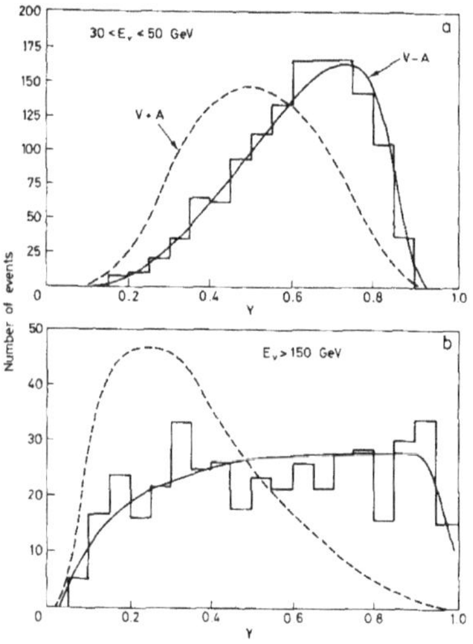

Fig. 7a and b. y distribution of observed neutrino events for a $30 \leq E_\nu \leq 50$ GeV and b $E_\nu > 150$ GeV; the solid and dashed curves represent the $V-A$ current and $V+A$ current, respectively, for charm production

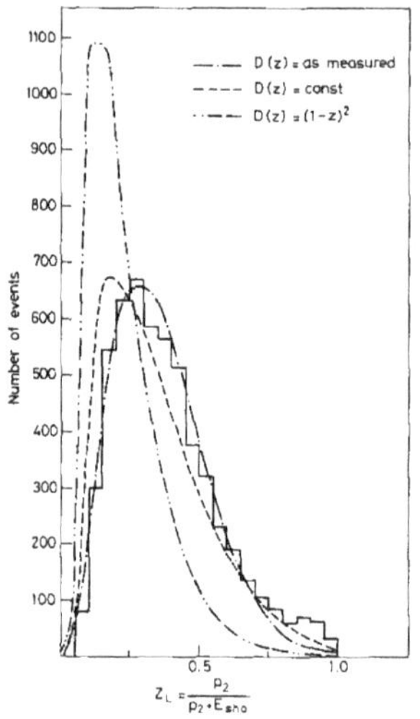

Fig. 8. Distribution of neutrino dimuon events in the variable z_L compared with Monte Carlo results

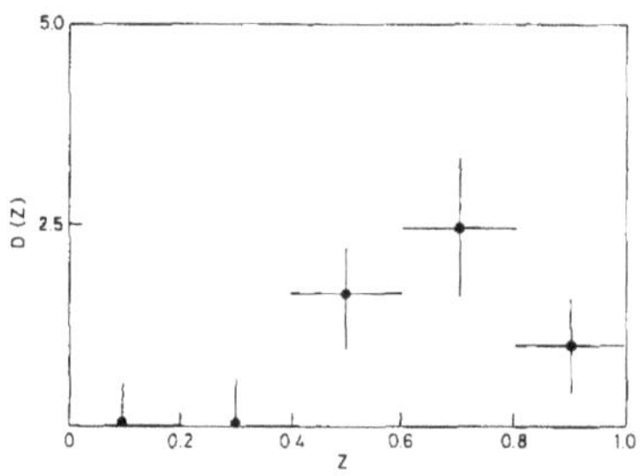

Fig. 9. Charm fragmentation function $D(z)$. Results of the fit in z_L and y_{had}

rameter β at the 95% confidence level:

E_ν (GeV)	30−50	50−100	100−150	≧150
β	≦0.30	≦0.10	≦0.10	≦0.34

For the combined data, and within the uncertainties of the fragmentation function, the limit $\beta < 0.07$ for the relative strength of the square of the right-handed current coupling constant can be deduced, at the 95% confidence level.

8. Charm Fragmentation

The fragmentation of heavy quarks into their appropriate heavy hadrons is expected to be harder than that of the light quarks (u, d, and s) into light hadrons [16]. It is expected that a charmed hadron coming from the fragmentation of a charm quark will carry most of the quark momentum. A first measurement of the charm fragmentation function $D(z)$, i.e. the distribution in the charm fragmentation variable $z = E_D/E_c$ was done by studying the observed distribution in $z_L = p_{\mu_2}/(p_{\mu_2} + E_{sho})$, as shown in Fig. 8. It is obvious that the data require a hard fragmentation function. $D(z)$ has been determined in a Monte Carlo fit to the two-dimensional distribution of dimuon data in z_L and $y_{had} = E_{sho}/E_{vis}$, leaving $D(z)$ as free parameters in 5 bins in z. The best fit to this distribution was obtained with the fragmentation function $D(z)$ as shown in Fig. 9. It should be noted that the data points $D(z)$ are correlated such that the detailed shape of $D(z)$ is not reliably obtained. For instance, the simple function $D(z) = \delta(z - 0.68)$ also gives an adequate fit to the data. On the other hand, the first moment of $D(z)$ is well determined. It is $\langle z \rangle = 0.68 \pm 0.05$, where the error is statistical. The systematic error from the following sources was estimated: uncertainties in *i)* the π- or K-decay background calculation, *ii)* the branching ratios of D mesons into K or K^* final states, *iii)* the charm quark mass used in the slow rescaling correction. These effects contribute about equally to the total systematic error of 0.06. We therefore give the result including systematic error, $\langle z \rangle = 0.68 \pm 0.08$. This compares well with the theoretical values at our average Q^2 of $20\,(\mathrm{GeV}/c)^2$ [16].

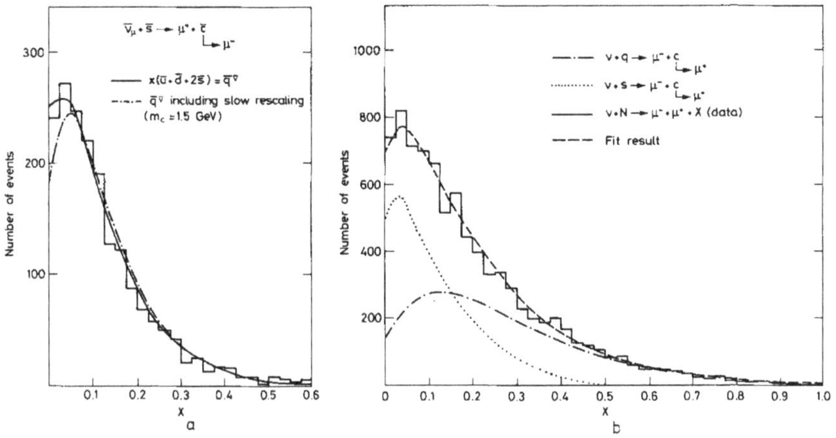

Fig. 10a and b. x_{vis} distribution for dimuon events. The histograms represent the data. **a** Antineutrino; the solid curve is the sea distribution obtained in our single-muon analysis [10], the dashed-dotted curve demonstrates the effect of slow rescaling. **b** Neutrino; the curves show the decomposition into 48% strange-sea contribution taken from the data of **a** (dotted curve) and 52% quark contribution (dashed-dotted curve). The dashed curve is the sum of both

9. The Strange-Sea Structure Function

The antineutrino dimuon production, in the GIM model, and also in the KM model if the mixing angles are not large, is dominated at the level of $\sim 90\%$ by $xs(x)$. The remaining smaller contribution is due to the non-strange sea $x[\bar{u}(x)+\bar{d}(x)]$. The observed x distribution is experimentally indistinguishable from the $x[\bar{u}(x)+\bar{d}(x)+2\bar{s}(x)]$ sea measured in single-muon production by antineutrinos at large y [10], so that the x distribution for antineutrino-induced dimuons in Fig. 10a is directly a measurement of the structure function of the strange sea $xs(x)$.

10. Dimuon to Single-Muon Production Ratios

Approximately 30% of the neutrino data and 60% of the antineutrino data were analysed also for the much more numerous single-muon events. Instead of dimuon cross-sections we give the ratios of dimuon to single-muon rates. This quantity is more directly obtained, and can be converted into absolute dimuon cross-sections on the basis of the single-muon absolute cross-sections [10]. Since the measured energy E_{vis} for the dimuon events, the sum of the two muon energies and the hadron energy, misses the outgoing neutrino energy (see Sect. 3), the measured energies are corrected by the energy-dependent factor shown in Fig. 6 to give the neutrino energy E_ν, at least on the average. All cross-section ratios given in the following are subject to a 10% scale error.

The raw ratios, corrected only for πK background, are shown in Figs. 11a and 12a. The acceptance corrections include reconstruction efficiencies, geometrical corrections (small), and the effect of the minimum muon-energy requirement. The dimuon acceptance – the same for neutrino and antineutrino events – is shown in Fig. 13 for two cases of the fragmentation function $D(z)=\delta(z-0.68)$ as well as for another function $D(z)=$constant, so that the effect of the fragmentation function may be judged. The acceptance-corrected results are shown in Figs. 11b and 12b. The solid error bars indicate the statistical errors only. The dotted error bars include the uncertainty of the fragmentation function. The ratio of neutrino to antineutrino $2\mu/1\mu$ production ratios ("double ratio") is shown in Fig. 14. In Fig. 15 our neutrino results are compared with bubble-chamber measurements [17, 18] of $\mu^- e^+$ production. The bubble-chamber data are of particular interest because the positrons are detected even at quite low energy (0.3 GeV) and the result is therefore much less affected by uncertainties in $D(z)$. It is not possible to compare our results with other counter experiments because data corrected for acceptance and for missing energy are not available.

11. Amount of Strange Sea and $\cos^2\theta_2$

The amount of the strange sea and $\cos^2\theta_2$ of the KM model can be deduced, subject to assumptions, from these results. Here we will find the momentum fraction of the strange sea, $S=\int_0^1 xs(x)\,dx$, or rather

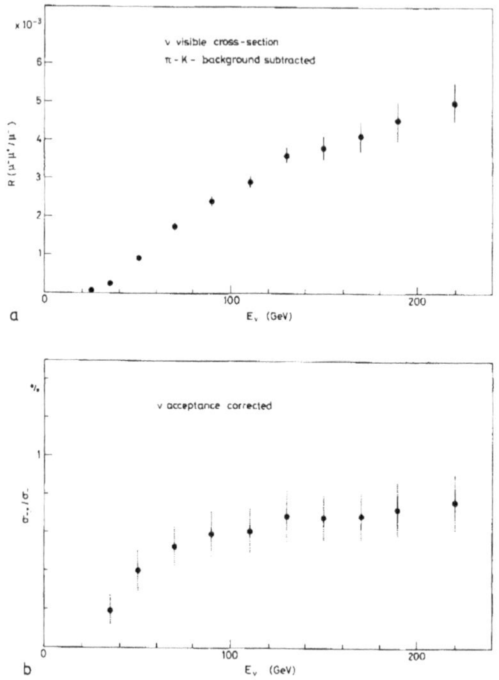

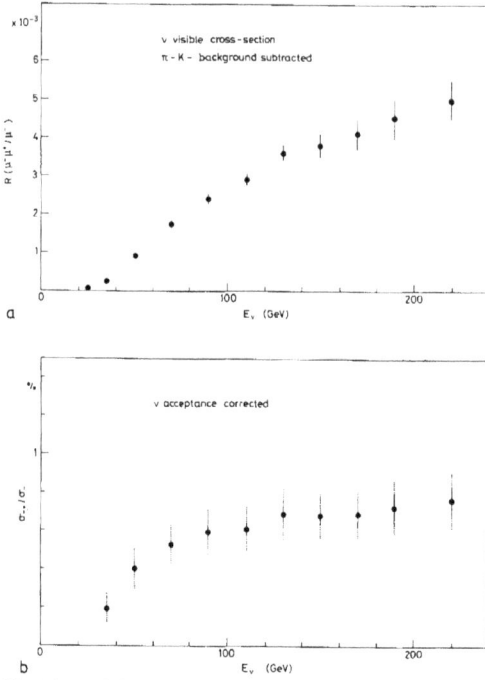

Fig. 11a and b. $R_2^\nu = \sigma^\nu(2\mu)/\sigma^\nu(1\mu)$ for neutrinos: **a** raw data, **b** acceptance corrected data. The solid error bars indicate statistical errors, the dotted error bars include systematic errors due to the uncertainty in the fragmentation function; scale error 10%

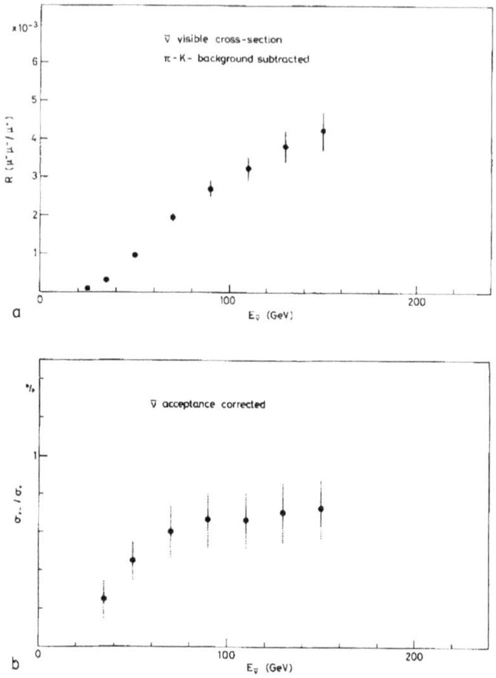

Fig. 12a and b. $R_2^{\bar\nu} = \sigma^{\bar\nu}(2\mu)/\sigma^{\bar\nu}(1\mu)$ for antineutrinos: **a** raw data, **b** acceptance corrected data. Errors as for Fig. 11

the quantity $S|U_{cs}^2|/U_{cd}^2$, from the shape of the x dependence of the neutrino dimuon production, and the quantity BU_{cd}^2 from the difference of the neutrino and antineutrino total cross-sections. Here B is the muonic branching ratio of the charmed particle. In the following analysis it is assumed that it is the same for neutrino- and antineutrino-produced charm. From these two quantities U_{cd}^2 and $|U_{cs}^2|S$ can be determined if B is known.

11.1. Determination of U_{cd}^2

From the neutrino and antineutrino dimuon to single-muon cross-section ratios, the quantity BU_{cd}^2 is determined as follows:

$$BU_{cd}^2 = \frac{(\sigma_{\mu^-\mu^+}^\nu/\sigma_{\mu^-}^\nu) - (R\sigma_{\mu^+\mu^-}^{\bar\nu}/\sigma_{\mu^+}^{\bar\nu})}{1-R}\,\frac{2}{3}. \qquad (5)$$

Here R is the ratio of antineutrino to neutrino total cross-sections, $R = \sigma_{\mu^+}^{\bar\nu}/\sigma_{\mu^-}^\nu = 0.48 \pm 0.02$ [10]. The dimuon to single-muon cross-section ratios, corrected for acceptance and slow rescaling, are shown in Figs. 16a and b. The results for $U_{cd}^2 B$ are given in Table 3. Within the statistical errors, the results are consistent with the average in the energy region $80 < E_\nu < 160$ GeV: $U_{cd}^2 B = (0.41 \pm 0.07) \times 10^{-2}$.

If slow rescaling were neglected the result would be $U_{cd}^2 B = (0.39 \pm 0.05) \times 10^{-2}$, which is very similar. A 10% increase (decrease) in the mass of the charm quark used in the slow rescaling correction increases (decreases) U_{cd}^2 by 2%. For the semileptonic branching ratio, a value of $B = (7.1 \pm 1.3)\%$ is obtained by

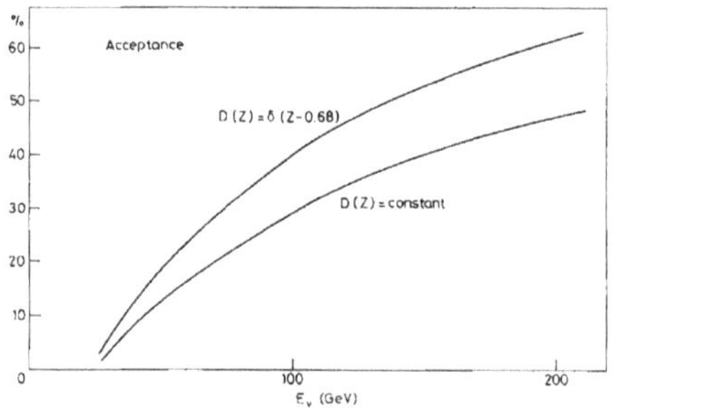

Fig. 13. Acceptance for neutrino dimuon events for two fragmentation functions. Solid line $D_c(z) = \delta(z - 0.68)$, dashed line $D_c(z) = \text{const}$

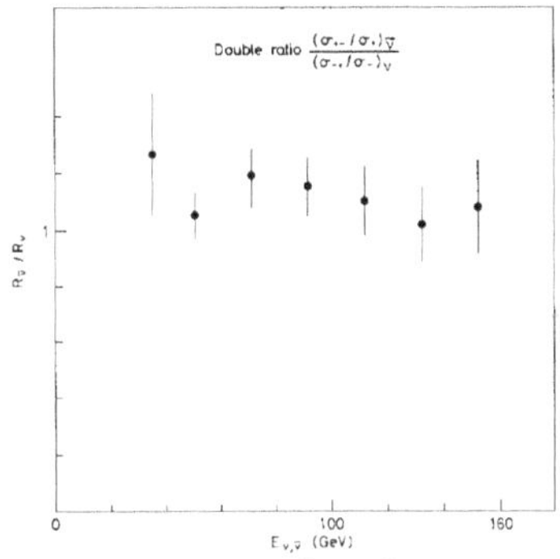

Fig. 14. Double ratio $[\sigma^{\bar{\nu}}(2\mu)/\sigma^{\bar{\nu}}(1\mu)]/[\sigma^{\nu}(2\mu)/\sigma^{\nu}(1\mu)]$ as a function of energy

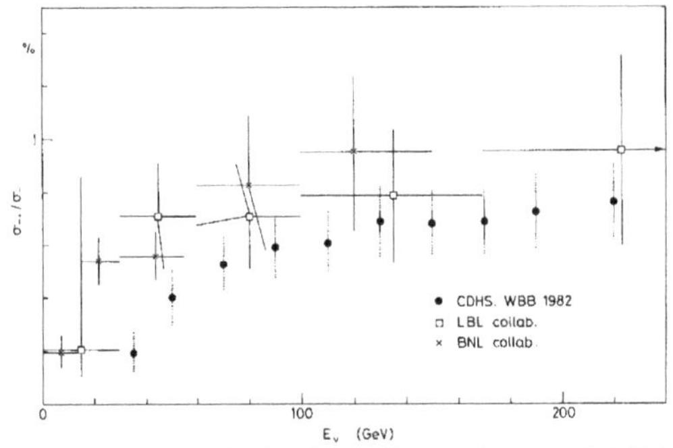

Fig. 15. $R_2^\nu=\sigma^\nu(2\mu)/\sigma^\nu(1\mu)$ from this experiment and bubble-chamber results

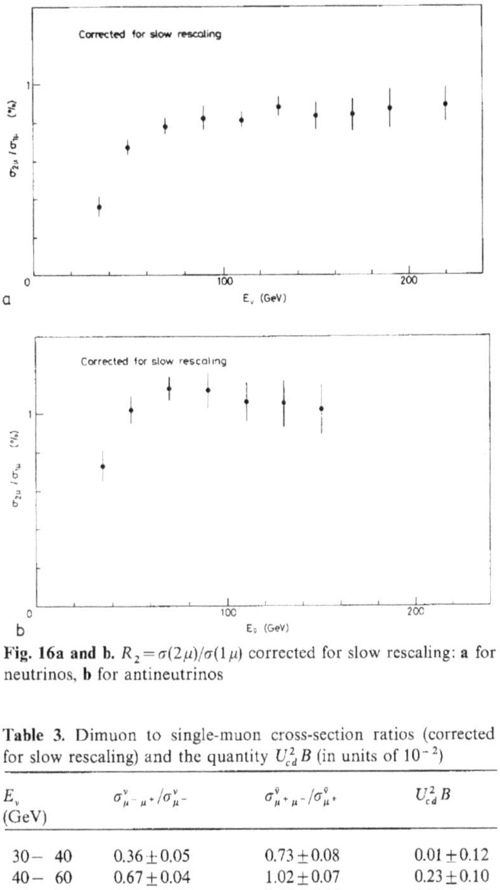

Fig. 16a and b. $R_2=\sigma(2\mu)/\sigma(1\mu)$ corrected for slow rescaling: a for neutrinos, b for antineutrinos

Table 3. Dimuon to single-muon cross-section ratios (corrected for slow rescaling) and the quantity $U_{cd}^2 B$ (in units of 10^{-2})

E_ν (GeV)	$\sigma^\nu_{\mu^-\mu^+}/\sigma^\nu_{\mu^-}$	$\sigma^{\bar{\nu}}_{\mu^+\mu^-}/\sigma^{\bar{\nu}}_{\mu^+}$	$U_{cd}^2 B$
30– 40	0.36±0.05	0.73±0.08	0.01±0.12
40– 60	0.67±0.04	1.02±0.07	0.23±0.10
60– 80	0.78±0.04	1.13±0.06	0.30±0.10
80–100	0.82±0.06	1.12±0.09	0.36±0.13
100–120	0.81±0.04	1.06±0.10	0.39±0.11
120–140	0.88±0.05	1.05±0.12	0.48±0.14
140–160	0.83±0.07	1.02±0.13	0.43±0.17
80–160	0.83±0.025	1.07±0.05	0.41±0.07

using the following experimental results: in an emulsion experiment on neutrino interactions in a 350 GeV WBB [19], it is found that above a visible energy of 30 GeV chiefly D mesons are produced, specifically a mixture of $32\pm11\%\, D^+$ and $68\pm11\%$ D^0 mesons. For this composition of D mesons, using the measured [15, 20, 21] average leptonic branching ratio for a 44%:56% mixture of $D^+:D^0$ in e^+e^- reactions - $B'=(8.2\pm1.2)\%$ - and the less precise separate semileptonic branching ratios for D^+ and D^0 [21] as constraints, the quoted value of B is obtained.

With this value $|U_{cd}|=0.24\pm0.03$. In the GIM model [12] U_{cd} is just the sine of the Cabibbo angle, and our experimental result is in good agreement with the accepted value $\sin\theta_c=0.230\pm0.003$ [22]. In the KM model [9], $U_{cd}=\sin\theta_1\cos\theta_2$, where $\sin\theta_1=0.228\pm0.011$ [13]. We therefore find $\cos\theta_2=1.05\pm0.14$. A 10% variation of the leptonic branching ratio B induces a 5% shift on U_{cd}, and a *difference* of 10% between the leptonic branching ratios of D^0 and $\bar{D}^0$ causes a 7.5% variation in U_{cd}. This, to our knowledge, is the first measurement of θ_2 and is clearly consistent with θ_2 small, although the accuracy leaves something to be desired.

11.2. The Amount of Strange Sea

The amount of the sea is determined from the *shape* of the neutrino and antineutrino x distributions. According to (1) and (2), and given the fact stated in Sect. 9 that the x dependence of $s(x)$ is experimentally consistent with the x dependence of $\bar{u}(x)+\bar{d}(x)$, it should be possible according to (1) to fit the neutrino x distribution with a mixture of $xs(x)$ as given

Table 4. The fraction of strange sea

E_ν (GeV)	r_{sea}	r_q	$\frac{\lvert U_{cs}^2\rvert}{U_{cd}^2}\frac{2S}{U+D}$	$\frac{\lvert U_{cs}^2\rvert}{U_{cd}^2}\frac{2S}{\bar{U}+\bar{D}}$	$\frac{2S}{\bar{U}+\bar{D}}$ for $\frac{U_{cd}^2}{\lvert U_{cs}^2\rvert}=0.056$
35–60	2.11	1.33	1.10±0.16	8.6±1.8	0.48±0.10
60–110	1.53	1.19	1.34±0.10	10.5±1.8	0.59±0.10
110–160	1.36	1.13	1.23±0.17	9.6±2.0	0.54±0.11
>160	1.24	1.10	1.36±0.20	10.6±2.2	0.59±0.12
>35	1.53	1.19	1.19±0.09	9.3±1.6	0.52±0.09

by the antineutrino x distribution and $x[u(x)+d(x)]\simeq 1/2[F_2(x)+xF_3(x)-2xs(x)]$.

Good fits are indeed obtained (see Fig. 10b), and give the results of Table 4, where r_{sea} and r_q are the slow rescaling factors for sea and quark distributions, respectively.

The experimental result for $2S/(U+D)$ is converted into the more interesting result for the ratio of strange to non-strange seas $2S/(\bar{U}+\bar{D})$ on the basis of the results obtained in the charged-current neutrino experiments [10], $\bar{U}+\bar{D}+2\bar{S}=0.070\pm0.005$ and $\int F_2(x)\,dx=0.438\pm0.022$, as well as the result of this experiment, $2S/(U+D)=0.052\pm0.004$ without the correction for slow rescaling.

With these values $(\bar{U}+\bar{D})/(U+D)=0.13\pm0.02$*. The final result, averaged over neutrino energy, is

$$\frac{|U_{cs}^2|}{U_{cd}^2}\frac{2S}{(\bar{U}+\bar{D})}=9.3\pm1.6. \tag{6}$$

If, consistent with the preceding result on θ_2, we assume θ_2 small and also assume θ_3 small in line with the results of the comparisons of β, muon, and K decays [13], so that $U_{cd}^2/|U_{cs}^2|=\tan^2\theta_c=0.056\pm0.005$, then the result for the amount of strange sea relative to the up-down sea becomes:

$$2S/(\bar{U}+\bar{D})=0.52\pm0.09. \tag{7}$$

A 10% variation in the value of m_c used in the slow rescaling correction changes this result by 5%, in the same direction. Subject to the above assumptions concerning the mixing angles, and after slow rescaling correction, the strange quark carries less of the nucleon momentum than the up or down antiquarks by a factor of about two. If the sea were flavour symmetric, we should have expected equal strange and up and down seas. The observation of non-symmetry, if not due to an inadequacy of the analysis, must then be attributed to some breakdown in the symmetry, perhaps to a difference in the strange quark and the up and down quark masses. From this result we can also obtain a limit on the coupling strength $|U_{cs}|$ of charmed and strange quarks. The maximum value for the strange-sea momentum fraction is reached for the symmetric case, $2S=\bar{U}+\bar{D}$. From (6) it follows that $|U_{cs}^2|\geqq(9.3\pm1.6)\,U_{cd}^2$, and therefore $|U_{cs}|>0.59$ at the 90% confidence level.

* This is the result of an iterative procedure based on $[|U_{cs}^2|/U_{cd}^2][2S/(U+D)]=0.92\pm0.06$ and $U_{cd}^2/|U_{cs}^2|=\tan^2\theta_c=0.056\pm0.005$

12. Conclusions

The main results of this work are the following:

i) Neutrino- and antineutrino-induced dimuon events show all the properties expected for a charged-current reaction and charm production and decay.

ii) The chirality of the weak charm-changing charged current is consistent with being $V-A$; a 95% confidence level upper limit of 0.07 is obtained for the square of the relative coupling strength of left-handed currents.

iii) The charm-fragmentation function $D(z)$ is peaked at large values of z with an average of $\langle z\rangle=0.68\pm0.08$.

iv) Two elements of the Kobayashi-Maskawa matrix have been determined, $U_{cd}=0.24\pm0.03$ and $|U_{cs}|>0.59$ at the 90% confidence level.

v) For the Kobayashi-Maskawa angle θ_2 a value of $\cos\theta_2=1.05\pm0.14$ is obtained.

vi) The sea of strange quarks carries 52±9% of the momentum of $\bar{u}$ or $\bar{d}$ quarks if the angles θ_2 and θ_3 are small.

References

1. A. Benvenuti et al.: Phys. Rev. Lett. **34**, 419 (1975)
A. Benvenuti et al.: Phys. Rev. Lett. **35**, 1199 (1975)
A. Benvenuti et al.: Phys. Rev. Lett. **41**, 1204 (1978)
2. B.C. Barish et al.: Phys. Rev. Lett. **36**, 939 (1976)
B.C. Barish et al.: Phys. Rev. Lett. **39**, 981 (1977)
3. M. Holder et al.: Phys. Lett. **69B**, 377 (1977)
4. J. Blietschau et al.: Phys. Lett. **58B**, 361 (1975)
J. Blietschau et al.: Phys. Lett. **60B**, 207 (1976)
B.C. Bosetti et al.: Phys. Lett. **73B**, 380 (1978)

5. J. von Krogh et al.: Phys. Rev. Lett. **36**, 710 (1976)
B.C. Bosetti et al.: Phys. Rev. Lett. **38**, 1248 (1977)
6. C. Baltay et al.: Phys. Rev. Lett. **39**, 62 (1977)
7. M. Jonker et al.: Phys. Lett. **107B**, 241 (1981)
8. M. Holder et al.: Nucl. Instrum. Methods **148**, 235 (1978)
9. M. Kobayashi, K. Maskawa: Progr. Theor. Phys. **49**, 652 (1973)
10. J.G.H. De Groot et al.: Z. Phys. C - Particles and Fields **1**, 143 (1979)
H. Abramowicz et al.: Neutrino and antineutrino charged-current inclusive scattering in iron in the energy range $20 < E_\nu < 300$ GeV. to be published in Z. Phys. C - Particles and Fields
11. J.G.H. De Groot et al.: Phys. Lett. **86B**, 103 (1979)
12. S.L. Glashow, J. Iliopoulos, L. Maiani: Phys. Rev. **D2**, 1285 (1970)
13. R.E. Shrock, L.L. Wang: Phys. Rev. Lett. **41**, 1692 (1978) and **42**, 1589 (1979)
14. R. Brock: Phys. Rev. Lett. **44**, 1027 (1980)
15. W. Bacino et al.: Phys. Rev. Lett. **43**, 1073 (1979)
16. J.D. Bjorken: Phys. Rev. **D17**, 171 (1978); M. Suzuki: Phys. Lett. **71B**, 139 (1977); H. Georgi, H.D. Politzer: Nucl. Phys. **B136**, 445 (1978)
17. H.C. Ballagh et al.: Phys. Rev. **D24**, 7 (1981)
18. C. Baltay: Recent results from neutrino experiment in heavy neon bubble chambers in Proc. 1979 JINR-CERN School of Physics, September 1979, p. 72. Budapest: Hungarian Academy of Sciences, 1980
19. N. Stanton et al.: E531 Collaboration in Neutrino 81, Proc. 11th Int. Conf. on Neutrino Physics and Astrophysics, Maui, Hawaii, 1981, p. 491. Honolulu: Dept. of Phys. and Astrophys., Univ. of Hawaii, 1981; E. Fisk: Rapporteur's talk, in Proc. Int. Symposium on Lepton and Photon Interactions at High Energy, Bonn, 1981, p. 703. Bonn: Universität Bonn, 1981
20. J.M. Feller et al.: Phys. Rev. Lett. **40**, 274 (1978)
21. R.H. Schindler: Ph.D. Thesis, SLAC report 219 (1979)
22. M. Roos: As quoted by K. Kleinknecht: In Proc. 17th Int. Conf. on High Energy Physics, London, England, 1974, p. III-23. Chilton, Didcot, Berks.: Rutherford High Energy Laboratory, 1975, and M. Nagel et al.: Nucl Phys. **B109**, 1 (1976). See also M. Roos: Nucl. Phys. **B77**, 420 (1974)

Z. Phys. C – Particles and Fields 17, 283–307 (1983)

Zeitschrift für Physik C Particles and Fields

Neutrino and Antineutrino Charged-Current Inclusive Scattering in Iron in the Energy Range $20 < E_\nu < 300$ GeV

H. Abramowicz[1], J.G.H. de Groot, J. Knobloch, J. May, P. Palazzi, A. Para[1], F. Ranjard, A. Savoy-Navarro[2], D. Schlatter, J. Steinberger, H. Taureg, W. von Rüden, H. Wahl, J. Wotschack

CERN, CH-1211 Geneva 23, Switzerland

P. Buchholz, F. Eisele, H.P. Klasen, K. Kleinknecht, H. Lierl, D. Pollmann, B. Pszola, B. Renk, H.J. Willutzki

Institut für Physik* der Universität, D-4600 Dortmund, Federal Republic of Germany

F. Dydak, T. Flottmann, C. Geweniger, J. Królikowski[1], K. Tittel

Institut für Hochenergiephysik* der Universität, D-6900 Heidelberg, Federal Republic of Germany

P. Bloch, B. Devaux, C. Guyot, J.P. Merlo, B. Peyaud, J. Rander, J. Rothberg[3], R. Turlay

DPhPE, CEN-Saclay, F-91190 Gif-sur-Yvette, France

J.T. He, T.Z. Ruan, W.M. Wu

Institute of High-Energy Physics, Beijing, China

Received 20 December 1982

Abstract. Inclusive charged-current interactions of high-energy neutrinos and antineutrinos have been studied with high statistics in a counter experiment at the CERN Super Proton Synchrotron. The energy dependence of the total cross-sections, the longitudinal structure function, and the nucleon structure functions F_2, xF_3, and $\bar{q}^{\bar{\nu}}$ are determined from these data. The analysis of the Q^2-dependence of the structure functions is used to test quantum chromodynamics, to determine the scale parameter Λ and the gluon distribution in the nucleon.

1. Introduction

The inclusive scattering of neutrinos and antineutrinos is of interest chiefly as a means of studying the structure of the nucleon and the strong interactions. The Callan-Gross relation [1], the Gross-Llewellyn Smith [2] sum rule, and the comparison of neutrino and charged lepton structure functions, permit tests of the quark parton model. The structure functions themselves show us how the quarks and gluons share the nucleon momentum, in particular, the small deviation from scaling permits a quantitative confrontation of quantum chromodynamics (QCD) with experiment.

We report here experimental results which are more extensive and precise than our previously published results [3–5]. In this work we were largely motivated by the interest in improving the comparison of the structure functions with QCD predictions. The results are based on the analysis of data obtained in 200 GeV neutrino and antineutrino narrow-band beams, in a 300 GeV narrow-band neutrino beam, and also in wide-band beams of both polarities. In total, we report here on 130,000 neutrino and 180,000 antineutrino charged-current events, after event selection, which can be compared with 23,000 and 6,200 events, respectively, in the previous publication [3].

Partial results based on the same data have already been published. These include the measurement of the longitudinal structure function [6], a limit on right-handed currents [7], the determination of the gluon distribution in the nucleon [8], and the comparison of the measured structure functions with QCD and non-asymptotically free theories of the strong interaction [9].

* Supported by the Bundesministerium für Forschung und Technologie, Bonn, Fed. Rep. Germany

1 On leave from the Institute of Experimental Physics, Warsaw University, Poland

2 On leave from the DPhPE, CEN-Saclay, France

3 On leave from the University of Washington, Seattle, USA

2. Phenomenology

The kinematic quantities for inclusive charged-current scattering are defined in the usual way:

$Q^2 = -(k-k')^2$

$\nu = (k-k')\cdot p/m_N$, where m_N = nucleon mass

$x = Q^2/2m_N\nu$

$y = m_N\nu/k\cdot p \simeq E_h/E_\nu$, where E_ν and E_h are the kinetic energy of the neutrino and of the hadron system in the laboratory system, respectively.

Assuming the standard $V-A$ theory of the weak interactions, the neutrino and antineutrino charged-current cross-sections can each be written in terms of three structure functions:

$$\frac{d^2\sigma^{\nu,\bar\nu}}{dx\,dy}=\frac{G^2 m_N E_\nu}{\pi}\,\frac{1}{(1+Q^2/m_W^2)^2}\cdot\left\{\left[1-y-\frac{m_N x y}{2E_\nu}\right]F_2^{\nu,\bar\nu}(x,Q^2)+\frac{y^2}{2}\,2xF_1^{\nu,\bar\nu}(x,Q^2)\pm\left[y-\frac{y^2}{2}\right]xF_3^{\nu,\bar\nu}(x,Q^2)\right\}. \qquad (1)$$

The propagator term $(1+Q^2/m_W^2)^{-2}$ is a small correction. With a vector boson mass of $m_W \approx 80$ GeV, as predicted in the Weinberg-Salam theory, and using the present experimental value for the electroweak angle, the correction of the cross-section is about 10% at the highest Q^2 of this experiment. It will be dropped in all formulae which follow.

In the quark parton model the structure functions are written in terms of the quark and antiquark momentum distributions in the nucleon. For isoscalar nuclei we have

$$2xF_1^{\nu}(x,Q^2)=2xF_1^{\bar\nu}(x,Q^2)=q(x,Q^2)+\bar q(x,Q^2)$$
= total momentum distribution of quarks and antiquarks. (2)

$$F_2^{\nu,\bar\nu}(x,Q^2)=2xF_1(x,Q^2)+F_L(x,Q^2) \qquad (3)$$

$$xF_3^{\nu,\bar\nu}(x,Q^2)=q(x,Q^2)-\bar q(x,Q^2)\pm 2x[s(x,Q^2)-c(x,Q^2)] \qquad (4)$$

$$xF_3(x,Q^2)=\tfrac{1}{2}(xF_3^{\nu}+xF_3^{\bar\nu})=q(x,Q^2)-\bar q(x,Q^2). \qquad (5)$$

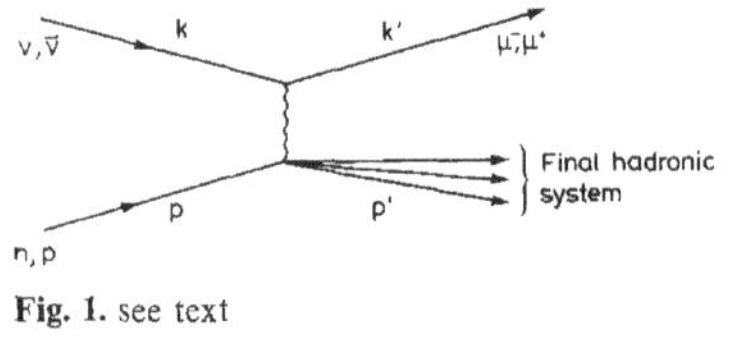

Fig. 1. see text

Here xF_3 is the momentum distribution of the valence quarks inside the nucleon, $F_L = F_2 - 2xF_1$ is the longitudinal structure function, and q and $\bar q$ are the sum of quark and antiquark momentum distributions, respectively:

$q = x(u+d+s+c)$

$\bar q = x(\bar u+\bar d+\bar s+\bar c)$;

u, d, s, and c are the up, down, strange, and charmed quark distribution in the proton as seen by the weak charged current.

If the transverse momentum of the quarks with respect to the nucleon momentum can be neglected, F_L is zero and the Callan-Gross relation [1] $F_2 = 2xF_1$ is valid. If F_L can be neglected, the neutrino and antineutrino cross-sections then have the simple form:

$$\frac{d^2\sigma^{\nu}}{dx\,dy}=\frac{G^2 m_N E_\nu}{\pi}\{q+x(s-c)+(1-y)^2[\bar q-x(\bar s-\bar c)]\}, \qquad (6)$$

$$\frac{d^2\sigma^{\bar\nu}}{dx\,dy}=\frac{G^2 m_N E_\nu}{\pi}\{\bar q+x(\bar s-\bar c)+(1-y)^2[q-x(s-c)]\}. \qquad (7)$$

The expressions (3) to (7) are valid up to terms $\sim Q^2/\nu^2$ which have been dropped for the sake of simplicity. The general formulae which have been used in the analysis are given in the Appendix. A combined analysis of neutrino and antineutrino differential cross-sections allows the separate measurement of the valence and the sea-quark distributions, as will be shown in Sect. 4. This is a unique feature of neutrino experiments and is due to the $V-A$ structure of charged-current weak interactions.

3. Experimental Procedure

3.1. Apparatus

The detector has been described elsewhere [10]. It consists of 19 toroidally magnetized iron modules, each composed of circular plates 3.75 m in diameter, with a total iron thickness of 75 cm per module, corresponding to a mass of 65 t. These modules serve simultaneously as neutrino target, hadron-shower calorimeter, and muon-spectrometer magnet. Scintillator sheets are sandwiched between the iron plates, to sample the hadron-shower energy and to trigger the detector. Between modules, drift chambers are inserted, each with three wire planes inclined at 60° to each other. The first seven modules have scintillator planes every five centimetres of iron, the next eight have them every fifteen centimetres; the last four modules serve as muon analysers

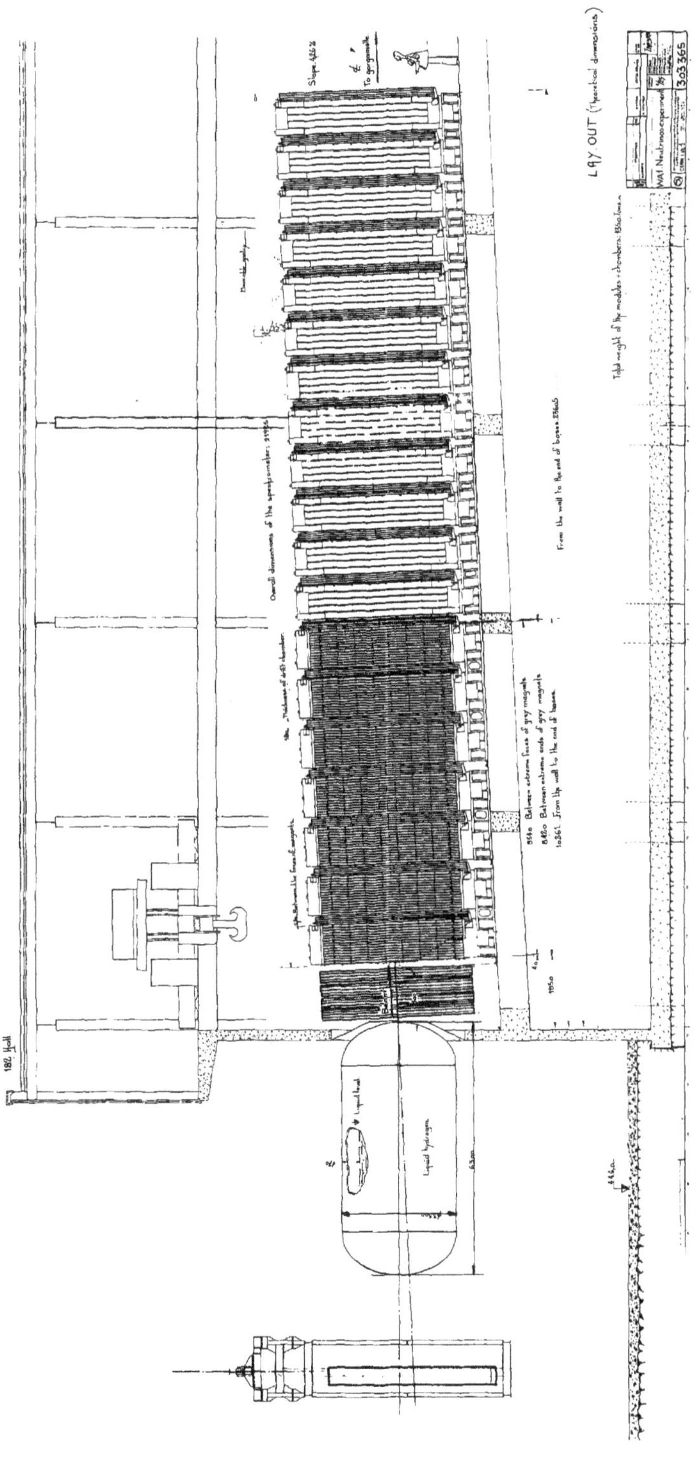

Fig. 2. Over-all view of the detector

only and are equipped with a single plane of scintillators for triggering purposes (see Fig. 2).

The hadron-energy response and the resolution of the detector have been measured by putting three modules into a hadron beam [11]. The resolution is approximately equal to $\Delta E/E \simeq 0.7/\sqrt{E}$ for modules with 5 cm sampling, and $1.35/\sqrt{E}$ for 15 cm sampling. The scintillator and phototube responses are continuously monitored by means of cosmic-ray muons which are recorded between bursts.

The drift-chamber resolution, including the uncertainty in the wire positions, is about 1 mm. However, the muon momentum error is dominated by multiple scattering in the iron; for the data reported here it is, on the average, $\Delta p_\mu/p_\mu \simeq 9\%$. We note that the sign of the magnetic field is chosen such that the muons are focused towards the axis, and in general, owing to the large diameter and high density of the detector, the muons are trapped and leave the apparatus only through the end. As a consequence the muon acceptance for charged-current events is nearly one and uniform except for muons with momenta below ~ 5 GeV, which do not traverse a sufficient number of drift chambers.

The fiducial mass of the detector is about 600 t, which gives rise to large event rates in intense neutrino beams. The rate of data taking is limited by the conversion time of the pulse-height circuits of ~ 2.5 μs. Pulse-height and drift-chamber circuits feed into fast buffers, 40 events deep, so that several events may be accepted even within the short 23 μs spill of narrow-band beams, and up to 40 events may be accepted within the millisecond spill of the wide-band beams.

The central area of the detector, where the magnet coils pass through, is not fully covered by scintillators. Furthermore, an area 30 cm in diameter (0.6% of the total area) is not magnetized, although it is filled with material of approximately the density of the iron. To ensure full shower containment and good muon momentum measurement, events which originate in a lozenge-shaped region (as shown in Fig. 3) are excluded from the analysis. The centre of the neutrino beam is directed ~ 45 cm below the centre of the modules so that we lose only a small fraction of all events with this cut and keep especially the highest energy neutrinos, which are in the beam centre.

The unique feature of this detector is the combination of calorimeter and muon spectrometer in one high-density unit, resulting in excellent muon acceptance. The hadron shower containment is complete in the fiducial region of about 600 t. These properties, combined with the ability to record large event numbers in intense neutrino beams, make this detector uniquely suitable for the systematic and precise measurement of inclusive neutrino cross-sections.

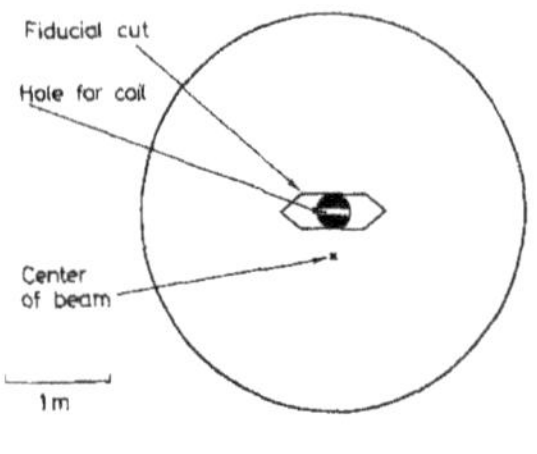

Fig. 3. Fiducial region of the calorimeter modules. The central region inside the lozenge-shaped area has been excluded from the analysis

Table 1. Characteristics of the five exposures used in this paper

Type of beam	Main characteristics	Number of protons on target	Number of charged-current events, after cuts
200 GeV νNBB	Positive, momentum-selected hadrons, $\Delta p/p = \pm 5\%$, $\Omega \simeq 15$ μsr	1.1×10^{18}	62,000
200 GeV $\bar{\nu} NBB$	Negative, momentum-selected hadrons, $\Delta p/p \simeq \pm 5\%$, $\Omega \simeq 15$ μsr	2.9×10^{18}	26,000
300 GeV νNBB	Positive, momentum-selected hadrons, $\Delta p/p \simeq \pm 5\%$, $\Omega \simeq 10$ μsr	2.4×10^{18}	32,000
350 GeV $\bar{\nu} WBB$	Horn-focused beam, negative particles focused, 350 GeV protons on target	4.3×10^{17}	155,000
350 GeV νWBB	Horn-focused beam, positive particles focused, 350 GeV protons on target	0.3×10^{17}	35,000

3.2. *Neutrino Beams*

The data were obtained in five different beams. Some of the characteristics of the exposures are given in Table 1.

The neutrino fluxes corresponding to these exposures are shown in Figs. 4a, b. The energy dependence of the narrow-band beam spectra are determined by the geometrical properties of the hadron beams and by the kaon-to-pion ratios, which were measured periodically with a differential Cherenkov counter in front of the decay tunnel. The K/π ratios were found to be the following:

a) 200 GeV ν (400 GeV/c p) $K^+/\pi^+ = 0.146 \pm 0.005$
b) 200 GeV $\bar{\nu}$ (400 GeV/c p) $K^-/\pi^- = 0.049 \pm 0.002$
c) 200 GeV $\bar{\nu}$ (450 GeV/c p) $K^-/\pi^- = 0.056 \pm 0.002$
d) 300 GeV ν (400 GeV/c p) $K^+/\pi^+ = 0.24 \pm 0.01$.

The wide-band beam spectra are not measured externally but are derived from the observed charged-current event rates and the total neutrino cross-sections as measured in the narrow-band beams. In the present work, the wide-band beam data are used specifically to determine the antiquark structure function (Sect. 4.2.3), where the large statistics obtained in the wide-band beam outweigh the disadvantage that the spectrum is not determined independently.

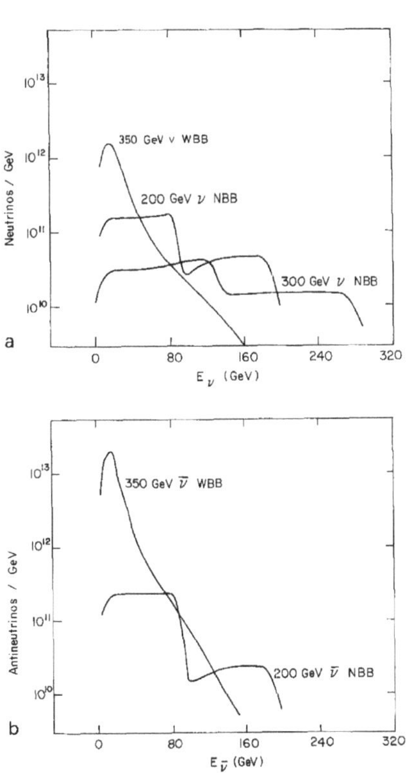

Fig. 4a and b. Absolute neutrino flux as a function of energy for the narrow-band and wide-band beams in which the present data have been obtained: **a** neutrino beams, **b** antineutrino beams

3.3. *Data Selection and Corrections*

3.3.1. Trigger Conditions

The trigger is based on the total scintillation pulse height observed in individual modules. It differs for wide- and narrow-band beams because of the very large difference in event rates in the two types of beam. The trigger conditions relevant for the selection of charged-current events are as follows:

i) Narrow-band beam trigger: At least three modules give a signal. The required signal level is sufficiently low, so that single muons will trigger with high efficiency. This trigger is effective for all charged-current events, provided a muon of at least 3 GeV is produced.

ii) Wide-band beam trigger: Events are selected if a total of approximately 7 GeV of energy, muonic or hadronic, is deposited anywhere in the apparatus. Since the energy loss of a muon is about 1 GeV per module, the trigger is satisfied even for single muons, provided they pass through more than seven modules.

3.3.2. Reconstruction

The reconstruction of a typical charged-current event by the off-line program is shown in Fig. 5. The hadronic shower energy is computed from the measured scintillator pulse heights on the basis of an algorithm designed to optimize the energy resolution and to reproduce data obtained with hadronic test beams in the modules [11]. The energy loss of the muon in the shower region is subtracted according to the observed muon momentum.

The muon-momentum reconstruction accounts for the energy loss due to ionization. Radiative losses in excess of ~ 1 GeV are taken into account using the observed pulse height in the scintillators along the track. The muon-track reconstruction has been cross-checked by studying a subset of several thousand events with the help of an interactive program and visual inspection. The efficiency of the muon reconstruction program was found to be 96 $\pm 1\%$ for events in the fiducial region and with muon momenta larger than 7 GeV. A fraction of the 4% of events which fail in the automatic recon-

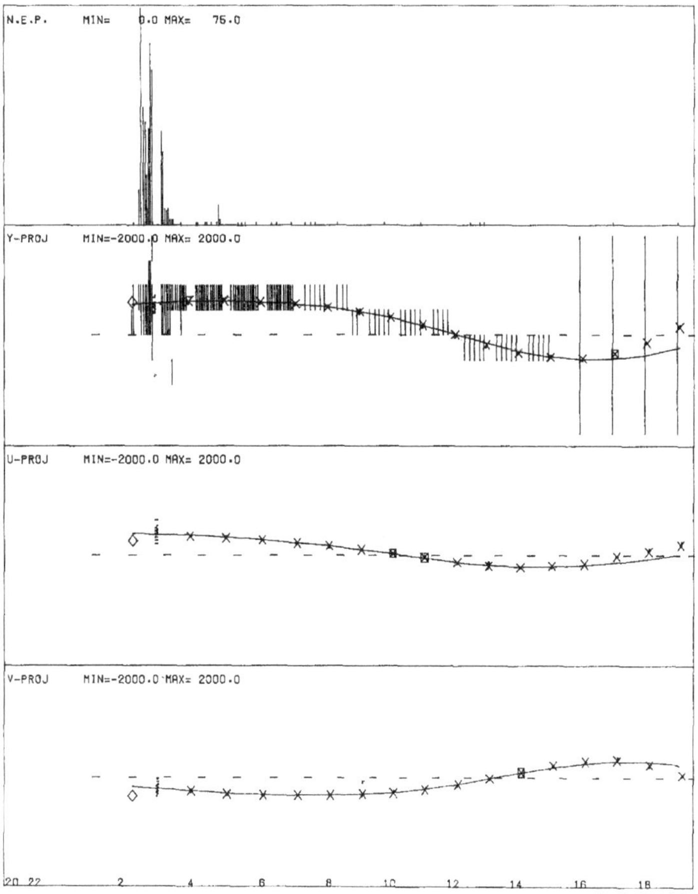

Fig. 5. Computer reconstruction of a charged-current neutrino interaction in the detector. The neutrino enters from the left. The top row gives the pulse height in the scintillator planes along the detector. The rows below show the measured coordinates in the three drift-chamber projections together with the fit curve for the muon reconstruction

struction have been reconstructed by hand and compared with the normal events. They show no significant difference in any kinematical quantity.

The energy scale for the muon- and hadron-energy reconstruction has been checked by means of the narrow-band beam events using the fact that the neutrino energy and the radial vertex position are correlated [10]. Events with small $y=E_h/E_\nu$ and large y serve to check the muon- and hadron-energy calibration, respectively.

3.3.3. Selection Criteria

Events are accepted provided the following conditions are met:

i) Track length: The muon has to pass through at least five drift chambers. As the hadron shower typically extends over one or two drift chambers only, the remaining event sample has a good muon reconstruction efficiency.

ii) Accuracy of track fit: The probability of the muon fit must be greater than 0.001.

iii) Fiducial volume: The origin of the event must be in the first 13 modules for the narrow-band beam (11 modules for the wide-band beam) within a radius of 1.6 m from the module centre and outside the central excluded region shown in Fig. 3. These fiducial requirements assure complete shower containment and adequate residual track length for the muon track measurement.

iv) Muon momentum: The reconstructed muon momentum must be greater than 7 GeV. This cut is applied only for the narrow-band beam data and is generally more severe than cut (i).

We note that these selection criteria are based on the event origin and the muon; the presence of a hadron shower is not required.

3.3.4. Corrections

The observed event numbers have to be corrected for experimental losses. The narrow-band beam data, which are used for the absolute cross-section measurement, are corrected for the reconstruction inefficiency and the measured dead-time of the detector. For the wide-band beam data a correction (<6%) is applied to the total event number in each energy bin to account for the trigger inefficiency for events with small hadron energy which originate at the end of the fiducial volume.

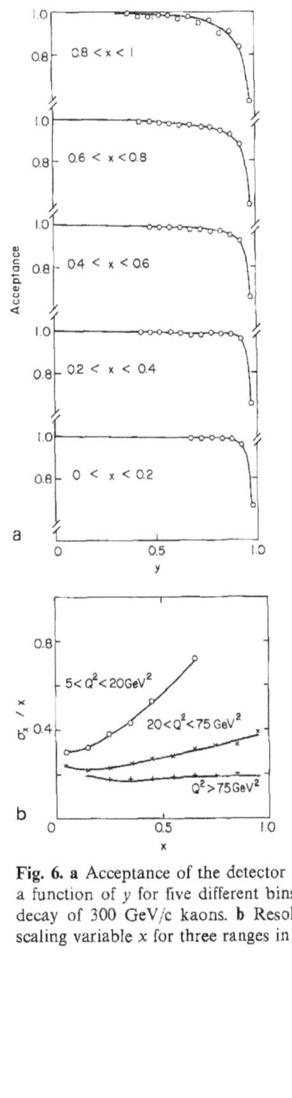

Fig. 6. **a** Acceptance of the detector for charged-current events as a function of y for five different bins in x for neutrinos from the decay of 300 GeV/c kaons. **b** Resolution of the detector in the scaling variable x for three ranges in Q^2

The observed event distributions are corrected for acceptance and experimental resolution with the help of a Monte Carlo simulation. In the case of the narrow-band beam, the Monte Carlo events were generated according to the known properties of the neutrino beam and the measured K/π ratios. For the wide-band beam, the energy and radial dependence of the neutrino flux was adjusted to match the observed event distributions. The measurement errors are simulated using the known resolution functions. The observed event numbers in any particular bin of x, y, and Q^2, or ν, are corrected for the effects of acceptance and resolution by multiplying by the ratio of generated-to-accepted Monte Carlo events for the same bin and beam condition. These corrections depend on the x, y, and Q^2 dependence of the cross-sections, which we want to determine. We have approached the true shape of the cross-sections iteratively by repeated comparison of the Monte Carlo simulation with the data.

The average acceptance as a function of y in different x-bins is shown in Fig. 6a for the example of high-energy neutrinos from the decay of 300 GeV/c kaons. The resolution in x for different ranges in Q^2 is shown in Fig. 6b.

4. Results

4.1. Energy Dependence of Total Cross-Section

Determination of the absolute cross-sections requires knowledge of the neutrino spectrum and flux. The narrow-band beam fluxes can be calculated from the hadron beam optics, the decay kinematics, and the absolute kaon and pion fluxes. The K/π flux ratios were given in Sect. 3.2, and the absolute fluxes were determined in two ways: i) on the basis of the absolute hadron flux, as measured with a beam-current transformer; ii) on the basis of the muon fluxes at different depths in the shield, measured by means of solid-state detectors, calibrated by track counting in nuclear emulsions. For the 200 GeV neutrino and anti-neutrino beams, the two methods are in agreement within 5–10%, and the resulting cross-sections are in agreement with those previously published [3]. However, both methods pose problems, and at 300 GeV the disagreement is greater. We hope to improve our understanding of the absolute neutrino fluxes in the future. For the present, we do not believe that we have made enough progress to publish new absolute cross-sections but content ourselves with normalizing the new results to the published cross-sections [3] in the common energy region. New absolute cross-section results will be postponed to a future experiment.

Although the absolute levels of the cross-sections are normalized to our old result, the energy variation of the cross-sections is new and independent, and for neutrinos extends to higher energies. The results are shown in Fig. 7 and tabulated in Table 2. The error bars include an estimate of the systematic point-to-point error. In addition, Table 2 gives also the over-all scale errors owing to the uncertainties of the absolute particle flux measurement and of the K/π ratios. For neutrinos, σ/E_ν shows a drop with energy below $E_\nu \simeq 70$ GeV. No other significant energy dependence is observed, either for neutrinos or antineutrinos, in agreement with other recent results [12] as well as with our previous result [3], but in disagreement with Blair et al. [13]. We note that the measured energy variations are in good agreement with the expectations based on the observed scaling violations as discussed in Sect. 5.

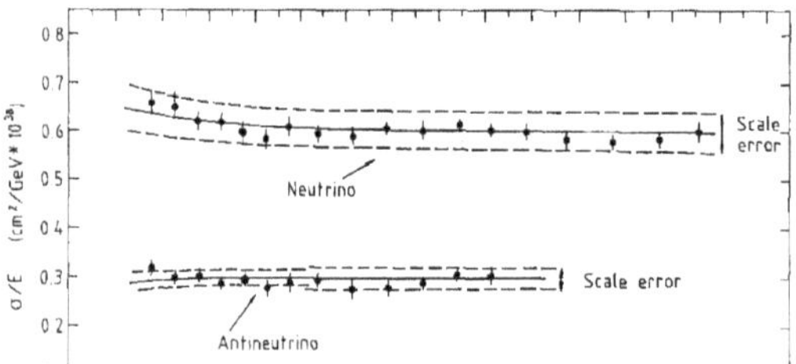

Fig. 7. Energy dependence of the total neutrino and antineutrino cross-sections. The error bars include an estimate of systematic point-to-point errors. The solid line shows the energy dependence as expected from the observed scaling violation of the structure functions. The dashed lines indicate the over-all scale errors

Table 2. Energy dependence of total neutrino and antineutrino cross-section from 200 and 300 GeV narrow-band beams. The systematic errors given are the estimated point-to-point errors due to energy calibration and resolution. The absolute normalization is fixed to yield $\sigma^\nu/E = 0.62$ and $\sigma^{\bar{\nu}}/E = 0.30$ in the energy range $30 < E_\nu < 90$ GeV, in agreement with the published results [3]. Both cross-section slopes have an additional over-all scale error

Neutrino energy bin	σ^ν/E_ν ($\times 10^{38}$ cm^{-2} GeV)		$\sigma^{\bar{\nu}}/E_{\bar{\nu}}$ ($\times 10^{38}$ cm^{-2} GeV)	
(GeV)	Value $\pm \sigma_{stat} \pm \sigma_{syst}$	Scale error	Value $\pm \sigma_{stat} \pm \sigma_{syst}$	Scale error
30–40	0.660 ± 0.015 ± 0.009	±6%	0.319 ± 0.009 ± 0.006	±5%
40–50	0.649 ± 0.013 ± 0.012		0.297 ± 0.008 ± 0.005	
50–60	0.618 ± 0.011 ± 0.011		0.303 ± 0.007 ± 0.006	
60–70	0.615 ± 0.010 ± 0.010		0.285 ± 0.006 ± 0.005	
70–80	0.599 ± 0.008 ± 0.004		0.297 ± 0.006 ± 0.002	
80–90	0.584 ± 0.009 ± 0.010		0.279 ± 0.007 ± 0.010	
90–100	0.611 ± 0.013 ± 0.007	±7.2%	0.287 ± 0.019 ± 0.008	±6.5%
100–115	0.594 ± 0.009 ± 0.005		0.294 ± 0.013 ± 0.007	
115–130	0.588 ± 0.009 ± 0.010		0.271 ± 0.011 ± 0.009	
130–145	0.605 ± 0.009 ± 0.006		0.274 ± 0.010 ± 0.006	
145–160	0.597 ± 0.008 ± 0.006		0.287 ± 0.010 ± 0.005	
160–175	0.610 ± 0.008 ± 0.004		0.302 ± 0.010 ± 0.006	
175–190	0.600 ± 0.009 ± 0.005		0.303 ± 0.018 ± 0.010	
190–205	0.595 ± 0.014 ± 0.009			
205–225	0.582 ± 0.012 ± 0.006			
225–245	0.572 ± 0.011 ± 0.008			
245–265	0.582 ± 0.012 ± 0.012			
265–285	0.595 ± 0.013 ± 0.015			

4.2. *Determination of Nucleon Structure Functions*

The first step in the determination of nucleon structure functions is the tabulation of the differential neutrino and antineutrino cross-sections in bins of x, y, and Q^2, or ν. The correction for acceptance and detector resolution is done with the help of a Monte Carlo simulation as described in Sect. 3. Bins are accepted only if the unsmearing correction, i.e. the ratio of the true population to the measured population in the bin, differs from one by less than 40%. This selection criterion eliminates essentially all bins with $x > 0.7$, where the majority of observed events has been shifted into this region from smaller values of x owing to resolution effects, and the highest y-bins where the acceptance is low owing to the muon momentum cut.

The differential cross-sections are determined for an isoscalar target, correcting for the small excess of neutrons in the iron nucleus according to the formulae given in the Appendix. Radiative effects are corrected according to the prescription of De Rujula et al. [14]. These corrections reduce the cross-sections for $x \lesssim 0.2$, increase them for large x, and are generally smaller than 10% except for very small x. Previously, published results [3] as well as preliminary results of the present experiment [15] have used an approximate parametrization for the radiative corrections due to Barlow and Wolfram [16] which differs substantially for small values of x from the present corrections.

The values of the structure functions are evaluated at the centre of the bins in x and Q^2. Fermi-motion effects have not been corrected since they are model-dependent. They mainly affect the shape of the structure functions at large x but have very little effect on the Q^2-dependence. Also no correction has been applied for the suppression of the strange sea due to the threshold effect in the transitions $s \to c$.

The expressions which have been used to obtain the structure functions from the differential cross-sections measurements are summarized in the Appendix and include all corrections.

4.2.1. Longitudinal Structure Function

The longitudinal structure function F_L is expected to be non-zero owing to the transverse momentum of the quarks with respect to the nucleon direction which, at high Q^2, can be calculated in perturbative QCD. F_L will also have a contribution if spin-zero constituents such as, for example, diquark systems contribute to neutrino scattering. From an experimental point of view, the longitudinal structure function gives rather small relative contributions to the differential cross-sections in most of the kinematic range so that its determination is subject to severe statistical and systematic errors.

The present analysis uses two different methods to determine the ratio $R=\sigma_L/\sigma_T=F_L/2xF_1$. The first method is based on the sum of neutrino and antineutrino differential cross-sections:

$$\frac{d^2\sigma^\nu}{dx\,dy}+\frac{d^2\sigma^{\bar\nu}}{dx\,dy}=\frac{G^2 m_N E_\nu}{\pi}\{[1+(1-y)^2]\,F_2 -y^2 F_L+2x(s-c)\,[1-(1-y)^2]\}. \tag{8}$$

The structure functions F_L and F_2 are separated on the basis of their y-dependence. The second method gives upper limits on R at large x and is based on the expression:

$$\frac{\pi}{G^2 m_N E_\nu}\left[\frac{d^2\sigma^{\bar\nu}}{dx\,dy}-(1-y)^2\frac{d^2\sigma^\nu}{dx\,dy}\right]\approx\{\bar q^{\bar\nu}+F_L[(1-y) -(1-y)^3]-2x(s-c)(1-y)^2\}, \tag{9}$$

which is approximately valid for $y\gtrsim 0.5$. Experimentally it is observed that the left-hand side is compatible with zero for $x\gtrsim 0.4$ which, in Sect. 4.2.3, is interpreted as the antiquark distribution being limited to small values of x. According to (9), this observation can be used to put an upper limit on F_L, keeping in mind that $\bar q^{\bar\nu}$ has to be larger or equal to zero. This second method is more sensitive and reliable since it is based on the magnitude of the observed cross-sections only.

4.2.1.1. Analysis of the y-Dependence. The analysis is based on the 200 GeV narrow-band beam data and has been reported in detail in [6]. The value of R is determined from the y-distribution for fixed bins in x and ν, i.e. using events from different neutrino energies. The result is free from assumptions about the nature of scaling violations in contrast to previous results which were obtained invoking either Bjorken scaling [17, 18] or a definite prescription of scaling violations which was not tested independently [3].

The results for R versus ν, averaged over x and for R versus x averaged over ν are shown in Figs. 8 and 9, respectively. As R does not seem to depend strongly on either ν or x, an average value of R for $\langle\nu\rangle=50$ GeV can be obtained by averaging the results in ν or x bins, giving $\langle R\rangle=0.10\pm0.025\pm0.06$, where statistical and systematic errors are given in turn. This result depends on the value used for the $s-c$ quark sea. To obtain the above result, we have assumed $x(s-c)=0.12\bar q$. A change to $x(s-c)=0.2x(\bar u+\bar d+\bar s)$ which is used throughout the rest of this paper, increases R by 0.02. The uncertainty in the amount of strange sea is not included in the systematic error of R.

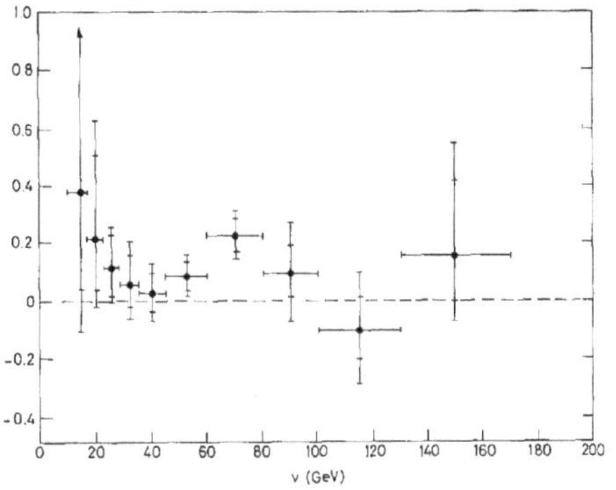

Fig. 8. $R=\sigma_L/\sigma_T$ averaged over x as a function of the energy transfer. The inner error bars are the statistical errors, the full error bars include an estimate of the systematic errors

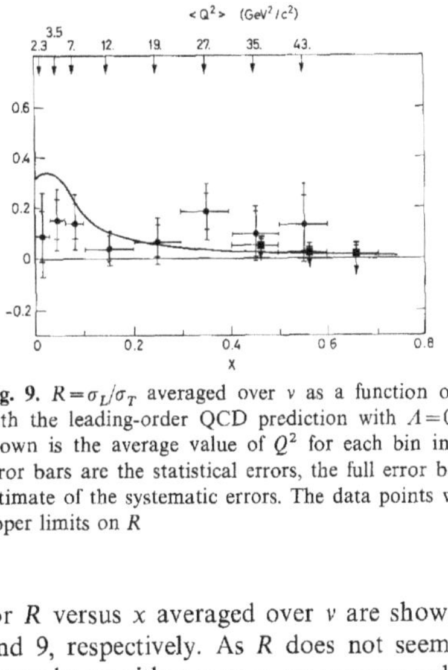

Fig. 9. $R=\sigma_L/\sigma_T$ averaged over ν as a function of x compared with the leading-order QCD prediction with $\Lambda=0.2$ GeV. Also shown is the average value of Q^2 for each bin in x. The inner error bars are the statistical errors, the full error bars include an estimate of the systematic errors. The data points with arrow are upper limits on R

4.2.1.2. Upper Limit on R at Large x. Both narrow-band and wide-band beam data are used to evaluate the left-hand side of (9) in the energy range

Table 3. Upper limits for $R=\sigma_L/\sigma_T$ at large x

x	0.35	0.45	0.55	0.65
$R=\sigma_L/\sigma_T \leqq$	0.152	0.058	0.022	0.017
σ_{stat}	0.019	0.019	0.024	0.033
σ_{syst}	0.042	0.022	0.025	0.035
$\langle Q^2\rangle$ [GeV^2/c^2]	30.0	37.0	39.0	37.0

$20 \leqq E_\nu \leqq 165$ GeV and for $y \geqq 0.44$. The results are summarized in Table 3 for four bins in x. Averaging over the x-range $0.4 \leqq x \leqq 0.7$, we obtain the result $R \leqq 0.039 \pm 0.014 \pm 0.025$ for $\langle Q^2\rangle = 38\,\text{GeV}^2/\text{c}^2$, where statistical and systematic errors are given in turn. The main systematic uncertainties are due to the errors in the cross-section ratio $\sigma^{\bar{\nu}}/\sigma^{\nu}$ mainly at small energies and to a smaller extent due to uncertainties in the hadron- and muon-energy calibration. Correction terms proportional to Q^2/ν^2, which have been omitted in (9) for the sake of simplicity, are important. They have been taken into account in order to derive these results according to the formulae given in the Appendix.

It should be noted that this method gives good upper limits on R only for the x-range where the antiquark contribution is small.

4.2.1.3. Discussion. In Fig. 10 our results, which correspond to an average value of $\langle\nu\rangle \simeq 50$ GeV, are compared with the SLAC-MIT results [20] corresponding to $\langle\nu\rangle$ of about 8 GeV, and the FNAL μp result [21]. The SLAC-MIT experiment measures values of R which are non-zero at large x, outside the given statistical errors, and are in contrast with the QCD expectation. This result has been interpreted as evidence for diquark contributions at large x [22]. The upper limits on R at large x and Q^2 from the present experiment, which are in agreement with the QCD prediction, do not exclude such a diquark contribution since it is expected to disappear very rapidly with Q^2. Comparing the SLAC results with the present analysis, there is an indication of a longitudinal contribution which decreases with Q^2 both at small x and at large x. The measured x-dependence of R is consistent with the QCD prediction. The experimental errors at small x are however still very large.

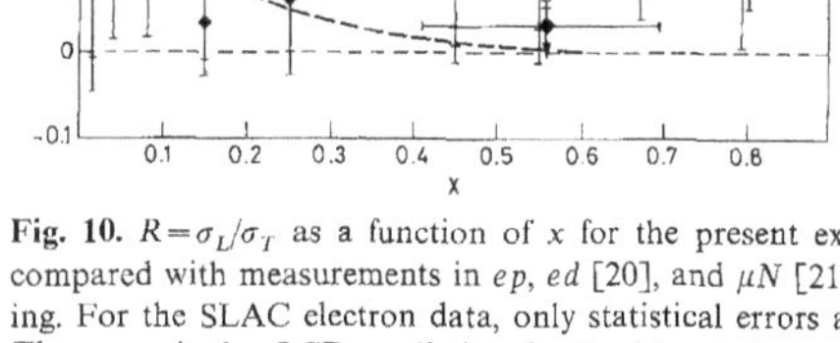

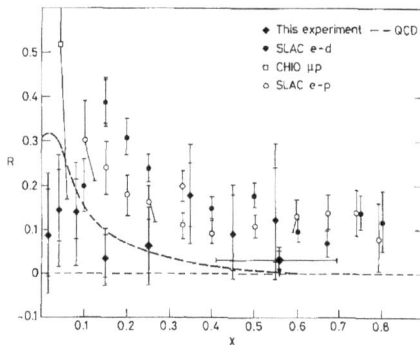

Fig. 10. $R=\sigma_L/\sigma_T$ as a function of x for the present experiment compared with measurements in ep, ed [20], and μN [21] scattering. For the SLAC electron data, only statistical errors are given. The curve is the QCD prediction for the kinematic range of this experiment neglecting the contribution of the charmed quark

4.2.2. The Structure Functions F_2, $2xF_1$, and xF_3

These structure functions are related to the differential cross-sections by the following approximate formulae:

$$F_2 = \frac{\left\{\frac{\pi}{G^2 m_N E_\nu}\left[\frac{d^2\sigma^\nu}{dx\,dy}+\frac{d^2\sigma^{\bar{\nu}}}{dx\,dy}\right] - 2x(s-c)\left[1-(1-y)^2\right]\right\}}{\left[1+(1-y)^2-y^2R/(1+R)\right]} \quad (10)$$

$$2xF_1 = F_2/(1+R) \quad (11)$$

$$xF_3 = \frac{\pi}{G^2 m_N E_\nu}\left[\frac{d^2\sigma^\nu}{dx\,dy}-\frac{d^2\sigma^{\bar{\nu}}}{dx\,dy}\right]/\{1-(1-y)^2\}. \quad (12)$$

Corrections for non-zero $R=\sigma_L/\sigma_T$ and for the difference $s-c$ have to be applied to obtain the structure function F_2. In Sect. 4.2 we have seen that the difference between F_2 and $2xF_1$, which is measured by R, is small and that the uncertainties at the present level of statistics are large compared to it. We therefore extract F_2 under the assumption that R = constant = 0.1. It should be noted that this correction is important only at large y, where the sum of neutrino and antineutrino cross-sections is proportional to $2xF_1$ rather than to F_2. The correction which involves the strange and charmed sea has been evaluated assuming $x[s(x)-c(x)]=0.2x(\bar{u}+\bar{d}+\bar{s})$, where we have used the result $2x\,s(x) \approx 0.4x(\bar{u}+\bar{d})$ obtained from the analysis of neutrino- and antineutrino-induced dimuon events [19], and assumed that the charm-quark component $x\,c(x)$ can be neglected.

The structure functions F_2 and xF_3 have been determined using 200 and 300 GeV narrow-band beam data in the hadron-energy range $E_h < 200$ GeV, where both neutrino and antineutrino data exist.

The results for the three structure functions F_2, $2xF_1$, and xF_3, after all corrections discussed in Sect. 4.2, are listed in Table 4, including an estimate of the systematic point-to-point errors and the magnitude of the correction to F_2 due to $R=0.1$.

Table 4. Structure functions F_2, $2xF_1$, xF_3, and F_+. The structure functions are evaluated at the bin centre with the assumptions $R = \sigma_L/\sigma_T = 0.1$ and $m_W = \infty$. No correction for Fermi motion has been applied. The errors given are statistical and systematic point-to-point errors. In addition we have an over-all scale error of $\pm 6\%$ for F_2 and F_+ and $\pm 8\%$ for xF_3. The column ΔR gives the change in F_2 if R is changed from zero to 0.1

Q^2 (GeV^2/c^2)	x	$F_2 \pm \sigma_{stat} \pm \sigma_{syst}$	ΔR	$2xF_1 \pm \sigma_{stat}$	x	$xF_3 \pm \sigma_{stat} \pm \sigma_{syst}$	$F_+ \pm \sigma_{stat} \pm \sigma_{syst}$
1.13	0.015	0.697 ± 0.077 ± 0.08	0.007	0.634 ± 0.070	0.015	0.160 ± 0.090 ± 0.08	
	0.045	0.754 ± 0.073 ± 0.08	0.001	0.689 ± 0.067			
	0.080	0.916 ± 0.081 ± 0.08	0.000	0.850 ± 0.075			
	0.150	0.940 ± 0.071 ± 0.05	0.000	0.915 ± 0.069			
1.42	0.015	0.838 ± 0.055 ± 0.05	0.007	0.762 ± 0.050	0.015	0.178 ± 0.077 ± 0.08	
	0.045	0.913 ± 0.062 ± 0.05	0.001	0.834 ± 0.057	0.045	0.345 ± 0.150 ± 0.05	
	0.080	0.983 ± 0.079 ± 0.05	0.000	0.908 ± 0.073			
	0.150	0.882 ± 0.057 ± 0.03	0.000	0.846 ± 0.055			
1.79	0.015	0.901 ± 0.060 ± 0.05	0.012	0.819 ± 0.054	0.015	0.249 ± 0.074 ± 0.07	
	0.045	0.950 ± 0.062 ± 0.05	0.003	0.867 ± 0.056	0.045	0.310 ± 0.128 ± 0.07	
	0.080	1.038 ± 0.070 ± 0.05	0.001	0.956 ± 0.065	0.080	0.853 ± 0.264 ± 0.07	
	0.150	0.934 ± 0.054 ± 0.05	0.000	0.887 ± 0.051			
	0.250	0.833 ± 0.071 ± 0.05	0.000	0.851 ± 0.072			
2.25	0.015	0.807 ± 0.056 ± 0.05	0.016	0.734 ± 0.051	0.015	0.311 ± 0.067 ± 0.07	
	0.045	1.025 ± 0.062 ± 0.05	0.005	0.935 ± 0.057	0.045	0.261 ± 0.105 ± 0.07	
	0.080	1.095 ± 0.067 ± 0.05	0.001	1.005 ± 0.061	0.080	0.524 ± 0.183 ± 0.07	
	0.150	0.910 ± 0.047 ± 0.05	0.000	0.856 ± 0.044	0.150	0.818 ± 0.244 ± 0.07	
	0.250	0.772 ± 0.058 ± 0.05	0.000	0.771 ± 0.058			
	0.350	0.577 ± 0.059 ± 0.05	0.000	0.625 ± 0.063			
2.84	0.015	1.065 ± 0.086 ± 0.04	0.018	0.969 ± 0.078	0.015	0.284 ± 0.109 ± 0.06	0.710 ± 0.055 ± 0.04
	0.045	0.838 ± 0.051 ± 0.04	0.004	0.764 ± 0.046	0.045	0.103 ± 0.072 ± 0.06	
	0.080	1.031 ± 0.059 ± 0.04	0.001	0.945 ± 0.054	0.080	0.473 ± 0.143 ± 0.06	
	0.150	0.946 ± 0.046 ± 0.04	0.000	0.884 ± 0.043	0.150	0.455 ± 0.195 ± 0.06	
	0.250	0.803 ± 0.054 ± 0.04	0.000	0.787 ± 0.053			
	0.350	0.612 ± 0.054 ± 0.04	0.000	0.641 ± 0.056			
	0.450	0.520 ± 0.053 ± 0.04	0.000	0.592 ± 0.060			
3.57	0.015	1.178 ± 0.099 ± 0.04	0.031	1.071 ± 0.090	0.015	0.278 ± 0.112 ± 0.06	0.746 ± 0.056 ± 0.04
	0.045	1.087 ± 0.058 ± 0.04	0.007	0.990 ± 0.053	0.045	0.390 ± 0.082 ± 0.06	
	0.080	1.105 ± 0.056 ± 0.03	0.003	1.011 ± 0.050	0.080	0.553 ± 0.117 ± 0.06	
	0.150	1.018 ± 0.043 ± 0.03	0.001	0.946 ± 0.040	0.150	0.468 ± 0.140 ± 0.06	
	0.250	0.762 ± 0.045 ± 0.03	0.000	0.735 ± 0.044	0.250	0.394 ± 0.289 ± 0.06	
	0.350	0.697 ± 0.053 ± 0.03	0.000	0.710 ± 0.053			
	0.450	0.507 ± 0.045 ± 0.03	0.000	0.553 ± 0.049			
	0.550	0.345 ± 0.039 ± 0.04	0.000	0.408 ± 0.046			
4.50	0.015	1.059 ± 0.097 ± 0.06	0.037	0.963 ± 0.088	0.015	0.211 ± 0.103 ± 0.06	0.631 ± 0.051 ± 0.06
	0.045	1.097 ± 0.057 ± 0.04	0.019	0.999 ± 0.052	0.045	0.282 ± 0.071 ± 0.06	
	0.080	1.000 ± 0.049 ± 0.03	0.006	0.914 ± 0.044	0.080	0.448 ± 0.080 ± 0.05	
	0.150	1.003 ± 0.038 ± 0.03	0.001	0.928 ± 0.035	0.150	0.716 ± 0.110 ± 0.05	
	0.250	0.854 ± 0.046 ± 0.03	0.000	0.814 ± 0.044	0.250	0.901 ± 0.230 ± 0.05	
	0.350	0.619 ± 0.043 ± 0.03	0.000	0.617 ± 0.043			
	0.450	0.446 ± 0.039 ± 0.03	0.000	0.470 ± 0.042			
	0.550	0.310 ± 0.033 ± 0.03	0.000	0.349 ± 0.037			

The data from the 300 GeV narrow-band beam exposure with hadron energies above 200 GeV, i.e. $y \gtrsim 0.66$, cannot be used to determine F_2 or xF_3 since no useful antineutrino data exist in this energy range. For this reason we use these data to determine the structure function

$$F_+ \equiv \tfrac{1}{2}[2xF_1 + xF_3^\nu] = x(u+d+2s),$$

which is obtained using

$$F_+ = \left(\frac{\pi}{G^2 m_N E_\nu}\frac{d^2\sigma^\nu}{dx\,dy} - \{x(\bar{u}+\bar{d}+2\bar{c}) \cdot [(1-y)^2 + R(1-y)]\}\right)/[1+R(1-y)]. \quad (13)$$

The correction term in braces $\{\ldots\}$, which is only

Table 4 (continued)

Q^2 (GeV^2/c^2)	x	$F_2 \pm \sigma_{stat} \pm \sigma_{syst}$	ΔR	$2xF_1 \pm \sigma_{stat}$	x	$xF_3 \pm \sigma_{stat} \pm \sigma_{syst}$	$F_+ \pm \sigma_{stat} \pm \sigma_{syst}$
5.66	0.015	1.078 ± 0.110 ± 0.08	0.057	0.980 ± 0.100	0.015	0.538 ± 0.108 ± 0.06	0.750 ± 0.065 ± 0.06
	0.045	1.147 ± 0.067 ± 0.04	0.034	1.044 ± 0.061	0.045	0.373 ± 0.074 0.06	
	0.080	1.142 ± 0.052 ± 0.03	0.010	1.043 ± 0.047	0.080	0.475 ± 0.073 0.06	
	0.150	1.073 ± 0.037 ± 0.03	0.003	0.989 ± 0.034	0.150	0.707 ± 0.082 0.06	
	0.250	0.789 ± 0.039 ± 0.03	0.000	0.745 ± 0.036	0.250	0.575 ± 0.141 0.04	
	0.350	0.658 ± 0.041 ± 0.03	0.000	0.644 ± 0.040	0.350	0.745 ± 0.224 0.04	
	0.450	0.440 ± 0.035 ± 0.02	0.000	0.451 ± 0.036	0.450	0.527 ± 0.274 0.04	
	0.550	0.237 ± 0.026 ± 0.02	0.000	0.256 ± 0.028			
	0.650	0.170 ± 0.033 ± 0.03	0.000	0.196 ± 0.037			
7.13	0.015				0.015		0.637 ± 0.076 ± 0.08
	0.045	1.149 ± 0.065 ± 0.04	0.031	1.046 ± 0.059	0.045	0.490 ± 0.074 0.06	0.814 ± 0.046 ± 0.03
	0.080	1.171 ± 0.052 ± 0.03	0.012	1.067 ± 0.048	0.080	0.601 ± 0.072 0.05	
	0.150	1.025 ± 0.034 ± 0.03	0.004	0.942 ± 0.031	0.150	0.593 ± 0.061 0.04	
	0.250	0.791 ± 0.036 ± 0.03	0.001	0.742 ± 0.033	0.250	0.618 ± 0.100 0.04	
	0.350	0.600 ± 0.036 ± 0.02	0.000	0.579 ± 0.035	0.350	0.450 ± 0.150 0.04	
	0.450	0.474 ± 0.036 ± 0.02	0.000	0.474 ± 0.035	0.450	0.664 ± 0.195 0.04	
	0.550	0.287 ± 0.026 ± 0.02	0.000	0.301 ± 0.027	0.550	0.185 ± 0.230 0.04	
	0.650	0.122 ± 0.023 ± 0.03	0.000	0.134 ± 0.025			
8.97	0.045	1.225 ± 0.075 ± 0.04	0.042	1.114 ± 0.068	0.045	0.610 ± 0.080 0.06	0.913 ± 0.051 ± 0.04
	0.080	1.086 ± 0.049 ± 0.03	0.014	0.990 ± 0.045	0.080	0.531 ± 0.060 0.04	
	0.150	0.987 ± 0.031 ± 0.03	0.006	0.905 ± 0.028	0.150	0.620 ± 0.047 0.04	
	0.250	0.838 ± 0.033 ± 0.03	0.001	0.780 ± 0.039	0.250	0.668 ± 0.077 0.04	
	0.350	0.573 ± 0.032 ± 0.02	0.000	0.546 ± 0.030	0.350	0.383 ± 0.100 0.04	
	0.450	0.424 ± 0.030 ± 0.02	0.000	0.416 ± 0.029	0.450	0.446 ± 0.139 0.03	
	0.550	0.247 ± 0.023 ± 0.02	0.000	0.251 ± 0.023	0.550	0.361 ± 0.114 0.03	
	0.650	0.171 ± 0.026 ± 0.02	0.000	0.182 ± 0.027	0.650	0.097 ± 0.158 0.03	
11.3	0.045	1.286 ± 0.088 ± 0.05	0.060	1.161 ± 0.080	0.045	0.580 ± 0.087 ± 0.08	0.897 ± 0.052 ± 0.05
	0.080	1.224 ± 0.058 ± 0.04	0.028	1.115 ± 0.053	0.080	0.572 ± 0.068 ± 0.06	
	0.150	1.080 ± 0.033 ± 0.03	0.009	0.987 ± 0.030	0.150	0.686 ± 0.047 ± 0.04	
	0.250	0.850 ± 0.031 ± 0.02	0.003	0.785 ± 0.029	0.250	0.681 ± 0.060 ± 0.03	
	0.350	0.591 ± 0.030 ± 0.02	0.001	0.558 ± 0.028	0.350	0.545 ± 0.083 ± 0.02	
	0.450	0.380 ± 0.025 ± 0.02	0.000	0.367 ± 0.024	0.450	0.369 ± 0.084 ± 0.02	
	0.550	0.230 ± 0.020 ± 0.02	0.000	0.227 ± 0.020	0.550	0.237 ± 0.114 ± 0.02	
	0.650	0.155 ± 0.042 ± 0.02	0.000	0.160 ± 0.043			
14.2	0.045	1.215 ± 0.130 ± 0.08	0.079	1.105 ± 0.119	0.045	0.390 ± 0.122 ± 0.10	0.835 ± 0.056 ± 0.05
	0.080	1.108 ± 0.060 ± 0.03	0.032	1.009 ± 0.054	0.080	0.560 ± 0.066 ± 0.04	0.849 ± 0.040 ± 0.03
	0.150	1.046 ± 0.030 ± 0.02	0.013	0.956 ± 0.028	0.150	0.725 ± 0.039 ± 0.03	
	0.250	0.772 ± 0.028 ± 0.02	0.004	0.713 ± 0.025	0.250	0.543 ± 0.046 ± 0.02	
	0.350	0.556 ± 0.025 ± 0.02	0.001	0.521 ± 0.024	0.350	0.500 ± 0.056 ± 0.02	
	0.450	0.356 ± 0.022 ± 0.015	0.001	0.340 ± 0.021	0.450	0.398 ± 0.067 ± 0.015	
	0.550	0.233 ± 0.015 ± 0.015	0.000	0.228 ± 0.018	0.550	0.357 ± 0.071 ± 0.015	
	0.650	0.133 ± 0.015 ± 0.01	0.000	0.133 ± 0.015	0.650	0.145 ± 0.079 ± 0.01	
17.9	0.045				0.045		0.809 ± 0.077 ± 0.06
	0.080	1.275 ± 0.068 ± 0.04	0.053	1.161 ± 0.062	0.080	0.584 ± 0.069 ± 0.05	0.920 ± 0.043 ± 0.04
	0.150	1.012 ± 0.030 ± 0.02	0.015	0.924 ± 0.028	0.150	0.648 ± 0.038 ± 0.03	
	0.250	0.797 ± 0.027 ± 0.02	0.006	0.733 ± 0.025	0.250	0.630 ± 0.040 ± 0.02	
	0.350	0.602 ± 0.025 ± 0.02	0.003	0.561 ± 0.024	0.350	0.476 ± 0.045 ± 0.02	
	0.450	0.366 ± 0.021 ± 0.015	0.001	0.346 ± 0.020	0.450	0.313 ± 0.048 ± 0.015	
	0.550	0.241 ± 0.018 ± 0.015	0.000	0.232 ± 0.017	0.550	0.262 ± 0.050 ± 0.015	
	0.650	0.106 ± 0.013 ± 0.01	0.000	0.105 ± 0.013	0.650	0.083 ± 0.062 ± 0.01	

present in the sea region and is always less than $\sim 5\%$ in the kinematic region $y > 0.66$, has been evaluated by an extrapolation of $x(\bar{u} + \bar{d} + 2\bar{s})$ from the lower hadron-energy region. The correction due to R is again evaluated using $R = 0.1$ and goes to zero for y approaching one. The structure function F_+ is listed in Table 4 for $\nu \gtrsim 100$ GeV using 200 and 300 GeV neutrino data. It should be noted, however,

Table 4 (continued)

Q^2 (GeV^2/c^2)	x	$F_2 \pm \sigma_{stat} \pm \sigma_{syst}$	ΔR	$2xF_1 \pm \sigma_{stat}$	x	$xF_3 \pm \sigma_{stat} \pm \sigma_{syst}$	$F_+ \pm \sigma_{stat} \pm \sigma_{syst}$
22.5	0.045				0.045		0.864 ± 0.117 ± 0.08
	0.080	1.207 ± 0.078 ± 0.06	0.060	1.098 ± 0.070	0.080	0.729 ± 0.075 ± 0.08	0.931 ± 0.050 ± 0.06
	0.150	0.998 ± 0.031 ± 0.02	0.023	0.910 ± 0.029	0.150	0.710 ± 0.036 ± 0.03	0.827 ± 0.026 ± 0.02
	0.250	0.764 ± 0.026 ± 0.02	0.010	0.701 ± 0.023	0.250	0.592 ± 0.034 ± 0.02	
	0.350	0.559 ± 0.023 ± 0.02	0.002	0.518 ± 0.022	0.350	0.509 ± 0.039 ± 0.02	
	0.450	0.396 ± 0.021 ± 0.015	0.001	0.371 ± 0.020	0.450	0.336 ± 0.038 ± 0.015	
	0.550	0.231 ± 0.016 ± 0.015	0.000	0.220 ± 0.015	0.550	0.296 ± 0.038 ± 0.015	
	0.650	0.110 ± 0.011 ± 0.01	0.000	0.106 ± 0.011	0.650	0.142 ± 0.038 ± 0.01	
28.4	0.080	1.128 ± 0.121 ± 0.08	0.078	1.026 ± 0.110	0.080	0.625 ± 0.110 ± 0.08	0.859 ± 0.056 ± 0.06
	0.150	1.071 ± 0.036 ± 0.03	0.035	0.976 ± 0.033	0.150	0.727 ± 0.039 ± 0.04	0.860 ± 0.026 ± 0.03
	0.250	0.802 ± 0.027 ± 0.02	0.013	0.735 ± 0.025	0.250	0.604 ± 0.034 ± 0.03	
	0.350	0.544 ± 0.022 ± 0.015	0.006	0.502 ± 0.020	0.350	0.456 ± 0.030 ± 0.015	
	0.450	0.335 ± 0.017 ± 0.015	0.002	0.312 ± 0.016	0.450	0.314 ± 0.028 ± 0.015	
	0.550	0.204 ± 0.014 ± 0.015	0.000	0.192 ± 0.013	0.550	0.210 ± 0.02 ± 0.015	
	0.650	0.098 ± 0.009 ± 0.01	0.000	0.094 ± 0.009	0.650	0.096 ± 0.023 ± 0.01	
35.7	0.150	1.117 ± 0.042 ± 0.03	0.047	1.018 ± 0.038	0.150	0.743 ± 0.042 ± 0.04	0.880 ± 0.029 ± 0.03
	0.250	0.731 ± 0.027 ± 0.02	0.018	0.668 ± 0.024	0.250	0.582 ± 0.031 ± 0.03	
	0.350	0.511 ± 0.021 ± 0.015	0.007	0.470 ± 0.019	0.350	0.474 ± 0.028 ± 0.015	
	0.450	0.328 ± 0.017 ± 0.01	0.003	0.305 ± 0.016	0.450	0.325 ± 0.025 ± 0.01	
	0.550	0.178 ± 0.013 ± 0.01	0.001	0.166 ± 0.012	0.550	0.169 ± 0.020 ± 0.01	
	0.650	0.074 ± 0.008 ± 0.01	0.000	0.070 ± 0.007	0.650	0.071 ± 0.015 ± 0.01	
45.0	0.150	1.039 ± 0.052 ± 0.04	0.060	0.946 ± 0.047	0.150	0.630 ± 0.049 ± 0.05	0.825 ± 0.031 ± 0.03
	0.250	0.809 ± 0.029 ± 0.02	0.025	0.735 ± 0.027	0.250	0.670 ± 0.032 ± 0.02	0.706 ± 0.023 ± 0.02
	0.350	0.497 ± 0.021 ± 0.015	0.010	0.456 ± 0.019	0.350	0.427 ± 0.025 ± 0.015	
	0.450	0.323 ± 0.016 ± 0.01	0.004	0.299 ± 0.015	0.450	0.306 ± 0.021 ± 0.01	
	0.550	0.172 ± 0.012 ± 0.01	0.002	0.160 ± 0.011	0.550	0.184 ± 0.017 ± 0.01	
	0.650	0.073 ± 0.008 ± 0.01	0.001	0.068 ± 0.007	0.650	0.067 ± 0.012 ± 0.01	
56.6	0.150	1.009 ± 0.085 ± 0.04	0.068	0.919 ± 0.078	0.150	0.654 ± 0.078 ± 0.05	0.828 ± 0.042 ± 0.04
	0.250	0.766 ± 0.030 ± 0.02	0.035	0.699 ± 0.027	0.250	0.620 ± 0.030 ± 0.02	0.665 ± 0.024 ± 0.02
	0.350	0.521 ± 0.023 ± 0.015	0.014	0.477 ± 0.021	0.350	0.433 ± 0.026 ± 0.015	0.458 ± 0.020 ± 0.015
	0.450	0.347 ± 0.018 ± 0.015	0.006	0.320 ± 0.017	0.450	0.295 ± 0.022 ± 0.015	
	0.550	0.162 ± 0.012 ± 0.01	0.002	0.150 ± 0.010	0.550	0.173 ± 0.015 ± 0.01	
	0.650	0.085 ± 0.008 ± 0.01	0.001	0.079 ± 0.007	0.650	0.092 ± 0.011 ± 0.01	
71.3	0.150				0.150		0.931 ± 0.064 ± 0.04
	0.250	0.687 ± 0.033 ± 0.02	0.041	0.626 ± 0.030	0.250	0.582 ± 0.031 ± 0.03	0.627 ± 0.026 ± 0.02
	0.350	0.499 ± 0.022 ± 0.015	0.021	0.456 ± 0.020	0.350	0.424 ± 0.023 ± 0.02	0.435 ± 0.018 ± 0.015
	0.450	0.312 ± 0.018 ± 0.015	0.009	0.287 ± 0.016	0.450	0.276 ± 0.020 ± 0.015	0.282 ± 0.016 ± 0.015
	0.550	0.176 ± 0.013 ± 0.015	0.003	0.163 ± 0.012	0.550	0.158 ± 0.015 ± 0.015	0.157 ± 0.011 ± 0.01
	0.650	0.072 ± 0.008 ± 0.01	0.001	0.067 ± 0.007	0.650	0.065 ± 0.009 ± 0.01	
89.7	0.150				0.150		1.014 ± 0.133 ± 0.06
	0.250	0.788 ± 0.063 ± 0.04	0.055	0.718 ± 0.057	0.250	0.712 ± 0.057 ± 0.05	0.685 ± 0.034 ± 0.02
	0.350	0.511 ± 0.026 ± 0.02	0.026	0.466 ± 0.024	0.350	0.407 ± 0.025 ± 0.03	0.433 ± 0.020 ± 0.015
	0.450	0.287 ± 0.017 ± 0.015	0.011	0.263 ± 0.016	0.450	0.265 ± 0.018 ± 0.015	0.257 ± 0.014
	0.550	0.133 ± 0.012 ± 0.015	0.003	0.122 ± 0.011	0.550	0.126 ± 0.013 ± 0.015	0.120 ± 0.010 ± 0.01
	0.650	0.065 ± 0.008 ± 0.01	0.000	0.060 ± 0.007	0.650	0.057 ± 0.009 ± 0.01	0.057 ± 0.007 ± 0.01
113.0	0.250				0.250		0.683 ± 0.051 ± 0.05
	0.350	0.491 ± 0.036 ± 0.03	0.031	0.448 ± 0.033	0.350	0.433 ± 0.033 ± 0.03	0.417 ± 0.023 ± 0.02
	0.450	0.316 ± 0.019 ± 0.02	0.016	0.289 ± 0.017	0.450	0.306 ± 0.018 ± 0.02	0.291 ± 0.016 ± 0.015
	0.550	0.143 ± 0.012 ± 0.015	0.005	0.131 ± 0.011	0.550	0.137 ± 0.012 ± 0.015	0.137 ± 0.010 ± 0.01
	0.650	0.080 ± 0.010 ± 0.01	0.002	0.074 ± 0.009	0.650	0.065 ± 0.010 ± 0.01	0.070 ± 0.007 ± 0.01

that F_+ is independent of the measurement of F_2 and xF_3 only for $\nu > 200$ GeV.

For x larger than 0.4 the contribution of the sea quarks disappears and the three structure functions $2xF_1 = q + \bar{q}$, $xF_3 = q - \bar{q}$, and $F_+ = q + s$ become progressively the same. The structure functions are displayed in Figs. 11–13 as functions of Q^2 and for all bins in x.

Table 4 (continued)

Q^2 (GeV^2/c^2)	x	$F_2 \pm \sigma_{stat} \pm \sigma_{syst}$	ΔR	$2xF_1 \pm \sigma_{stat}$	x	$xF_3 \pm \sigma_{stat} \pm \sigma_{syst}$	$F_+ \pm \sigma_{stat} \pm \sigma_{syst}$
142.2	0.250				0.250		0.707 ± 0.126 ± 0.06
	0.350				0.350		0.414 ± 0.031 ± 0.02
	0.450	0.296 ± 0.026 ± 0.02	0.019	0.270 ± 0.024	0.450	0.285 ± 0.024 ± 0.02	0.246 ± 0.017 ± 0.015
	0.550	0.147 ± 0.014 ± 0.02	0.008	0.135 ± 0.013	0.550	0.141 ± 0.013 ± 0.02	0.135 ± 0.011 ± 0.01
	0.650	0.084 ± 0.009 ± 0.01	0.004	0.077 ± 0.009	0.650	0.075 ± 0.009 ± 0.01	0.074 ± 0.007 ± 0.01
179.0	0.350				0.350		0.384 ± 0.058 ± 0.04
	0.450				0.450		0.257 ± 0.023 ± 0.02
	0.550	0.164 ± 0.020 ± 0.02	0.010	0.150 ± 0.019	0.550	0.157 ± 0.018 ± 0.02	0.136 ± 0.013 ± 0.015
	0.650	0.068 ± 0.011 ± 0.01	0.003	0.062 ± 0.010	0.650	0.068 ± 0.010 ± 0.01	0.060 ± 0.007 ± 0.01
225.3	0.450				0.450		0.264 ± 0.044 ± 0.02
	0.550				0.550		0.140 ± 0.018 ± 0.02
	0.650				0.650		0.064 ± 0.011 ± 0.015
283.7	0.550				0.550		0.086 ± 0.033 ± 0.02
	0.650				0.650		0.069 ± 0.014 ± 0.02

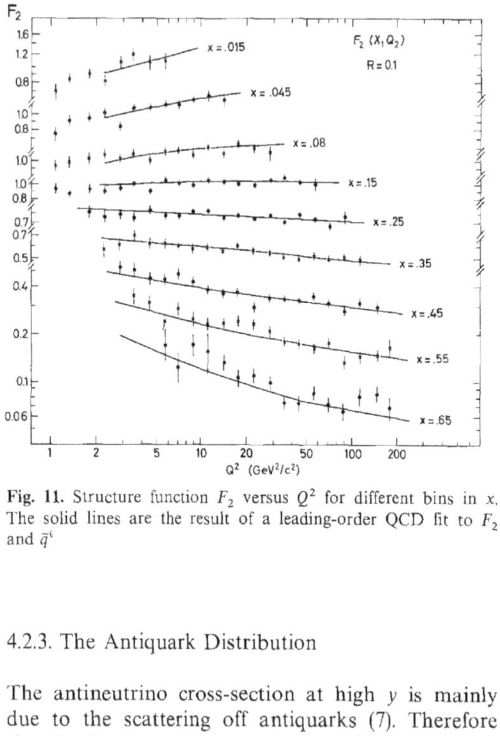

Fig. 11. Structure function F_2 versus Q^2 for different bins in x. The solid lines are the result of a leading-order QCD fit to F_2 and $\bar{q}^{\bar{\nu}}$

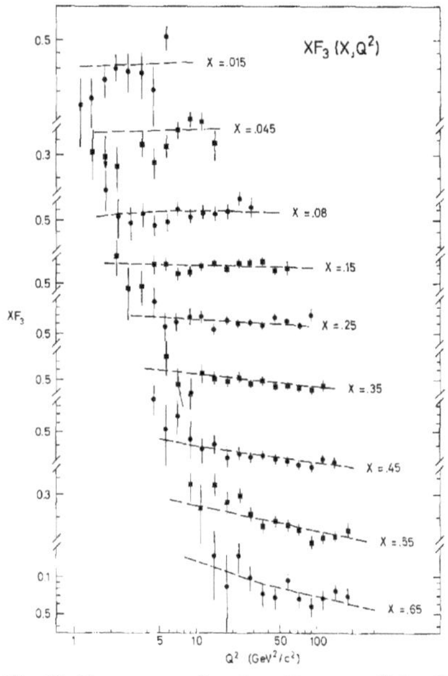

Fig. 12. The structure function xF_3 versus Q^2 for different bins in x. The dashed lines are the result of a leading-order QCD fit to the data

4.2.3. The Antiquark Distribution

The antineutrino cross-section at high y is mainly due to the scattering off antiquarks (7). Therefore the distribution of sea quarks in the specific combination $\bar{q}^{\bar{\nu}}(x, Q^2) = x(\bar{u} + \bar{d} + 2\bar{s})$ is directly measurable. In the narrow-band beam this measurement suffers from statistics owing to low $\bar{\nu}$-flux and the restricted y-range. We have therefore added results from about 155,000 $\bar{\nu}$ and 35,000 ν events with $E_\nu > 20$ GeV which have been recorded in wide-band beams. The evaluation of differential cross-sections is done in the same way as for the narrow-band beam data except that the energy spectrum is obtained from the data themselves, counting all events in a given energy bin. The normalization for the antineutrino is obtained using a linearly rising total

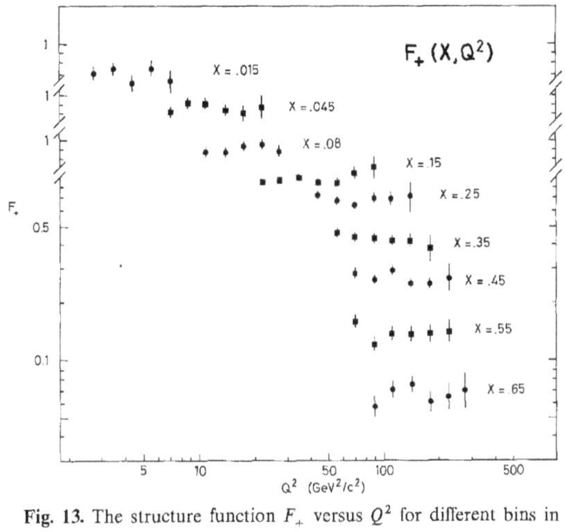

Fig. 13. The structure function F_+ versus Q^2 for different bins in x

cross-section with the slope $\sigma^{\bar{\nu}}/E=0.30$ (10^{-38} cm^2/GeV). For the neutrino wide-band beam data we use $\sigma/E_\nu=0.62$ (10^{-38} cm^2/GeV) for $E_\nu>70$ GeV and a rise of 11% down from 70 GeV to 20 GeV, in agreement with the results of Sect. 4.2. The differential cross-sections are in good agreement with those from the narrow-band beam. The results of wide-band and narrow-band beams have been averaged. Below $E_\nu \simeq 100$ GeV the wide-band beam data dominate; above, the narrow-band beam data are dominant.

The sea distribution is obtained using $\bar{\nu}$ and ν differential cross-sections for $y>0.5$ according to the expression:

$$\bar{q}^{\bar{\nu}}=x(\bar{u}+\bar{d}+2\bar{s})=\left\{\frac{\pi}{G^2 m_N E_\nu}\left[\frac{d^2\sigma^{\bar{\nu}}}{dx\,dy}-(1-y)^2\frac{d^2\sigma^{\nu}}{dx\,dy}\right]\right.$$
$$+2x(s-c)\left[(1-y)^2-(1-y)^4\right]$$
$$\left.-F_L\left[(1-y)-(1-y)^3\right]\right\}/\left[1-(1-y)^4\right]. \qquad (14)$$

The term proportional to $d^2\sigma^\nu/dx\,dy$ subtracts the amount of scattering due to quarks. It is zero at $y=1$ and amounts to about 50% at $y=0.5$.

The method is illustrated in Fig. 14, which shows $d\sigma^{\bar{\nu}}/dx$ for two energy bins and for four bins in y. The neutrino differential cross-section weighted by $(1-y)^2$ is also shown.

The evaluation of the antiquark distribution requires assumptions about the amount of strange sea and the magnitude of the longitudinal structure function. We use $R=0.1$ and $2s/(\bar{u}+\bar{d})=0.4$, i.e. the same assumption as for the determination of F_2 and F_+. Experimentally the antiquark structure function is best obtained in bins of x and ν. It is listed in Table 5a together with the estimate of the systematic point-to-point error and the correction due to $R=0.1$. The results, translated into bins of x and Q^2, are shown in Fig. 15 together with the result of a

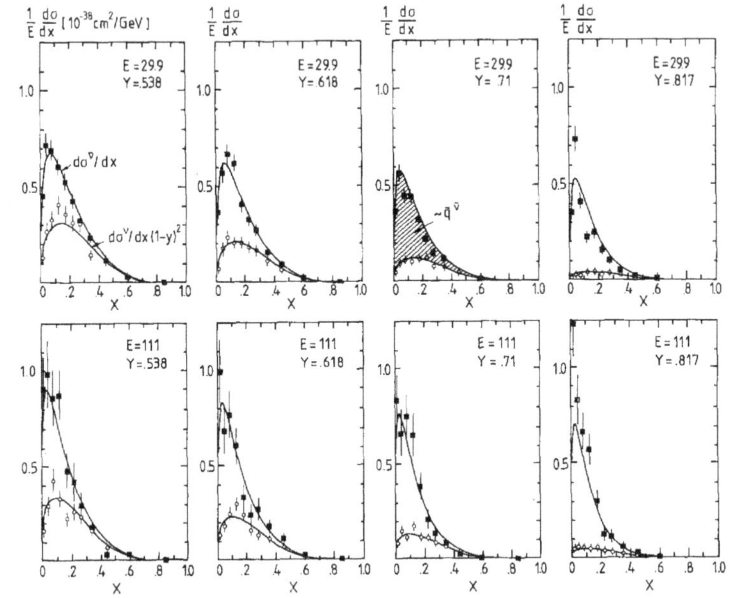

Fig. 14. The differential antineutrino cross-section as a function of x for four bins in y and two neutrino energies. Also shown is the neutrino cross-section multiplied by $(1-y)^2$. The curves are parametrizations of the data which have been used in the Monte Carlo simulation to determine the unsmearing corrections

Table 5a. Antiquark distribution $\bar{q}^{\bar{\nu}}(x, E_h)$ evaluated assuming $R=\sigma_L/\sigma_T=0.1$. The estimate of systematic errors is given in Table 5b

x	E_h	12.6	15.8	19.9	25.1	31.6	39.8	50.1	63.1	79.4	100.0	125.9	158.5
x = 0.015	QBAR	0.123	0.163	0.186	0.266	0.207	0.324	0.331	0.360	0.412	0.425	0.335	0.400
	ERR	0.044	0.019	0.020	0.023	0.023	0.021	0.024	0.028	0.070	0.044	0.048	0.093
	Q_2	0.354	0.446	0.562	0.707	0.890	1.120	1.410	1.776	2.235	2.814	3.543	4.460
	ΔR	0.018	0.016	0.015	0.016	0.012	0.018	0.021	0.018	0.029	0.019	0.013	0.017
x = 0.045	QBAR	0.214	0.219	0.276	0.317	0.286	0.327	0.374	0.391	0.290	0.407	0.319	0.364
	ERR	0.076	0.030	0.032	0.038	0.032	0.029	0.030	0.037	0.081	0.057	0.061	0.071
	Q_2	1.063	1.338	1.684	2.121	2.670	3.361	4.231	5.327	6.706	8.442	10.628	13.380
	ΔR	0.034	0.027	0.027	0.031	0.019	0.028	0.028	0.030	0.040	0.032	0.022	0.014
x = 0.08	QBAR	0.180	0.230	0.230	0.241	0.262	0.289	0.323	0.280	0.381	0.343	0.252	0.278
	ERR	0.074	0.029	0.029	0.031	0.032	0.024	0.025	0.028	0.068	0.046	0.039	0.052
	Q_2	1.889	2.995	2.995	3.770	4.746	5.975	7.522	9.469	11.921	15.008	18.894	23.786
	ΔR	0.042	0.031	0.031	0.031	0.025	0.028	0.029	0.028	0.034	0.030	0.018	0.013
x = 0.125	QBAR	0.161	0.181	0.209	0.151	0.174	0.195	0.211	0.182	0.251	0.198	0.205	0.167
	ERR	0.072	0.028	0.027	0.031	0.025	0.019	0.020	0.023	0.066	0.028	0.036	0.055
	Q_2	2.952	3.717	4.679	5.890	7.416	9.336	11.753	14.796	18.627	23.450	29.522	37.166
	ΔR	0.046	0.033	0.030	0.033	0.023	0.025	0.026	0.026	0.041	0.020	0.023	0.021
x = 0.175	QBAR	0.138	0.103	0.118	0.129	0.104	0.101	0.123	0.146	0.146	0.147	0.153	0.124
	σ_{stat}	0.057	0.027	0.026	0.028	0.021	0.018	0.019	0.022	0.047	0.031	0.036	0.034
	Q_2	4.133	5.203	6.550	8.247	10.381	13.070	16.454	20.714	26.078	32.830	41.331	52.032
	ΔR	0.034	0.031	0.029	0.030	0.017	0.023	0.025	0.026	0.028	0.025	0.023	0.009
x = 0.225	QBAR	0.015	0.073	0.060	0.087	0.077	0.076	0.078	0.063	0.046	0.043	0.042	0.052
	ERR	0.081	0.024	0.026	0.027	0.020	0.018	0.015	0.018	0.037	0.015	0.018	0.022
	Q_2	5.314	6.690	8.422	10.603	13.348	16.804	21.155	26.633	33.529	42.210	53.139	66.898
	ΔR	0.046	0.027	0.027	0.026	0.017	0.022	0.019	0.021	0.024	0.011	0.011	0.008
x = 0.275	QBAR	0.052	0.028	0.054	0.049	0.078	0.044	0.040	0.031	0.013	0.043	-0.004	0.034
	ERR	0.062	0.023	0.021	0.021	0.019	0.015	0.014	0.014	0.033	0.019	0.018	0.031
	Q_2	6.495	8.177	10.294	12.959	16.314	20.538	25.856	32.551	40.979	51.590	64.948	81.765
	ΔR	0.031	0.021	0.019	0.018	0.015	0.017	0.017	0.016	0.019	0.013	0.014	0.010
x = 0.35	QBAR	-0.039	0.019	0.026	0.009	0.005	0.011	0.004	0.019	-0.007	0.013	0.014	0.010
	ERR	0.038	0.013	0.013	0.014	0.010	0.008	0.008	0.008	0.019	0.011	0.006	0.007
	Q_2	8.266	10.406	13.101	16.493	20.764	26.140	32.908	41.429	52.156	65.660	82.661	104.064
	ΔR	0.024	0.016	0.015	0.006	0.009	0.013	0.012	0.011	0.014	0.010	0.006	0.004
x = 0.45	QBAR	0.003	-0.006	-0.002	-0.007	0.002	-0.006	-0.002	-0.002	-0.016	0.007	-0.008	0.000
	ERR	0.029	0.011	0.009	0.012	0.007	0.007	0.005	0.005	0.013	0.008	0.013	0.000
	Q_2	10.628	13.380	16.844	21.205	26.696	33.608	42.310	53.265	67.057	86.500	106.279	0.000
	ΔR	0.013	0.010	0.009	0.010	0.006	0.009	0.007	0.006	0.008	0.005	0.007	0.000
x = 0.60	QBAR	-0.009	-0.002	-0.002	0.001	-0.001	-0.001	-0.001	0.001	0.002	0.002	-0.001	0.002
	ERR	0.010	0.003	0.003	0.003	0.002	0.002	0.001	0.001	0.003	0.003	0.003	0.003
	Q_2	14.171	17.840	22.459	28.274	35.595	44.811	56.414	71.021	89.410	116.220	141.705	178.396
	ΔR	0.005	0.003	0.002	0.002	0.001	0.002	0.001	0.001	0.001	0.001	0.001	0.001

Row 1 (QBAR): Value of structure function
Row 2 (ERR) : Statistical error on $\bar{q}^{\bar{\nu}}$
Row 3 (Q_2) : Value of Q_2 corresponding to centre of bin
Row 4 (ΔR) : Effect of Callan-Gross violation $\Delta R = \bar{q}(R = 0.0) - \bar{q}(R = 0.1)$

QCD fit which will be described in Sect. 5. Finally, in order to allow an easier comparison with the other structure functions, $\bar{q}^{\bar{\nu}}(x, Q^2)$ is tabulated in the same bins as those used for F_2 and xF_3 in Table 5b.

4.2.4. Discussion of Systematic Errors

Systematic errors are twofold: i) errors which can be absorbed in an overall scale error, and ii) errors which change the shape and/or the Q^2-slope of the structure functions. The scale error is estimated to be $\pm 6\%$ for F_2 and $\pm 8\%$ for $\bar{q}^{\bar{\nu}}$ and xF_3, mainly due to the error in absolute cross-section measurements. The shapes of the structure functions are affected by the uncertainties in $\sigma^{\bar{\nu}}/\sigma^{\nu}$ and by effects caused by the apparatus, such as errors in the hadron- and muon-energy calibration and in the unfolding of acceptance and resolution effects. The cross-

Table 5b. The structure function $\bar{q}^{\bar{\nu}}(x, Q^2)$ evaluated under the assumption $R=\sigma_L/\sigma_T=0.1$. Statistical and systematic point-to-point errors are given separately. In addition, $\bar{q}^{\bar{\nu}}$ has an over-all scale error of $\pm 8\%$

Q^2 (GeV^2/c^2)	x	$\bar{q}^{\bar{\nu}} \pm \sigma_{stat} \pm \sigma_{syst}$
0.450	0.015	0.146 ± 0.028 ± 0.03
0.566	0.015	0.195 ± 0.033 ± 0.03
0.713	0.015	0.300 ± 0.035 ± 0.03
0.897	0.015	0.311 ± 0.037 ± 0.02
	0.045	0.431 ± 0.151 ± 0.08
1.13	0.015	0.323 ± 0.041 ± 0.02
	0.015	0.248 ± 0.074 ± 0.08
1.42	0.015	0.339 ± 0.038 ± 0.02
	0.045	0.252 ± 0.046 ± 0.05
1.79	0.015	0.348 ± 0.035 ± 0.02
	0.045	0.241 ± 0.038 ± 0.02
	0.08	0.281 ± 0.178 ± 0.06
2.25	0.015	0.307 ± 0.037 ± 0.02
	0.045	0.271 ± 0.036 ± 0.02
	0.08	0.182 ± 0.066 ± 0.05
2.84	0.015	0.438 ± 0.058 ± 0.02
	0.045	0.357 ± 0.038 ± 0.02
	0.08	0.212 ± 0.040 ± 0.03
	0.15	0.236 ± 0.142
3.57	0.015	0.456 ± 0.067 ± 0.02
	0.045	0.370 ± 0.034 ± 0.02
	0.08	0.238 ± 0.034 ± 0.02
	0.15	0.111 ± 0.062 ± 0.02
4.50	0.015	0.472 ± 0.089 ± 0.03
	0.045	0.395 ± 0.034 ± 0.02
	0.08	0.251 ± 0.030 ± 0.02
	0.15	0.179 ± 0.028 ± 0.02
5.66	0.015	0.259 ± 0.082 ± 0.05
	0.045	0.342 ± 0.035 ± 0.02
	0.08	0.288 ± 0.026 ± 0.02
	0.15	0.174 ± 0.022 ± 0.02
	0.25	0.012 ± 0.113 ± 0.02
7.13	0.045	0.269 ± 0.034 ± 0.02
	0.08	0.278 ± 0.025 ± 0.02
	0.15	0.153 ± 0.021 ± 0.01
	0.25	0.064 ± 0.037 ± 0.02
8.97	0.045	0.284 ± 0.057 ± 0.03
	0.08	0.297 ± 0.027 ± 0.02
	0.15	0.141 ± 0.016 ± 0.01
	0.25	0.045 ± 0.021 ± 0.01
	0.35	-0.047 ± 0.047 ± 0.03
11.30	0.045	0.343 ± 0.066 ± 0.04
	0.08	0.292 ± 0.030 ± 0.02
	0.15	0.179 ± 0.016 ± 0.01
	0.25	0.051 ± 0.018 ± 0.01
	0.35	0.019 ± 0.021 ± 0.01
	0.45	-0.037 ± 0.051 ± 0.01
14.22	0.045	0.417 ± 0.104 ± 0.05
	0.08	0.284 ± 0.038 ± 0.02
	0.15	0.143 ± 0.014 ± 0.01
	0.25	0.070 ± 0.014 ± 0.005
	0.35	0.022 ± 0.015 ± 0.01
	0.45	0.013 ± 0.015 ± 0.01
	0.55	-0.006 ± 0.042 ± 0.01
17.90	0.08	0.346 ± 0.051 ± 0.03
	0.15	0.165 ± 0.015 ± 0.01
	0.25	0.058 ± 0.011 ± 0.005
	0.35	0.030 ± 0.013 ± 0.005
	0.45	-0.007 ± 0.012 ± 0.005
	0.55	-0.013 ± 0.014 ± 0.005
22.54	0.08	0.166 ± 0.047 ± 0.05
	0.15	0.125 ± 0.017 ± 0.01
	0.25	0.065 ± 0.010 ± 0.005
	0.35	0.015 ± 0.009 ± 0.005
	0.45	-0.001 ± 0.010 ± 0.003
	0.55	-0.007 ± 0.008 ± 0.003
28.37	0.15	0.152 ± 0.024 ± 0.02
	0.25	0.077 ± 0.011 ± 0.005
	0.35	0.017 ± 0.008 ± 0.005
	0.45	0.004 ± 0.007 ± 0.003
	0.55	-0.008 ± 0.007 ± 0.003
35.72	0.15	0.174 ± 0.027 ± 0.03
	0.25	0.055 ± 0.012 ± 0.005
	0.35	0.013 ± 0.007 ± 0.003
	0.45	-0.002 ± 0.005 ± 0.003
	0.55	-0.000 ± 0.005 ± 0.003
44.96	0.25	0.044 ± 0.013 ± 0.005
	0.35	0.010 ± 0.007 ± 0.003
	0.45	0.006 ± 0.005 ± 0.002
	0.55	0.001 ± 0.005 ± 0.002
56.61	0.25	0.042 ± 0.014 ± 0.005
	0.35	0.005 ± 0.006 ± 0.003
	0.45	0.004 ± 0.005 ± 0.002
	0.55	0.001 ± 0.005 ± 0.002
71.26	0.25	0.011 ± 0.012 ± 0.01
	0.35	0.011 ± 0.007 ± 0.003
	0.45	0.002 ± 0.008 ± 0.002
	0.55	-0.002 ± 0.005 ± 0.002
89.72	0.35	0.034 ± 0.013 ± 0.006
	0.45	-0.003 ± 0.007 ± 0.003
	0.55	0.001 ± 0.004 ± 0.003
112.9	0.35	0.015 ± 0.014 ± 0.01
	0.45	-0.003 ± 0.005 ± 0.003
	0.55	-0.000 ± 0.008 ± 0.003
142.2	0.45	-0.002 ± 0.007 ± 0.004
	0.55	-0.002 ± 0.003 ± 0.003

section ratio $\sigma^{\bar{\nu}}/\sigma^{\nu}$ is well known except for energies below $E_\nu \simeq 50$ GeV [3, 12]. This uncertainty gives an error contribution mainly to $\bar{q}^{\bar{\nu}}(x, \nu)$ at low values of ν. The hadron- and muon-energy calibrations have been varied within the estimated uncertainties, and the uncertainty due to resolutions and acceptance effects has been estimated as 15% of the unsmearing correction. All error contributions have been added in quadrature. The results are given in Tables 4 and 5. The systematic errors are always smaller than or at most equal to the statistical errors. They mainly affect the x-dependence of the structure functions but have a smaller effect on their Q^2-dependence.

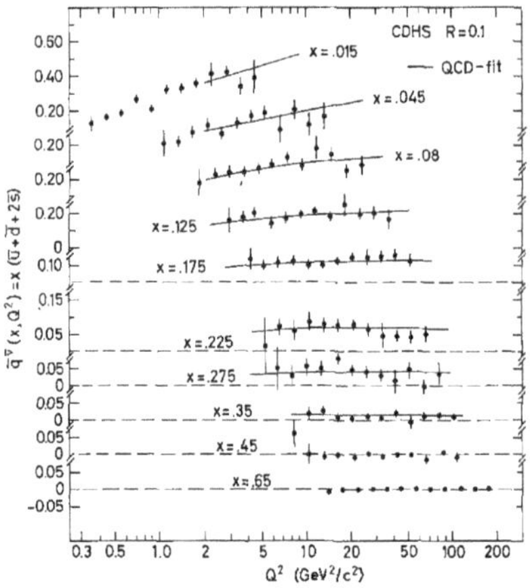

Fig. 15. The structure function $\bar{q}^{\bar{\nu}}$ versus Q^2 for different bins in x. The solid lines are the result of a QCD fit to F_2 and $\bar{q}^{\bar{\nu}}$

4.2.5. Dependence of the Structure Functions on Physics Assumptions

The structure function $xF_3(x,Q^2)$ is obtained from the differential cross-sections without further assumptions. Unfortunately it is statistically poorly determined owing to the limited statistics of the narrow-band beam data and the fact that it is due to the *difference* of cross-sections. The structure functions F_2, $2xF_1$, F_+, and $\bar{q}^{\bar{\nu}}$, on the other hand, have been extracted under specific assumptions about $R = \sigma_L/\sigma_T$ and the amount of strange and charmed sea. The magnitude of the correction due to R can be seen in Tables 4 and 5a, where the entry ΔR gives the change in the structure function going from $R=0.1$ to $R=0$. The effect on the measurement of F_2 is generally small compared to the statistical error except in the highest bins of Q^2 for each x bin, since the correction is only substantial at large y. The structure function $2xF_1$, on the other hand, is directly measured at large y only, and therefore suffers from substantial uncertainties in most of our kinematic range. The structure function F_+ is evaluated for high hadron energies ($E_h > 100$ GeV) only, i.e. for $y \gtrsim 0.5$, where the uncertainty due to R is small. This structure function is the most reliable at large x and high values of Q^2. The antiquark distribution finally depends very strongly on R, especially at large values of x. This leaves a substantial uncertainty for $x<0.4$ where the error in R is still very large.

The effect of the strange-sea correction is very small for $\bar{q}^{\bar{\nu}}$ ($\lesssim 1\%$). The effect on the structure function F_2 is given in Table 6 for two values of $2x(s-c)/\bar{q}(x)$. Here also, the correction is rather small. It should be kept in mind, however, that neutrino interactions with strange quarks lead predominantly to charmed quarks in the final state so that the contribution of $xs(x)$ is kinematically suppressed by a charm mass threshold effect. The magnitude of this suppression can be estimated with the slow rescaling model [23], to be about 0.2 for low hadron energies and 0.8 for the highest energies of this experiment. This leads to scaling violations for the structure functions F_2 and $\bar{q}^{\bar{\nu}}$ at small values of x, which account for up to 30% of the observed slopes in Q^2 at small x-values.

Table 6. Dependence of F_2 on the magnitude of the strange- and charmed-sea correction. The ratio of F_2 for the two values $2(s-c)/(\bar{u}+\bar{d})=0.4$ and $2(s-c)/(\bar{u}+\bar{d})=0.2$ is given as a function of x and Q^2

Q^2	x					
	0.015	0.045	0.08	0.15	0.25	0.35
1.4	0.95	0.98	0.99	1	1	
3.5	0.93	0.96	0.98	0.99	1	
9.0		0.95	0.96	0.97	0.99	1
18			0.95	0.97	0.99	1
36				0.97	0.99	1
72					0.99	1

The structure functions have been evaluated assuming an infinite mass for the intermediate vector boson. For a vector boson mass in the range of 80 GeV, the propagator has however a significant effect on the Q^2-slopes at large x.

The structure functions tabulated in Tables 4 and 5 are evaluated under reasonable physics assumptions, given the magnitude of the experimental errors. Any refined analysis has, however, to keep in mind the uncertainties in $R(x,Q^2)$, the effects of the charm mass theshold, and the propagator effect.

5. Interpretation of Structure-Function Measurements

5.1. The Shape of Structure Functions

The measurements of $2xF_1$, xF_3, F_+, and $\bar{q}^{\bar{\nu}}$, as described in the previous section, form a consistent set of structure functions for an isoscalar target. Their x-dependences are shown in Fig. 16 for a fixed bin in Q^2, together with empirical fits to the data which fulfil the expected quark parton model relations between these structure functions. The measurements of $2xF_1$ and xF_3 agree at large values of x, and their difference at small x is well

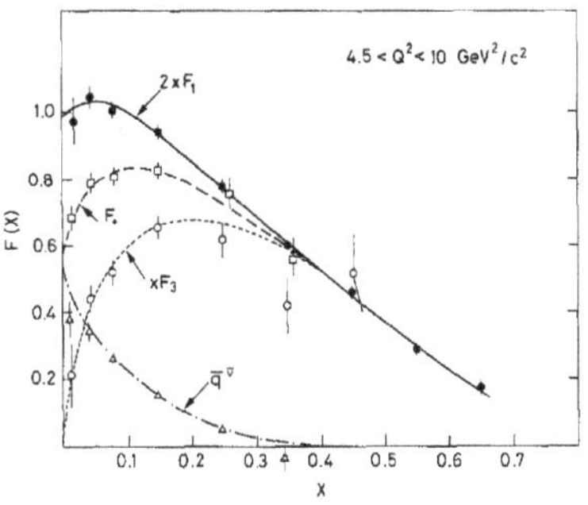

Fig. 16. Comparison of the structure functions $2xF_1$, F_+, xF_3, and $\bar{q}^{\bar{\nu}}$ for fixed Q^2 as a function of x. The curves are empirical fits to the data which fulfil the quark parton model relations between these structure functions: $2xF_1 = q + \bar{q}$, $xF_3 = q - \bar{q}$, $F_+ = q + xs$, $\bar{q}^{\bar{\nu}} = \bar{q} + xs$ with $2s/\bar{q} = 0.4$

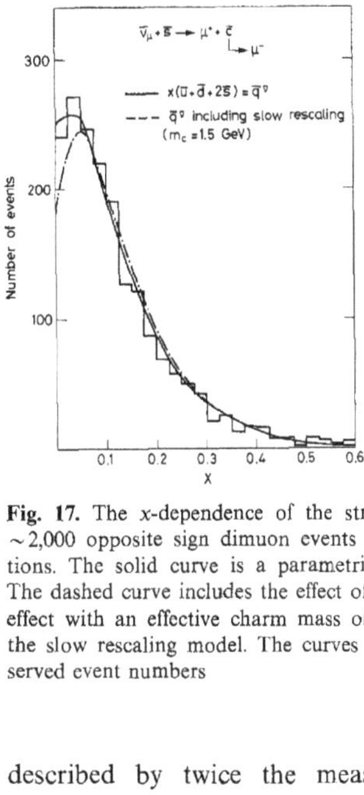

Fig. 17. The x-dependence of the strange sea as measured by ~2,000 opposite sign dimuon events from antineutrino interactions. The solid curve is a parametrization of $\bar{q}^{\bar{\nu}} = x(\bar{u} + \bar{d} + 2\bar{s})$. The dashed curve includes the effect of the charm mass threshold effect with an effective charm mass of 1.5 GeV as calculated in the slow rescaling model. The curves are normalized to the observed event numbers

described by twice the measured antiquark distribution if the effect of the strange sea is taken into account. An important aspect of the data is that the antiquark contribution disappears for $x \gtrsim 0.4$, so that $xF_3 \approx 2xF_1$ at large x. Quantitatively, we find $\bar{q}^{\bar{\nu}}/q \lesssim 0.000 \pm 0.005$ for $x > 0.5$ and an average value of $Q^2 = 33\,\mathrm{GeV}^2/c^2$ [7]. Finally, the shape of the strange sea $xs(x)$ has also been measured using ~2,000 antineutrino-induced opposite-sign dimuon events [19]. Figure 17 shows $xs(x)$ for an average hadronic energy $\nu = 50\,\mathrm{GeV}$ compared to the shape of $\bar{q}^{\bar{\nu}} = x(\bar{u} + \bar{d} + 2\bar{s})$, and to $\bar{q}^{\bar{\nu}}$ including the slow rescaling correction. The distributions do not differ by more than the experimental error. The small difference between the two curves shows that the shape of the effective strange-sea structure function is affected very little by the charm threshold discussed in Sect. 4.2.5.

5.2. Comparison of Neutrino and Charged-Lepton Structure Functions

In the quark parton model, the structure function F_2^{lN} observed in electron or muon inelastic scattering, and the structure function $F_2^{\nu N}$ observed in neutrino scattering, are related outside the sea region by

$$F_2^{\nu N}(x, Q^2) = \tfrac{18}{5} F_2^{lN}(x, Q^2). \tag{15}$$

In the sea region, neutrino and muon experiments measure different contributions of the strange and charmed quarks. For muon experiments the strange sea is suppressed by its quark charge $-1/3$, whereas in neutrino scattering it is suppressed by the charm threshold effect in the transition $s \to c$. The two effects nearly cancel in the present kinematic range such that the QPM relation (15) should be reasonably well satisfied also at small x. In Fig. 18a our measurements of $F_2^{\nu N}$ are compared with the measurements of $F_2^{\mu N}$ in muon-iron scattering obtained by the European Muon Collaboration (EMC) [24] and with F_2^{ed} obtained in electron-deuteron scattering at SLAC [20]. The EMC data are most easily compared with our data, since the same target material is used and the kinematic range is almost the same. The structure functions $F_2^{\mu N}$ and $F_2^{\nu N}$ agree well in shape within the given statistical and systematic errors, and the normalization agrees with the QPM prediction 18/5. For the SLAC data, on the other hand, we find $F_2^{\nu N}/F_2^{ed} = 1.46 \pm 0.12$ for $x > 0.4$, averaging over the whole Q^2-range including the flux normalization errors on both experiments where $9/5 = 1.8$ is expected. This difference may be related to the uncertainty in R. The agreement with expectation is substantially improved if the value of R is set to zero for the SLAC data at large x and Q^2.

The Q^2-variation $d \ln F_2 / d \ln Q^2$ of the structure functions $F_2^{\nu N}$ and (18/5) $F_2^{\mu N}$ of [24] is shown in Fig. 18b as a function of x. The measurements of the slopes have been obtained for each value of x by linear fits in $\ln Q^2$ over the whole available Q^2-range.

Neutrino and muon data show pronounced scaling violations which agree in shape and magnitude. The pattern of scaling violations is well described by leading-order QCD with $\Lambda_{\mathrm{LO}} \simeq 0.2\,\mathrm{GeV}$, as described below.

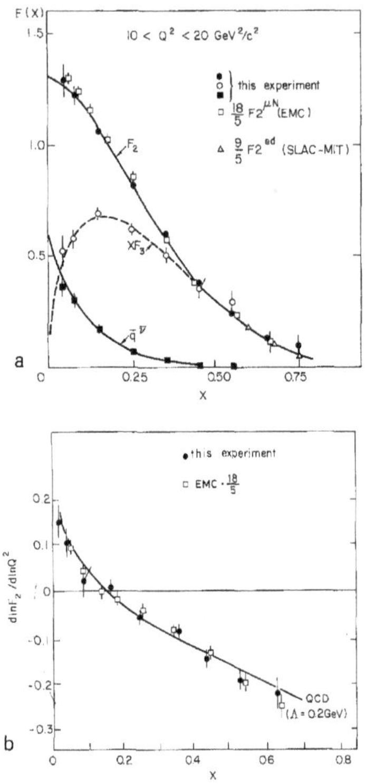

Fig. 18. a The structure functions F_2, xF_3, and $\bar{q}^{\bar{v}}$ as measured in this experiment for fixed Q^2 as a function of x. The data on F_2 are compared with the measurements of $F_2^{\mu N}$ by EMC [24] and of F_2^{ed} by SLAC-MIT [20] multiplied by the quark parton model factors. **b** The Q^2 dependence of the structure function F_2 versus x as measured by this experiment compared with the results of the EMC muon experiment [24]. The data points are obtained by linear fits of $\ln F_2$ versus $\ln Q^2$ in the whole Q^2 range of the experiments. The solid line gives the QCD expectation for $\Lambda=0.2$ GeV

5.3. *Confrontation of Structure Function Measurements with QCD*

Perturbative QCD predicts the Q^2-evolution of the nucleon structure functions, although the functions themselves are at present not calculable. The evolution equations as given by Altarelli and Parisi [25] in leading order are

$$\frac{dxF_3(x,Q^2)}{d\ln Q^2}=\frac{\alpha_s(Q^2)}{2\pi}\int_x^1\left[Pqq\left(\frac{x}{z}\right)zF_3(z,Q^2)\right]\frac{x\,dz}{z^2} \quad (16a)$$

$$\frac{dF_2(x,Q^2)}{d\ln Q^2}=\frac{\alpha_s(Q^2)}{2\pi}\int_x^1\left[Pqq\left(\frac{x}{z}\right)F_2(z,Q^2)\right.$$
$$\left.+2N_F Pgq\left(\frac{x}{z}\right)G(z,Q^2)\right]\frac{x\,dz}{z^2} \quad (16b)$$

$$\frac{d\bar{q}^{\bar{v}}(x,Q^2)}{d\ln Q^2}=\frac{\alpha_s(Q^2)}{2\pi}\int_x^1\left[Pqq\left(\frac{x}{z}\right)\bar{q}^{\bar{v}}(z,Q^2)\right.$$
$$\left.+N_F Pgq\left(\frac{x}{z}\right)G(z,Q^2)\right]\frac{x\,dz}{z^2} \quad (16c)$$

$$\frac{dG(x,Q^2)}{d\ln Q^2}=\frac{\alpha_s(Q^2)}{2\pi}\int_x^1\left[Pqg\left(\frac{x}{z}\right)F_2(z,Q^2)\right.$$
$$\left.+Pgg\left(\frac{x}{z}\right)G(z,Q^2)\right]\frac{x\,dz}{z^2} \quad (16d)$$

In the above equations the P_{ij} are splitting functions given by QCD, $G(x,Q^2)$ is the gluon structure function, N_f is the number of active flavours taken to be four in our range of Q^2, and $\alpha_s=12\pi/[(33-2N_f)\ln(Q^2/\Lambda^2)]$ is the strong coupling constant. The scale parameter Λ is not given by the theory.

The comparison of the inclusive neutrino and antineutrino scattering results with the QCD equations permits conclusions on several distinct points:

i) Tests of the validity of the QCD predictions
ii) The determination of Λ
iii) A determination of the gluon distribution.

All conclusions are based on the measured Q^2-dependence of the structure functions. For tests of the validity of QCD, and for the systematically most correct determination of Λ, the evolution of xF_3, (16a), is the most useful, since it involves no other structure functions, and is free from uncertainties in $R(x,Q^2)$. The gluon distribution, as well as a statistically superior Λ value, are best determined using both the F_2 and the $\bar{q}^{\bar{v}}$ structure functions.

5.3.1. Slopes of Structure Functions

The predictions of the Altarelli-Parisi equations for the Q^2 evolution of F_2 and $\bar{q}^{\bar{v}}$ are compared directly with the measured slopes of the structure functions in Figs. 19a and b for $Q_0^2=4.5\,\text{GeV}^2/c^2$. The data points are obtained by linear fits in $\ln\ln Q^2$ for each value of x. The full curves correspond to the best QCD fit to F_2 and $\bar{q}^{\bar{v}}$ as described below. They are in good agreement with the data. On the other hand, non-asymptotically free theories of the strong interaction with scalar or vector gluons are not able to describe the observed scaling violations as shown by the dotted and dashed curves in Fig. 19a and described in detail in another paper [9].

5.3.2. Fitting Procedure

The quantitative confrontation of the data with QCD is based on the numerical integration of the

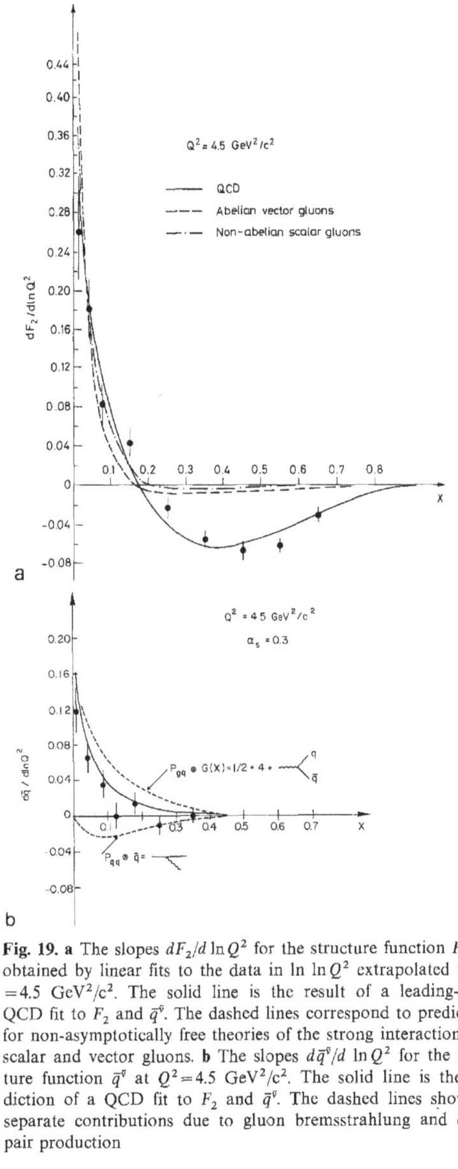

Fig. 19. a The slopes $dF_2/d\ln Q^2$ for the structure function $F_2^{\nu N}$ as obtained by linear fits to the data in $\ln\ln Q^2$ extrapolated to $Q^2 = 4.5\ \mathrm{GeV}^2/c^2$. The solid line is the result of a leading-order QCD fit to F_2 and $\bar{q}^{\bar{\nu}}$. The dashed lines correspond to predictions for non-asymptotically free theories of the strong interaction with scalar and vector gluons. **b** The slopes $d\bar{q}^{\bar{\nu}}/d\ln Q^2$ for the structure function $\bar{q}^{\bar{\nu}}$ at $Q^2=4.5\ \mathrm{GeV}^2/c^2$. The solid line is the prediction of a QCD fit to F_2 and $\bar{q}^{\bar{\nu}}$. The dashed lines show the separate contributions due to gluon bremsstrahlung and quark pair production

Altarelli-Parisi Eqs. (16a) to (16d) using leading- or second-order expressions for $\alpha_s(Q^2)$ and the splitting functions. The structure functions for a starting value $Q^2=Q_0^2$ have been parametrized in the following way:

$$
\begin{aligned}
xF_3(x,Q_0^2)&=a_3(1+b_3x)(1-x)^{c_3}\\
F_2(x,Q_0^2)&=a_2(1+b_2x)(1-x)^{c_2}\\
G(x,Q_0^2)&=a_g(1+b_gx)(1-x)^{c_g}\\
\bar{q}^{\bar{\nu}}(x,Q_0^2)&=a_q(1-x)^{c_q}.
\end{aligned}
\qquad (17)
$$

The parameter a_3 has been fixed by imposing the Gross-Llewellyn Smith sum rule [2] $\int_0^1 F_3\,dx=3$ or $\int_0^1 F_3\,dx=3[1-\alpha_s(Q^2)/\pi]$ in leading and second order, respectively, and a_g by the momentum sum rule $\int_0^1 G\,dx+\int_0^1 F_2\,dx=1$. The shape parameters a_i, b_i, c_i, d_i and the scale parameter Λ are then determined by least squares fits to the data*. This method allows the use of all available data in the whole (x, Q^2) range and needs no data at low invariant hadron mass W. In addition, it is rather insensitive to the contributions of unmeasured kinematic regions. We have verified that the above parametrizations are sufficiently general by varying Q_0^2. We note, however, that the effects due to b_i being non-zero are quite significant for the quality of the fits and the results.

Quantumchromodynamic fits have been performed to the non-singlet structure function xF_3, and to the singlet structure functions F_2 and $\bar{q}^{\bar{\nu}}$. These fits use only data with $Q^2>2\ \mathrm{GeV}^2/c^2$ and $W^2>11\ \mathrm{GeV}^2$. The low Q^2 cut is justified *a posteriori* by the small value of Λ obtained in these fits. The W cut was imposed to avoid the kinematic region where higher twist contributions may be expected to be important and where non-leading corrections are substantial. Target mass corrections are applied according to the prescription of [26]. Finally, we include a propagator term with a mass $m_W = 80\ \mathrm{GeV}$.

5.3.3. Analysis of the Non-Singlet Structure Function xF_3

The Q^2 evolution of xF_3 as given by (15a) is most easily obtained as it does not involve the gluon distribution. The result of the second-order fit is compared with the measurements in Fig. 12. The best value for Λ is $\Lambda_{\overline{MS}}=0.2^{+0.2}_{-0.1}$ GeV including our estimate of the systematic uncertainties. This result is not only independent of assumptions about the shape of the gluon distribution, it is also not affected by the uncertainties on the value of R, the amount of the strange sea and the threshold effects due to charm production. This result is therefore most significant from a systematic point of view.

* We have used the programs developed by Abbott and Barnett, for the numerical solution of the Altarelli-Parisi equation. The second-order calculations for the non-singlet case have been verified using the program of Lopez and Yndurain

5.3.4. Combined Analysis of F_2 and $\bar{q}^{\bar{\nu}}$

The QCD predictions for the Q^2-evolution of F_2 and $\bar{q}^{\bar{\nu}}$ involve the unknown gluon structure function $G(x, Q_0^2)$ and the scale parameter Λ. The analysis of F_2 alone is unable to determine Λ and $G(x, Q_0^2)$ simultaneously since Λ is very strongly correlated to the width of the gluon distribution. If the functional form of the gluon distribution is not constrained, no separation between the effects of Λ and the gluon distribution is possible with F_2 alone. The additional measurement of $\bar{q}^{\bar{\nu}}$, however, provides the means for the separation of the two. This can be seen in two ways:

i) Λ can be determined from the Q^2-evolution of F_2 at large x, i.e. $x \gtrsim 0.3$ after subtracting the sea-quark contribution to get the "non-singlet" structure function

$$F_{NS}(x, Q^2) \equiv F_2(x, Q^2) - 2[\bar{q}^{\bar{\nu}} - x\, s(x, Q^2)] \approx F_2(x, Q^2) - 1.7\, \bar{q}^{\bar{\nu}}. \tag{18}$$

For $x > 0.3$ the correction is small and well known. Note also that R is well determined in this x-range and is practically zero such that we are justified in analysing F_{NS} as a non-singlet structure function.

ii) The gluon distribution can be determined by a simultaneous analysis of F_2 and $\bar{q}^{\bar{\nu}}$ including the sea region, leaving both $G(x, Q_0^2)$ and Λ as unknowns. In this analysis we use mainly the fact that $\bar{q}^{\bar{\nu}}$ is very small at large x. A broad gluon distribution leads to a rise of the antiquark distribution at large x with Q^2 due to the convolution $Pgq \otimes G(x, Q^2)$ in (16c) such that the QCD prediction will be above the measured $\bar{q}^{\bar{\nu}}$ even if we start with a vanishing $\bar{q}$-distribution at small Q^2 (say $Q^2 = 2\,\text{GeV}^2/c^2$). From the point of systematic uncertainties, the magnitude of $\bar{q}^{\bar{\nu}}$ at large x is well measured and more reliable than the measurement of scaling violations. It should be noted that the values of $\bar{q}^{\bar{\nu}}$ from Table 5 which are given for $R = 0.1$ for all x, should be corrected to $R = 0$, at large x, in agreement with the results of Sect. 4.2.1.2.

5.3.4.1. Non-Singlet Analysis of F_2 at Large x. The structure function F_2 is most accurately determined from a statistical point of view. If the analysis is restricted to large x where R is well bounded and the sea contribution small, the uncertainties due to $R = \sigma_L/\sigma_T$ and the charm threshold are negligible. We choose $x > 0.3$ and subtract the small sea-quark contribution according to (18) using our measurement of $\bar{q}^{\bar{\nu}}$. The structure functions F_2 and $\bar{q}^{\bar{\nu}}$ are evaluated using the QCD prediction for R in agreement with the upper limits derived in Sect. 4.2.1.2.

The results of leading and second-order fits are summarized in Table 7. We find $\Lambda_{\overline{MS}} = 0.30 \pm 0.075\,\text{GeV}$ in agreement with the results of the non-singlet fits to xF_3 and the combined singlet fit to F_2 and $\bar{q}^{\bar{\nu}}$.

The dependence of Λ on various systematic effects and different cuts is summarized in Table 8. Whereas target mass corrections and Fermi-motion correction lead to small variations only, the weak propagator has a substantial influence on the result. Previous results which have been reported for the same data [27] had been obtained using a value of $R = 0.1$ which reduces the value of Λ by 0.1 GeV.

Including our estimate of the systematic uncertainties, the non-singlet analysis of F_2 at large x gives the result $\Lambda_{\overline{MS}} = 0.30 \pm 0.15\,\text{GeV}$.

Table 7. Results of QCD fits to the structure function measurements with $Q^2 > 2\,\text{GeV}^2/c^2$ and $W^2 > 11\,\text{GeV}^2$. The errors on Λ are statistical only

Structure function	Λ (GeV)	$F_i(x, Q_0^2)$	χ^2/DF
$F_{NS}(x > 0.03)$ $R = R_{QCD}$	$\Lambda_{LO} = 0.275 \pm 0.08$ $\Lambda_{\overline{MS}} = 0.30 \pm 0.08$	$F_{NS} = \frac{3}{B} x^{0.374}(1-x)^{3.31}(1+5.86x)$	48/49
F_2 and $\bar{q}^{\bar{\nu}}$ all x $R = 0.1$	$\Lambda_{LO} = 0.18 \pm 0.02$	$F_2 = 1.10(1+3.7x)(1-x)^{3.19}$ $\bar{q}^{\bar{\nu}} = 0.52(1-x)^{8.54}$ $G = 2.62(1+3.5x)(1-x)^{5.90}$ $Q_0^2 = 5\,\text{GeV}^2/c^2$	209/196
$F_2(x > 0.03)$ $\bar{q}^{\bar{\nu}}(x > 0.3)$ $R = R_{QCD}$ Corrected for slow rescaling	$\Lambda_{LO} = 0.29 \pm 0.03$	$F_2 = 1.18(1+3.27x)(1-x)^{3.12}$ $\bar{q}^{\bar{\nu}} = 0.53(1-x)^{7.12}$ $G = 1.75(1+8.9x)(1-x)^{6.03}$ $Q_0^2 = 5\,\text{GeV}^2/c^2$	136/130

Table 8. Systematic dependences of Λ_{LO} for the QCD fit to $F_{NS}(x>0.3$ and $W^2>11$ GeV$^2)$

Fit		Λ_{LO} (GeV)	χ^2/DF
Standard fit	$R=R_{QCD}$	0.275 ± 0.09	48/49
	$R=0.0$	0.295	47.9/49
	$R=0.1$	0.21	53/49
	$R=0.2$	0.10	58.4/49
$m_W=\infty$		0.37	48/49
No target mass correction		0.28	48/49
No sea quark correction		0.21	49/49

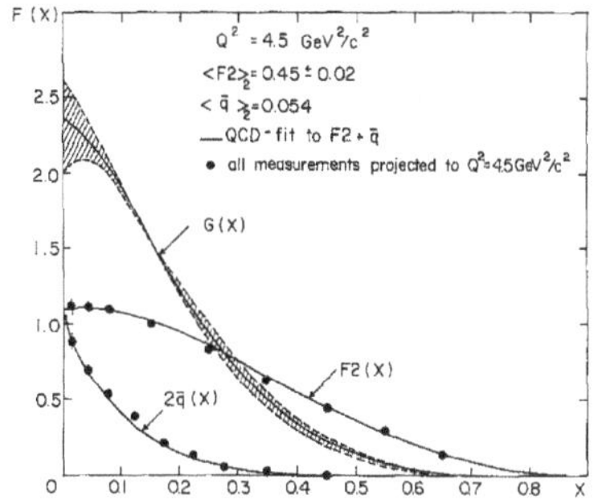

Fig. 20. The structure functions F_2, $\bar{q}^{\bar{\nu}}$, and $G(x)$ for fixed Q^2 as obtained from the QCD fit to F_2 and $\bar{q}^{\bar{\nu}}$. Also shown are the $\pm1\sigma$ bands for $G(x)$ and the measurements of F_2 and $\bar{q}^{\bar{\nu}}$ projected to $Q^2=4.5$ GeV^2/c^2 along the QCD fit lines

5.3.4.2. Determination of the Gluon Distribution by an Analysis of F_2 and $\bar{q}^{\bar{\nu}}$. The combined analysis of F_2 and $\bar{q}^{\bar{\nu}}$ allows a simultaneous determination of Λ and the gluon distribution as discussed in detail in [8]. The result of the leading-order QCD fit to the tabulated structure functions is given in Table 7. It is compared with the measured structure functions in Figs. 11 and 15 and with the measured slopes in Fig. 19. The structure functions $G(x)$, $F_2(x)$, and $\bar{q}^{\bar{\nu}}(x)$ as given by this fit are shown in Fig. 20 for a fixed value of Q^2.

The gluon distribution can only be determined if the observed scaling violations are due to QCD effects. Unfortunately the Q^2-dependence of F_2 and $\bar{q}^{\bar{\nu}}$ is subject to substantial uncertainties due to the charm threshold effect for the strange sea contribution and the poor knowledge of R at small values of x. The magnitude of these effects is estimated in Fig. 21 for the slopes of F_2 using the QCD prediction for $R(x, Q^2)$ and the slow rescaling model for the charm threshold effect. The small x region ($x \lesssim 0.3$) is seriously affected, whereas the large x region shows only a weak dependence. The slopes of $\bar{q}^{\bar{\nu}}$ at small x are affected even more severely by these uncertainties. These effects have been studied in [8] using different assumptions about the strange- and charmed-sea distribution and about R with the result that the gluon distribution was only moderately affected by these effects. Meanwhile we have obtained upper limits on R at large x, such that the uncertainties of the $\bar{q}^{\bar{\nu}}$ measurement and the slopes of F_2 at large x are reduced and we can hope to get a more reliable estimate of $G(x, Q_0^2)$.

We have therefore repeated the singlet analysis using R equal to the QCD prediction which at large x agrees well with the upper limits of Sect. 4.2.1.2. In order to eliminate the regions with large uncertainties due to R at small x, we use F_2 for $0.03<x<0.7$ and $\bar{q}^{\bar{\nu}}$ for $0.3<x<0.7$ only. Finally we correct F_2 for the charm threshold effect using the slow rescaling model [23] with an effective mass of the charm quark $m_c=1.5\,\text{GeV/c}^2$. The result of this fit is also given in Table 7. The gluon distribution comes out slightly broader compared to the fits with $R=0.1$ and Λ_{LO} increases to 0.29 GeV.

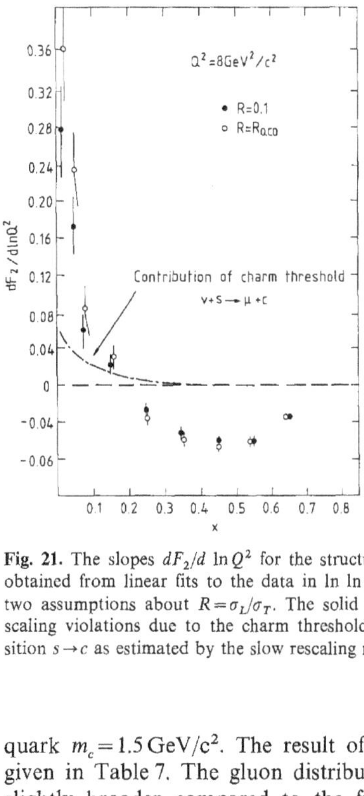

Fig. 21. The slopes $dF_2/d\ln Q^2$ for the structure function $F_2^{\nu N}$ as obtained from linear fits to the data in ln lnQ^2 at fixed Q^2 and two assumptions about $R=\sigma_L/\sigma_T$. The solid curve indicates the scaling violations due to the charm threshold effect in the transition $s\to c$ as estimated by the slow rescaling model

Clearly, the shape of the gluon distribution depends on the functional form which is chosen for the reference value Q_0^2. Our parametrization (17) involves only two free parameters. It gives nevertheless substantial freedom for the shape of the distribution. Within the given parametrization, the gluon distribution is well determined because the integral of $G(x, Q_0^2)$ is given by the energy-momentum sum rule and the width is constrained by the measured width of the antiquark distribution.

The effect of second-order QCD corrections has been studied by several authors [28]. The effect on the gluon distribution is small and comparable to the experimental uncertainties.

We conclude that the combined analysis of F_2 and $\bar{q}^{\bar{\nu}}$ has provided, for the first time, a determination of the gluon distribution.

5.3.5. Conclusions

The Q^2-dependence of all structure functions at high values of the invariant hadron mass ($W^2 > 11\,\mathrm{GeV}^2$) is consistently described by QCD with a value of Λ around 0.25 GeV. It is incompatible with the predictions of non-asymptotically-free theories of the strong interactions as discussed in detail in [9].

The magnitude of non-perturbative contributions, on the other hand, cannot be estimated reliably. If there is just one higher twist term, then the scaling violations at high W are dominated by QCD effects and our determination of Λ and the gluon distribution remain unaffected. In principle, however, all scaling violations could be explained by more complicated higher twist contributions, provided they mimic the x-dependence of QCD [29]. Except for this improbable possibility, we have achieved a reliable determination of Λ and the gluon distribution. The results summarized in Tables 7 and 8 can be combined to obtain a best value of Λ from the present experiment. Including our estimate of systematic uncertainties we find $\Lambda_{\overline{\mathrm{MS}}} = 0.25^{+0.15}_{-0.1}$ GeV.

6. General Conclusions

We list here the more important results of this experiment:

i) The neutrino and antineutrino cross-sections are proportional to E_ν at energies above ~ 70 GeV. For the neutrino cross-section, σ^ν/E shows a significant rise towards lower energies in agreement with the pattern of the observed scaling violations.

ii) The complete and consistent set of structure functions F_2, xF_3, $\bar{q}^{\bar{\nu}}$ and $x\,s(x)$ has been determined for an isoscalar target in the Q^2-range $1 < Q^2 < 200\,\mathrm{GeV}^2/c^2$. It is more precise and comprehensive than measurements previously available and provides a reliable basis for the study of parton dynamics in hard scattering processes. The measurement of F_2 agrees well with similar measurements in charged-lepton scattering, in agreement with the quark parton model.

iii) A deviation from the Callan-Gross relation is observed at small x, whereas the longitudinal structure function is negligible at large x in agreement with the QCD expectation.

iv) Clear scaling violations are observed for all structure functions. The observed Q^2-dependence for $W^2 > 11\,\mathrm{GeV}^2$ is in good agreement with QCD but at variance with non-asymptotically free theories of the strong interaction.

v) The combined analysis of F_2 and $\bar{q}^{\bar{\nu}}$ has, for the first time, allowed a determination of the gluon structure function.

vi) The value of Λ obtained from data at $W^2 > 11\,\mathrm{GeV}^2$ is $\Lambda_{\overline{\mathrm{MS}}} = 0.25^{+0.15}_{-0.10}$, including an estimate of systematic errors.

Acknowledgements. We thank our technical staff, from all the participating institutions, who have contributed so much to the construction and running of the detector. We also thank H. Wind for important contributions to the muon reconstruction program. We thank M. Barnett and C. Lopez for helpful discussions concerning the QCD comparisons.

Appendix

In the following we summarize the formulae and assumptions which have been used to determine the structure functions from the measured differential cross-sections.

1. Correction of the Differential Cross-Sections for the Excess of Neutrons in Iron (Isoscalar Correction)

The correction is evaluated in the framework of the quark parton model assuming $\bar{u} = \bar{d}$. The cross-sections per nucleon in iron are given by the expressions:

$$\left.\frac{d^2\sigma^{\nu N}}{dx\,dy}\right|_{\text{iron}} = \left.\frac{d^2\sigma^{\nu N}}{dx\,dy}\right|_{I=0} + \frac{N-Z}{N+Z}\,x[u_v - d_v]\,\frac{G^2 m_N E_\nu}{\pi} \tag{A1}$$

$$\left.\frac{d^2\sigma^{\bar{\nu} N}}{dx\,dy}\right|_{\text{iron}} = \left.\frac{d^2\sigma^{\bar{\nu} N}}{dx\,dy}\right|_{I=0} - \frac{N-Z}{N+Z}\,x[u_v - d_v]\,(1-y)^2\,\frac{G^2 m_N E_\nu}{\pi} \tag{A2}$$

Here Z and N are the number of protons and neutrons in iron, respectively, and the effects of the longitudinal structure function has been neglected.

The second term on the right-hand side of (A1) and (A2) are the isoscalar corrections. They are evaluated using

$$x(u_v - d_v) = \frac{1 - d_v/u_v}{1 + d_v/u_v} x F_3, \qquad (A3)$$

where xF_3 is the valence quark distribution as measured in the present experiment. To evaluate the structure functions F_2, xF_3, F_+, $\bar{q}^{\bar{\nu}}$, and $R = \sigma_L/\sigma_T$, we have used the simple assumption $d_v/u_v = 0.5$ such that $[(N-Z)/(N+Z)]\, x(u_v - d_v) = \delta x F_3$ with $\delta = (N-Z)/3(N+Z) \simeq 0.023$. The corrections are generally so small that this approximation is adequate. The evaluation of the upper limit on R in Sect. 4.2.1, however, is more sensitive to small corrections. For this determination we used $d_v/u_v = 0.57\ (1-x)$ which we have obtained from a comparison of neutrino and antineutrino interactions in hydrogen [30].

2. Relations Between Structure Functions and Differential Cross-Sections for Isoscalar Targets

The structure functions have been evaluated from the differential cross-sections using the following equations:

$$\frac{\pi}{G^2 m_N E_\nu} \frac{d^2(\sigma^\nu + \sigma^{\bar{\nu}})}{dx\,dy} = F_2[1 + (1-y)^2 - y^2(R(1 + Q^2/\nu^2)/(1+R) - Q^2/2\nu^2)] + 2x(s-c)[1-(1-y)^2] \qquad (A4)$$

$$\frac{\pi}{G^2 m_N E_\nu} \frac{d^2(\sigma^\nu - \sigma^{\bar{\nu}})}{dx\,dy} = x F_3[1 - (1-y)^2] \qquad (A5)$$

$$\frac{\pi}{G^2 m_N E_\nu} \left[\frac{d^2\sigma^{\bar{\nu}}}{dx\,dy} - \frac{d^2\sigma^\nu}{dx\,dy}(1-y)^2\right] = \tfrac{1}{2}[2xF_1 - xF_3 + 2x(s-c)][1-(1-y)^4] + (F_2 - 2xF_1)[(1-y)-(1-y)^3] - \frac{Q^2}{4\nu^2} F_2(2y^3 - y^4) - 2x(s-c)[(1-y^2)-(1-y)^4]. \qquad (A6)$$

Equation (A6) is used to determine the antiquark distribution which is defined by

$$\tfrac{1}{2}[2xF_1 - xF_3 + 2x(s-c)] = \bar{q}^{\bar{\nu}} + \frac{Q^2}{4\nu^2} x F_3. \qquad (A7)$$

This definition ensures that $\bar{q}^{\bar{\nu}}$ is always larger than or equal to zero owing to the inequality $2xF_1 \geqq \sqrt{1 + (Q^2/\nu^2)}\, x F_3$ [2].

The upper limit on $R = \sigma_L/\sigma_T$ from Sect. 4.2.1 is obtained from the experimental quantity:

$$\left[\frac{(d^2\sigma^{\bar{\nu}}/dx\,dy) - (1-y)^2(d^2\sigma^\nu/dx\,dy)}{(d^2\sigma^\nu/dx\,dy) - (1-y)^2(d^2\sigma^{\bar{\nu}}/dx\,dy)}\right] \geqq \frac{[(1-y)-(1-y)^3][R - (Q^2/2\nu^2)]}{1 + [(1-y)-(1-y)^3](Q^2/2\nu^2)}. \qquad (A8)$$

The right-hand side has been obtained from Eqs. (A6) and (A7) putting $\bar{q}^{\bar{\nu}}$ equal to zero.

Assumptions that Enter Into the Structure Function Evaluation. The structure functions F_2, $2xF_1$, and $\bar{q}^{\bar{\nu}}$ can only be obtained with assumptions about $R = \sigma_L/\sigma_T$ and the difference between strange and charmed sea.

To get the values of Tables 4 and 5, we used $R = 0.1$ and $x(s-c) = 0.4\, x(\bar{u} + \bar{d})$ as explained in Sect. 4.2.5. Other assumptions can easily be applied using the correction columns of Tables 4 and 5a and Table 6.

References

1. C.G. Callan, D.J. Gross: Phys. Rev. Lett. **22**, 156 (1969)
2. D.J. Gross, C.H. Llewellyn Smith: Nucl. Phys. **B14**, 337 (1969)
3. J.G.H. de Groot et al.: Z. Phys. C - Particles and Fields **1**, 143 (1979)
4. J.G.H. de Groot et al.: Phys. Lett. **82B**, 292 (1979)
5. J.G.H. de Groot et al.: Phys. Lett. **82B**, 456 (1979)
6. H. Abramowicz et al.: Phys. Lett. **107B**, 141 (1981)
7. H. Abramowicz et al.: Z. Phys. C - Particles and Fields **12**, 225 (1982)
8. H. Abramowicz et al.: Z. Phys. C - Particles and Fields **12**, 289 (1982)
9. H. Abramowicz et al.: Z. Phys. C - Particles and Fields **13**, 199 (1982)
10. M. Holder et al.: Nucl. Instrum. Methods **148**, 235 (1978)
11. H. Abramowicz et al.: Nucl. Instrum. Methods **180**, 429 (1981)
12. For a recent compilation of total cross-section measurements, see, for example, H. Wahl, Proc. 16th Rencontre de Moriond, Les Arcs, 1981, p. 233. Dreux: Frontières 1981
13. R. Blair et al. CCFRR Collaboration: Proc. Neutrino '81 Conf., Hawaii, 1981, p. 311. (High-Energy Physics Group, Univ. of Hawaii, Honolulu, 1981)
14. A. de Rújula et al.: Nucl. Phys. **B154**, 394 (1979)
15. F. Eisele: Proc. Int. Conf. on Neutrino Physics and Astrophysics, Erice, 1980, p. 143. ed. L. Fiorini. New York: Plenum Press 1982
16. R. Barlow, S. Wolfram: Phys. Rev. **D20**, 2198 (1979)
17. M. Jonker et al.: Phys. Lett. **109B**, 133 (1982)
18. S.M. Heagy et al.: Phys. Rev. **D23**, 1045 (1981)
19. H. Abramowicz et al.: Z. Phys. C - Particles and Fields **15**, 19 (1982)
20. A. Bodek et al.: Phys. Rev. **D20**, 1471 (1979)
21. B.A. Gordon et al.: Phys. Rev. **D20**, 2645 (1979)
22. L.F. Abbott et al.: Phys. Rev. **D22**, 582 (1980)
23. R.M. Barnett: Phys. Rev. **D14**, 70 (1976)
24. J.J. Aubert et al.: Phys. Lett. **105B**, 322 (1981)
25. G. Altarelli, G. Parisi: Nucl. Phys. **B126**, 298 (1979)
26. R. Barbieri et al.: Nucl. Phys. **B117**, 50 (1976); H. Georgi, H.D. Politzer: Phys. Rev. **D14**, 1829 (1976)
27. F. Eisele et al. CDHS Collaboration: Proc. Neutrino '81 Conf., Hawaii, 1981, p. 297. (High-Energy Physics Group, Univ. of Hawaii, Honolulu, 1981)
28. L. Baulieu, C. Kounnas: Preprint TH.3266-CERN (1982); D. Duke: Private communication
29. F. Eisele et al.: Phys. Rev. **D26**, 41 (1982)
30. B. Pszola CDHS Collaboration: Proc. Neutrino '82 Conf., Balatonfüred, 1982 Vol. II, p. 133. (Eds. A. Frenkel, L. Jenik (1982))

Volume 231, number 4 PHYSICS LETTERS B 16 November 1989

DETERMINATION OF THE NUMBER OF LIGHT NEUTRINO SPECIES

ALEPH Collaboration

D. DECAMP, B. DESCHIZEAUX, J.-P. LEES, M.-N. MINARD
Laboratoire de Physique des Particules (LAPP), F-74019 Annecy-le-Vieux Cedex, France

J.M. CRESPO, M. DELFINO, E. FERNANDEZ [1], M. MARTINEZ, R. MIQUEL, M.L. MIR, S. ORTEU, A. PACHECO, J.A. PERLAS, E. TUBAU
Laboratorio de Fisica de Altas Energias, Universidad Autonoma de Barcelona, E-08193 Bellaterra (Barcelona), Spain [2]

M.G. CATANESI, M. DE PALMA, A. FARILLA, G. IASELLI, G. MAGGI, A. MASTROGIACOMO, S. NATALI, S. NUZZO, A. RANIERI, G. RASO, F. ROMANO, F. RUGGIERI, G. SELVAGGI, L. SILVESTRIS, P. TEMPESTA, G. ZITO
INFN, Sezione di Bari e Dipartimento di Fisica dell' Università, I-70126 Bari, Italy

Y. CHEN, D. HUANG, J. LIN, T. RUAN, T. WANG, W. WU, Y. XIE, D. XU, R. XU, J. ZHANG, W. ZHAO
Institute of High-Energy Physics, Academia Sinica, Beijing, P.R. China

H. ALBRECHT [3], F. BIRD, E. BLUCHER, T. CHARITY, H. DREVERMANN, Ll. GARRIDO, C. GRAB, R. HAGELBERG, S. HAYWOOD, B. JOST, M. KASEMANN, G. KELLNER, J. KNOBLOCH, A. LACOURT, I. LEHRAUS, T. LOHSE, D. LÜKE [3], A. MARCHIORO, P. MATO, J. MAY, V. MERTENS, A. MINTEN, A. MIOTTO, P. PALAZZI, M. PEPE-ALTARELLI, F. RANJARD, J. RICHSTEIN [4], A. ROTH, J. ROTHBERG [5], H. ROTSCHEIDT, W. VON RÜDEN, D. SCHLATTER, R. ST.DENIS, M. TAKASHIMA, M. TALBY, H. TAUREG, W. TEJESSY, H. WACHSMUTH, S. WHEELER, W. WIEDENMANN, W. WITZELING, J. WOTSCHACK
European Organization for Nuclear Research (CERN), CH-1211 Geneva 23, Switzerland

Z. AJALTOUNI, M. BARDADIN-OTWINOWSKA, A. FALVARD, P. GAY, P. HENRARD, J. JOUSSET, B. MICHEL, J-C. MONTRET, D. PALLIN, P. PERRET, J. PRAT, J. PRORIOL, F. PRULHIÈRE
Laboratoire de Physique Corpusculaire, Université Blaise Pascal, Clermont-Ferrand, F-63177 Aubière, France

H. BERTELSEN, F. HANSEN, J.R. HANSEN, J.D. HANSEN, P.H. HANSEN, A. LINDAHL, B. MADSEN, R. MØLLERUD, B.S. NILSSON, G. PETERSEN
Niels Bohr Institute, DK-2100 Copenhagen, Denmark [6]

E. SIMOPOULOU, A. VAYAKI
Nuclear Research Center Demokritos (NRCD), Athens, Greece

J. BADIER, D. BERNARD, A. BLONDEL, G. BONNEAUD, J. BOUROTTE, F. BRAEMS, J.C. BRIENT, M.A. CIOCCI, G. FOUQUE, R. GUIRLET, P. MINÉ, A. ROUGÉ, M. RUMPF, H. VIDEAU, I. VIDEAU [1], D. ZWIERSKI
Laboratoire de Physique Nucléaire Hautes Energies, École Polytechnique, F-91128 Palaiseau Cedex, France

0370-2693/89/$ 03.50 © Elsevier Science Publishers B.V.
(North-Holland Physics Publishing Division)

Volume 231, number 4 PHYSICS LETTERS B 16 November 1989

D.J. CANDLIN
Department of Physics, University of Edinburgh, Edinburgh EH9 3JZ, UK [7]

A. CONTI, G. PARRINI
Dipartimento di Fisica, Università di Firenze, I-50125 Florence, Italy

M. CORDEN, C. GEORGIOPOULOS, J.H. GOLDMAN, M. IKEDA, D. LEVINTHAL [8], J. LANNUTTI, M. MERMIKIDES, L. SAWYER
High-Energy Particle Physics Laboratory, Florida State University, Tallahassee, FL 32306, USA [9,10,11]

A. ANTONELLI, R. BALDINI, G. BENCIVENNI, G. BOLOGNA, F. BOSSI, P. CAMPANA, G. CAPON, V. CHIARELLA, G. DE NINNO, B. D'ETTORRE-PIAZZOLI, G. FELICI, P. LAURELLI, G. MANNOCCHI, F. MURTAS, G.P. MURTAS, G. NICOLETTI, P. PICCHI, P. ZOGRAFOU
Laboratori Nazionali dell' INFN (LNF-INFN), I-00044 Frascati, Italy

B. ALTOON, O. BOYLE, A.J. FLAVELL, A.W. HALLEY, I. TEN HAVE, J.A. HEARNS, I.S. HUGHES, J.G. LYNCH, D.J. MARTIN, R. O'NEILL, C. RAINE, J.M. SCARR, K. SMITH [1], A.S. THOMPSON
Department of Natural Philosophy, University of Glasgow, Glasgow G12 8QQ, UK [7]

B. BRANDL, O. BRAUN, R. GEIGES, C. GEWENIGER [1], P. HANKE, V. HEPP, E.E. KLUGE, Y. MAUMARY, M. PANTER, A. PUTZER, B. RENSCH, A. STAHL, K. TITTEL, M. WUNSCH
Institut für Hochenergiephysik, Universität Heidelberg, D-6900 Heidelberg, FRG [12]

G.J. BARBER, A.T. BELK, R. BEUSELINCK, D.M. BINNIE, W. CAMERON [1], M. CATTANEO, P.J. DORNAN, S. DUGEAY, R.W. FORTY, D.N. GENTRY, J.F. HASSARD, D.G. MILLER, D.R. PRICE, J.K. SEDGBEER, I.R. TOMALIN, G. TAYLOR
Department of Physics, Imperial College, London SW7 2BZ, UK [7]

P. GIRTLER, D. KUHN, G. RUDOLPH
Institut für Experimentalphysik, Universität Innsbruck, A-6020 Innsbruck, Austria

T.J. BRODBECK, C. BOWDERY [1], A.J. FINCH, F. FOSTER, G. HUGHES, N.R. KEEMER, M. NUTTALL, B.S. ROWLINGSON, T. SLOAN, S.W. SNOW
Department of Physics, University of Lancaster, Lancaster LA1 4YB, UK [7]

T. BARCZEWSKI, L.A.T. BAUERDICK, K. KLEINKNECHT, D. POLLMANN [13], B. RENK, S. ROEHN, H.-G. SANDER, M. SCHMELLING, F. STEEG
Institut für Physik, Universität Mainz, D-6500 Mainz, FRG [12]

J.-P. ALBANESE, J.-J. AUBERT, C. BENCHOUK, A. BONISSENT, F. ETIENNE, R. NACASCH, P. PAYRE, B. PIETRZYK [1], Z. QIAN
Centre de Physique des Particules, Faculté des Sciences de Luminy, F-13288 Marseille, France

W. BLUM, P. CATTANEO, M. COMIN, B. DEHNING, G. COWAN, H. DIETL, M. FERNANDEZ-BOSMAN, D. HAUFF, A. JAHN, E. LANGE, G. LÜTJENS, G. LUTZ, W. MÄNNER, H.-G. MOSER, Y. PAN, R. RICHTER, A. SCHWARZ, R. SETTLES, U. STIEGLER, U. STIERLIN, G. STIMPFL [14], J. THOMAS, G. WALTERMANN
Max-Planck-Institut für Physik und Astrophysik, Werner-Heisenberg-Institut für Physik, D-8000 Munich, FRG [12]

Volume 231, number 4 PHYSICS LETTERS B 16 November 1989

J. BOUCROT, O. CALLOT, A. CORDIER, M. DAVIER, G. DE BOUARD, G. GANIS, J.-F. GRIVAZ, Ph. HEUSSE, P. JANOT, V. JOURNÉ, D.W. KIM, J. LEFRANÇOIS, D. LLOYD-OWEN, A.-M. LUTZ, P. MAROTTE, J.-J. VEILLET

Laboratoire de l'Accélérateur Linéaire, Université de Paris-Sud, F-91405 Orsay Cedex, France

S.R. AMENDOLIA, G. BAGLIESI, G. BATIGNANI, L. BOSISIO, U. BOTTIGLI, C. BRADASCHIA, I. FERRANTE, F. FIDECARO, L. FOÀ [1], E. FOCARDI, F. FORTI, A. GIASSI, M.A. GIORGI, F. LIGABUE, A. LUSIANI, E.B. MANNELLI, P.S. MARROCCHESI, A. MESSINEO, F. PALLA, G. SANGUINETTI, S. SCAPELLATO, J. STEINBERGER, R. TENCHINI, G. TONELLI, G. TRIGGIANI

Dipartimento di Fisica dell' Università, INFN Sezione di Pisa, e Scuola Normale Superiore, I-56010 Pisa, Italy

J.M. CARTER, M.G. GREEN, A.K. McKEMEY, P.V. MARCH, T. MEDCALF, M.R. SAICH, J. STRONG [1], R.M. THOMAS, T. WILDISH

Department of Physics, Royal Holloway & Bedford New College, University of London, Surrey TW20 0EX, UK [7]

D.R. BOTTERILL, R.W. CLIFFT, T.R. EDGECOCK, M. EDWARDS, S.M. FISHER, D.L. HILL, T.J. JONES, G. McPHERSON, M. MORRISSEY, P.R. NORTON, D.P. SALMON, G.J. TAPPERN, J.C. THOMPSON, J. HARVEY

High-Energy Physics Division, Rutherford Appleton Laboratory, Chilton, Didcot, Oxon OX11 0QX, UK [7]

B. BLOCH-DEVAUX, P. COLAS, C. KLOPFENSTEIN, E. LANÇON, E. LOCCI, S. LOUCATOS, L. MIRABITO, E. MONNIER, P. PEREZ, F. PERRIER, B. PIGNARD, J. RANDER, J.-F. RENARDY, A. ROUSSARIE, J.-P. SCHULLER, R. TURLAY

Département de Physique des Particules Elémentaires, CEN-Saclay, F-91191 Gif-sur-Yvette Cedex, France

J.G. ASHMAN, C.N. BOOTH, F. COMBLEY, M. DINSDALE, J. MARTIN, D. PARKER, L.F. THOMPSON

Department of Physics, University of Sheffield, Sheffield S3 7RH, UK [7]

S. BRANDT, H. BURKHARDT, C. GRUPEN, H. MEINHARD, E. NEUGEBAUER, U. SCHÄFER, H. SEYWERD, K. STUPPERICH

Fachbereich Physik, Universität Siegen, D-5900 Siegen, FRG [12]

B. GOBBO, F. LIELLO, E. MILOTTI, F. RAGUSA [15], L. ROLANDI [1]

Dipartimento di Fisica, Università di Trieste e INFN Sezione di Trieste, I-34127 Trieste, Italy

L. BELLANTONI, J.F. BOUDREAU, D. CINABRO, J.S. CONWAY, D.F. COWEN, Z. FENG, J.L. HARTON, J. HILGART, R.C. JARED [16], R.P. JOHNSON, B.W. LECLAIRE, Y.B. PAN, T. PARKER, J.R. PATER, Y. SAADI, V. SHARMA, J.A. WEAR, F.V. WEBER, SAU LAN WU, S.T. XUE and G. ZOBERNIG

Department of Physics, University of Wisconsin, Madison, WI 53706, USA [17]

Received 12 October 1989

The cross-section for $e^+e^- \to$ hadrons in the vicinity of the Z boson peak has been measured with the ALEPH detector at the CERN Large Electron Positron collider, LEP. Measurements of the Z mass, $M_Z = (91.174 \pm 0.070)$ GeV, the Z width $\Gamma_Z = (2.68 \pm 0.15)$ GeV, and of the peak hadronic cross-section, $\sigma^{peak}_{had} = (29.3 \pm 1.2)$ nb, are presented. Within the constraints of the standard electroweak model, the number of light neutrino species is found to be $N_\nu = 3.27 \pm 0.30$. This result rules out the possibility of a fourth type of light neutrino at 98% CL.

1. Introduction

The Z boson was discovered in 1983 at the $p\bar{p}$ collider at CERN [1,2]. Detailed studies of its properties can now be performed in e^+e^- collisions at LEP. In the standard electroweak model [3], the Z boson is expected to decay with comparable probability into all species of fermions that are kinematically allowed. The decay rate of the Z into light, neutral, penetrating particles such as neutrinos, that would otherwise escape detection, can be measured through an increase in the total width Γ_Z. Detection of additional neutrino species would put in evidence additional fermion families, even if the masses of their charged partners are inaccessible at presently available energies. One additional species of neutrino would result in an increase of 6.6% in Γ_Z. Furthermore the peak cross-section to any detectable final state f is very sensitive to this change and would decrease by 13% for one more neutrino species. This cross-section can be expressed in terms of Γ_Z and of the partial widths, Γ_{ee}, Γ_ν, Γ_f, of the Z into e^+e^-, neutrinos, and the final state f as

$$\sigma_f^{peak} = \frac{12\pi}{M_Z^2} \frac{\Gamma_{ee}\Gamma_f}{\Gamma_Z^2} (1-\delta_{rad}) \equiv \sigma_f^0 (1-\delta_{rad}) , \qquad (1)$$

with

$$\Gamma_Z = N_\nu \Gamma_\nu + 3\Gamma_{ee} + \Gamma_{had} , \qquad (2)$$

where N_ν is the number of light neutrino species. In the particular case where f comprised mostly hadronic final states, uncertainties in the overall scale of partial widths, such as those related to the lack of knowledge of the top quark mass and more generally to electroweak radiative effects, as well as uncertainties in the ratio of hadronic to leptonic partial widths, largely cancel in this formula. The QED initial state radiative correction, δ_{rad}, is quite large, but has been calculated to an accuracy believed to be better than 0.5% by several authors [4].

Cross-sections are measured by taking the ratio of the number of selected Z decays, hereafter referred to as hadronic events, to the number of small angle e^+e^- events from the well calculable Bhabha scattering process, hereafter referred to as luminosity events.

2. Description of the ALEPH detector

The data presented here have been collected with the ALEPH detector during the first three weeks of running at LEP, from 20 September to 9 October 1989. Guided by earlier measurements of the Z mass by the CDF [5] and MarkII [6] Collaborations, data were collected mainly at the Z peak and in the near vicinity of it. Integrated luminosities of 64 nb^{-1} at the peak and 88 nb^{-1} on the sides of the peak were recorded.

A detailed description of the ALEPH detector is in preparation [7]. The principal components relevant for this measurement are:

– The Inner Tracking Chamber, ITC, and 8-layer cylindrical drift chamber with sense wires parallel to the beam axis from 13 cm to 29 cm in radius. Tracks with polar angles from 14° to 166° traverse all 8 layers.

– The large cylindrical Time Projection Chamber, TPC, extending from an inner radius of 31 cm to an outer radius of 180 cm over a length of 4.4 m. Up to 21 space coordinates are recorded for tracks with polar angles from 47° to 133°. Requiring 4 coordinates, tracks are reconstructed down to 15°.

– The Electromagnetic Calorimeter, ECAL, a lead wire-chamber sandwich. The cathode readout is subdivided into a total of 73 728 projective towers. Each

[1] Present address: CERN, CH-1211 Geneva 23, Switzerland.
[2] Supported by CAICYT, Spain.
[3] Permanent address: DESY, D-2000 Hamburg 52, FRG.
[4] Present address: Lecroy, Geneva, Switzerland.
[5] On leave of absence from University of Washington, Seattle, WA 98195, USA.
[6] Supported by the Danish Natural Science Research Council.
[7] Supported by the UK Science and Engineering Research Council.
[8] Supported by SLOAN fellowship, contract BR 2703.
[9] Supported by the US Department of Energy, contract DE-FG05-87ER40319.
[10] Supported by the NSF, contract PHY-8451274.
[11] Supported by the US Department of Energy, contract DE-FC0S-85ER250000.
[12] Supported by the Bundesministerium für Forschung und Technologie.
[13] Permanent address: Institut für Physik, Universität Dortmund, D-4600 Dortmund 50, FRG.
[14] Present address: FSU, Tallahassee, FL 32306, USA.
[15] Present address: INFN, Sezione di Milano, I-20133 Milan, Italy.
[16] Permanent address: LBL, Berkeley, CA 94720, USA.
[17] Supported by the US Department of Energy, contract DE-AC02-76ER00881.

tower of about $1^\circ \times 1^\circ$ solid angle is read out in three stacks of 10, 23 and 12 layers (respectively 4, 9 and 9 radiation lengths). The signals from the 45 wire planes of each of the 36 modules are also read out. The two endcaps cover polar angles from 11° to 40° and 140° to 169° and the barrel covers polar angles from 40° to 140°.

- The superconducting, solenoidal coil, providing a magnetic field of 1.5 T.
- The Hadron Calorimeter, HCAL, comprised of 23 layers of streamer tubes interleaved in the iron of the magnet return yoke, read out by a total of 4608 projective towers. Signals from each of the tubes are also read out. The modules are rotated in azimuth by $\sim 2^\circ$ with respect to the electromagnetic calorimeter so that inactive zones do not align in the two calorimeters. The two endcaps and the barrel of the hadron calorimeter cover polar angles down to 6°.
- The Small Angle Tracking chamber, SATR, with 9 planes of drift tubes, covering angles from 40 to 90 mrads, for precise measurement of small angle electron tracks.
- The Luminosity Calorimeter, LCAL, similar in its construction and read-out to ECAL, extending from 50 to 180 mrads, providing energy and position measurement of the showers produced by luminosity events.

Hadronic events were triggered by two independent first level triggers: (i) an ECAL-based trigger, requiring a total energy of 6 GeV deposited in the ECAL barrel or 3 GeV in either of the ECAL endcaps or 1 GeV in both in coincidence; (ii) an ITC–HCAL coincidence, for penetrating charged particles, requiring 6 ITC wire planes and 4 to 8 planes of HCAL tubes in the same azimuthal region.

Luminosity events were also triggered in two ways: (i) a coincidence of 15 GeV deposited in LCAL on one side with 10 GeV deposited on the other side, without requiring azimuthal correlation; (ii) a single arm requirement of 32 GeV deposited in either side of LCAL. In addition, prescaled single arm triggers with 10 and 15 GeV thresholds were recorded to provide an estimate of the beam related background.

All types of events were processed simultaneously through the same trigger system (with the same deadtime), through the same data acquisition and reconstruction programs. In order to ensure that the events were counted during the same life-time, the list of enabled triggers and the status of each of the relevant subdetectors was recorded with each event. Events were accepted only when both ECAL and LCAL were running and when both ECAL and LCAL triggers were enabled. A possible bias could come from the few data acquisition failures that occurred during the run. A careful investigation of all the events before, during, and after each failure revealed a possible but small loss of 0.4% of the hadronic events. This check could be done for the events that were mishandled because their trigger pattern was always recorded and the trigger patterns of hadronic and luminosity events are unique enough to allow an estimate of the losses in each category.

3. Event selection

To provide a check, two independent event selections were used. The first one selected hadronic Z decays only, and was based on TPC tracks. The second one selected decays of the Z into hadrons as well as τ pairs, and was based on calorimetric energy. The event samples overlapped to the extent of 95%. The efficiencies of both methods were very close to unity and systematic uncertainties in these efficiencies were less than 1%. Altogether 3112 events were retained in the track selection and 3320 events in the calorimetric selection.

3.1. Selection with TPC tracks

Hadronic Z decays (a typical event is shown in fig. 1) were selected on the basis of charged tracks only, requiring at least 5 charged tracks. The energy sum of all charged tracks was required to be at least 10% of the centre-of-mass energy. Tracks were required to have a polar angle larger than 18.2°, to be reconstructed from at least 4 TPC coordinates, and to originate from a 2 cm radius 20 cm long cylinder around the nominal beam position. The performance of the TPC was in remarkable agreement with expectations. In order to estimate the acceptance, a complete simulation of the $e^+e^- \to$hadrons process was performed, including initial state radiation effects and hadronization. The properties that are relevant for acceptance calculation, total charged-particle energy, track multiplicity, sphericity distribution and polar

Volume 231, number 4 PHYSICS LETTERS B 16 November 1989

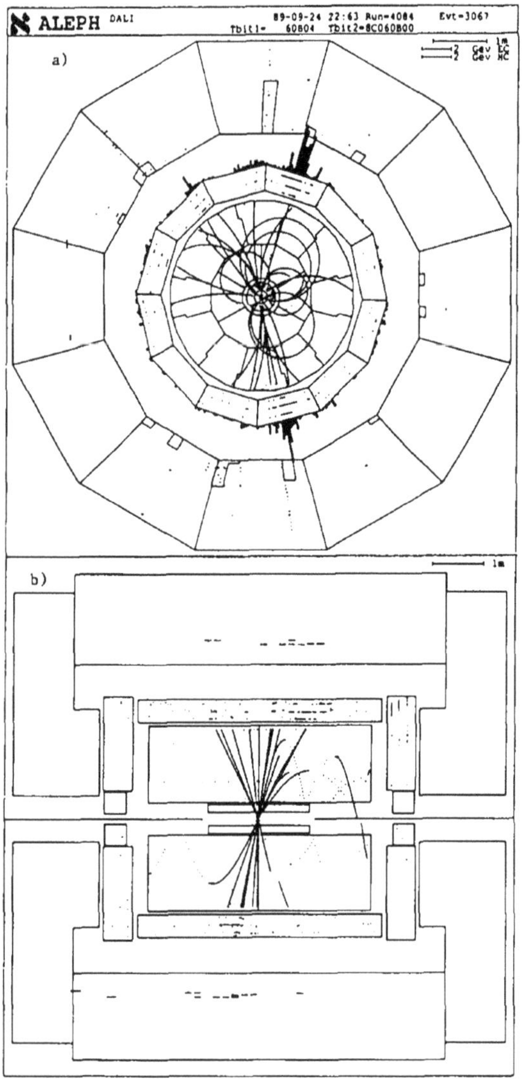

Fig. 1. A hadronic Z decay in the ALEPH detector: (a) x–y view; (b) r–z view.

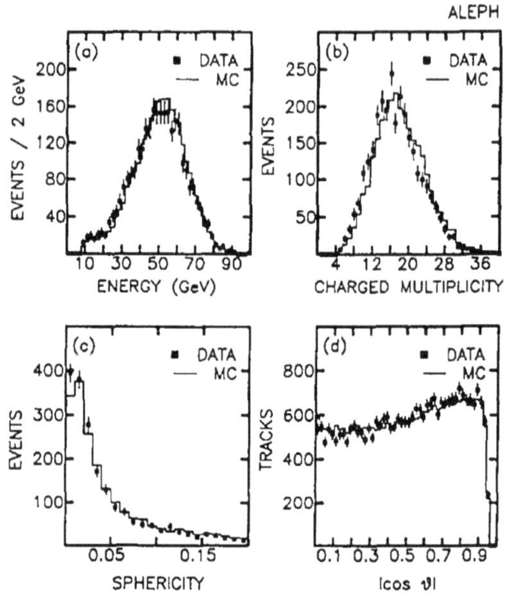

Fig. 2. Properties of charged tracks in hadronic events and comparison with simulations. In each plot, the solid points represent data and the lines represent the simulation normalized to the data: (a) distribution of the charged-track energy sum per event; (b) charged-track multiplicity distribution; (c) sphericity distribution for events where the sphericity axis had a polar angle such that $|\cos \vartheta_{sph}| < 0.8$; and (d) polar angle distribution for charged tracks.

angle distribution, are in good agreement with the simulation, as shown in fig. 2. The efficiency of this selection method is 0.975 ± 0.006, on the peak; the error corresponds to assigning a conservative 20% energy scale uncertainty near the cut. Uncertainties in hadronization models were reduced to a very small level by using the measured sphericity distribution for the acceptance calculation. Due to initial state radiation, this efficiency varies slightly on the side of the Z peak by up to -0.003.

Contamination of the sample of hadronic events by τ pairs from Z decays was estimated to be (5.1 ± 1.5) events, and in fact three events compatible with that hypothesis were found in the sample. Contamination by beam–gas interactions was estimated from the number of events found passing the selection cuts except for the longitudinal vertex position: about one event is expected. Finally the background from $e^+e^- \to e^+e^- +$ hadrons ("two-photon" events) was calculated to be about 15 pb, representing a contamination of 0.5×10^{-3} to the peak cross-section.

3.2. Selection using the calorimeters

The aim of this method was to select hadronic and τ events. The basic requirement was that the total calorimetric energy be above 20 GeV, as well as either ≥6 GeV in the ECAL barrel or at least 1.5 GeV in each ECAL endcap. These requirements reduce both two-photon events and muon-pair events to a negligible level. Large-angle e^+e^- events were rejected on the basis of their characteristic tight energy clusters in the electromagnetic calorimeter. For the few events with no tracks at all, cuts were applied to eliminate cosmic rays: a timing cut and a cut on the minimal number (2) of clusters above 3 GeV in ECAL. On the basis of Monte Carlo simulation as well as the scanning of events the resulting selection efficiency is 0.994 ± 0.005 for hadronic events and 0.60 ± 0.05 for τ events, giving a combined $\sigma_{had} + \sigma_\tau$ efficiency of 0.974 ± 0.006. Contamination by events other than hadronic or τ was estimated to be less than 0.4%.

The two event samples were compared event by event. Differences were well understood given the different characteristics of the two selections.

3.3. Trigger efficiency

The trigger efficiency was measured by counting events where one or both of the ECAL and ITC–HCAL triggers occurred; it was found to be 100% for the ECAL trigger and 87% for the ITC–HCAL trigger, giving an overall efficiency of 100%.

4. Determination of the luminosity

Luminosity events were selected on the basis of the energy deposited in the LCAL towers. Neighboring towers containing more than 50 MeV were joined into clusters, giving energy and position of the shower. Events were required to have a shower reconstructed on each side of LCAL.

In order to minimize the dependence of the acceptance upon precise knowledge of the beam parameters, an asymmetric selection was performed: on one side (e.g., the e^+ side), showers were required to have more than half of their energy deposited in a fiducial volume (fig. 3) excluding the towers situated at the edge of the detector; the total energy deposit on that side was also required to be larger than 55% of the beam energy; on the other side (e^- side), only a total energy deposition of more than 44% of beam energy was required. The respective roles of the e^+ and e^- sides were interchanged in every other event. Finally, the difference in azimuth, $\Delta\phi$, between the e^+ and the e^- was required to be larger than 170°.

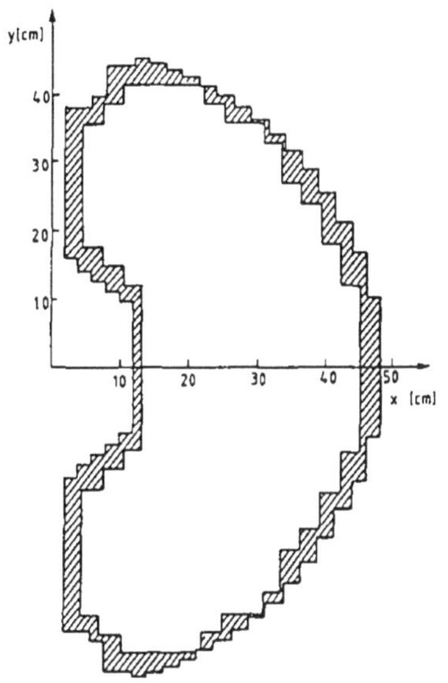

Fig. 3. End view of the Luminosity Calorimeter, showing the tower limits; the shaded area represent the towers excluded from the fiducial area.

The accepted cross-section was calculated using a first order event generator [8] with a full simulation of the detector. The value obtained for the cross-section for Bhabha scattering into the region within these cuts was found to be $(31.12 \pm 0.45_{exp} \pm 0.31_{th})$ nb, at a centre-of-mass energy of 91.0 GeV and for a Z mass of 91.0 GeV. the first error represents the effect of uncertainties in the simulation, calibration, and positioning of the apparatus; the second one represents the possible error due to neglecting higher order radiative effects, and the uncertainty in the photon vacuum polarization [9]. Properties of the luminosity events are shown in fig. 4, and compared with the simulation. The optimum energy resolution has not yet been obtained, but the acceptance is quite insensitive to this. Other properties are in good agreement with expectations. The displacements of the beam

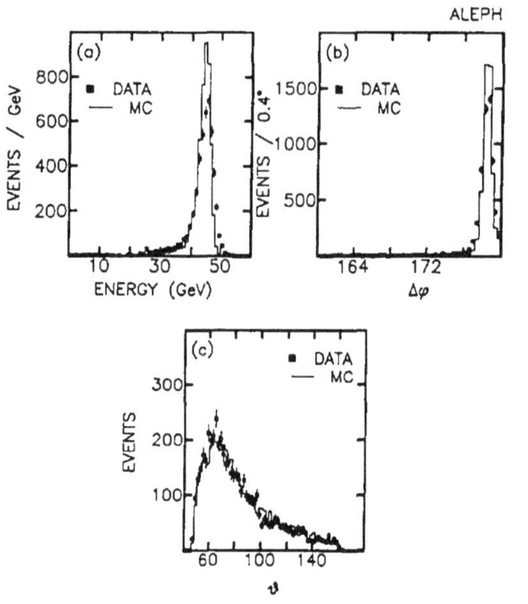

Fig. 4. Properties of luminosity events passing selection criteria and comparison with simulations. In each plot, the solid points represent data and the lines represent the simulation normalized to the data: (a) shower energy distribution for events passing the tight fiducial cut; (b) azimuthal separation $\Delta\phi$ of the two opposite clusters; (c) polar angle distribution.

were measured using luminosity events themselves and corrected for.

The beam-related background contamination was estimated from single arm, prescaled triggers. These triggers were combined into artificial double arm events. The number of such combinations passing the selection cuts was normalized to the number of real coincidences with $\Delta\phi < 90°$. This background subtraction was performed for each run and was of the order of 1% or smaller.

The contamination by physics sources such as $e^+e^- \to \gamma\gamma$ or $e^+e^- \to e^+e^- f\bar{f}$ has been estimated to be less than 2×10^{-3}. The interference of the Z exchange diagram with the purely QED contribution has been taken into account when determining the resonance parameters.

The trigger efficiency was measured for each data taking period by comparing the number of events passing the selection criteria that set the single arm trigger, the coincidence trigger, or both. Inefficiencies in the trigger were traced down to faulty electronic channels. Efficiencies vary with the run, ranging from 0.98 to 1.00 with an average value of 0.997 ± 0.002.

Table 1
Summary of systematic errors in the luminosity measurement.

transverse and longitudinal shower profile	±0.005
energy scale	±0.002
energy resolution and cell-to-cell calibration	±0.007
external alignment and beam parameters	±0.002
internal alignment and inner radius	±0.010
description of material	±0.005
higher order radiative effects	±0.01
total uncertainty:	±0.02

A summary of the luminosity systematic errors is given in table 1. A relatively small normalization error was possible mainly as a result of the excellent spatial resolution of the calorimeter (~300 μm for electrons near the tower boundaries), the good background conditions delivered by the machine, the availability of a redundant set of triggers and progress in the theoretical calculations [10]. The relative normalization uncertainty of cross-section measurements at different energies comes mostly from differences in background conditions and trigger efficiency. These uncertainties were taken into account in the statistical error.

5. Determination of the Z resonance parameters

The number of Z events, of luminosity events, together with the cross-sections are given in table 2 for the two event selection methods.

Two different fits were performed to the data. In the first fit, the basic parameters of the Z resonance, its mass M_Z, width Γ_Z, and QED corrected peak cross-section

$$\sigma_f^0 \equiv \frac{12\pi}{M_Z^2}\frac{\Gamma_{ee}\Gamma_f}{\Gamma_Z^2}$$

are extracted with little model dependence. In the second fit, the constraints from the standard model are applied to determine M_Z and N_ν.

The three parameter fit was performed using com-

Table 2
Event numbers and cross-section as a function of centre-of-mass energy. The overall systematic error of ±2% in the cross-sections is not included.

Energy GeV	Selection from TPC tracks			Selection by calorimeters		
	N_{had}	N_{lumi}	σ_{had} (nb)	$N_{had}+N_\tau$	N_{lumi}	$\sigma_{had}+\sigma_\tau$ (nb)
89.263	120	443	9.00±0.92	134	450	9.89±0.97
90.265	406	715	18.43±1.14	445	736	19.62±1.17
91.020	656	668	31.29±1.72	678	669	32.28±1.76
91.266	1156	1295	28.16±1.14	1243	1309	29.96±1.19
92.260	258	377	21.11±1.72	268	374	22.10±1.78
92.519	125	247	15.52±1.71	142	260	16.75±1.76
93.264	391	883	13.36±0.82	410	889	13.92±0.84

puter programs by Burgers [11] and Borelli et al. [12], folding a Breit–Wigner resonance (with s-dependent width) with the second order exponentiated initial state radiation spectrum. Results from fitting with the two programs were in good agreement with each other. These approximate programs agree adequately with complete electroweak calculations (see D.Y. Bardin et al., in ref. [4] and ref. [13]) for centre-of-mass energies within ±2.5 GeV of the peak. These fits yield the values for the mass and width of the Z boson shown in table 3.

The data points and the result of the fit are shown in fig. 5 for the track selected events.

The error in M_Z does not yet include the uncertainty in the mean e^+e^- collision energy. This error was determined by the LEP division [14] on the basis of measurements of uncertainties in the magnetic field integrals and in the orbit positions to be 5×10^{-4} or 45 MeV. Including this uncertainty, we find

$$M_Z = (91.174 \pm 0.055_{exp} \pm 0.045_{LEP})\ \text{GeV}\,. \qquad (3)$$

The value agrees with the two previous best measurements, refs [5,6], but the uncertainty is smaller by a factor of 2.

M_Z is effectively uncorrelated with the other two parameters. The correlation between Γ_Z and σ^0 for the track selected data sample is shown in fig. 6.

In the standard model, Γ_{ee}, Γ_{had} and Γ_ν are calculable with a small uncertainty of about ±1% due to (i) electroweak radiative effects involving unknown

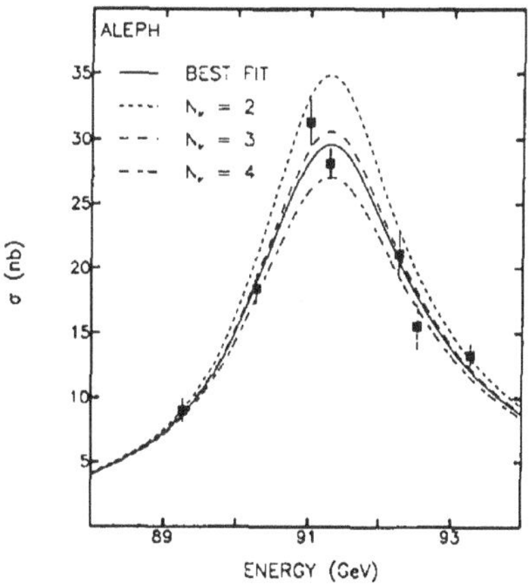

Fig. 5. The cross-section for $e^+e^-\to$hadrons as a function of centre-of-mass energy and result of the three parameter fit.

Table 3

	M_Z (GeV)	Γ_Z (GeV)	σ^0 (nb)	σ^{peak} (nb)
hadronic events	91.178±0.055	2.66±0.16	39.1±1.6	29.3±1.2
hadronic+τ events	91.170±0.054	2.70±0.15	40.9±1.7	30.5±1.3
combined	91.174±0.054	2.68±0.15		

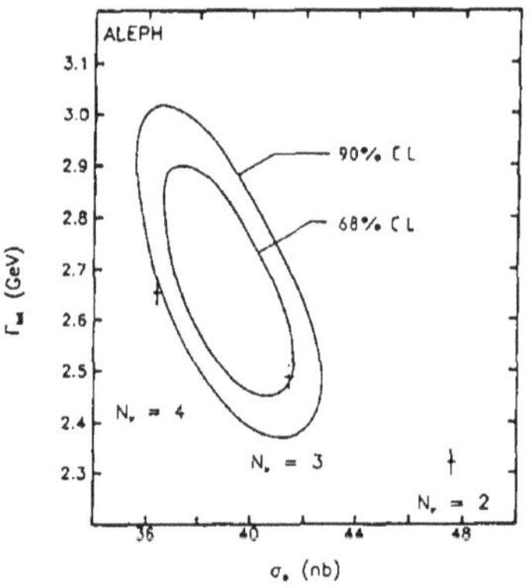

Fig. 6. The total width versus the peak hadronic cross-section, with 68% and 90% CL experimental contours. The standard model prediction for 2, 3 or 4 species of neutrinos is also shown, with its theoretical error.

Table 4
Standard model partial widths of the Z in MeV for the measured value of M_Z; $\alpha_s = 0.12 \pm 0.02$ and $\sin^2\vartheta_w = 0.230 \pm 0.006$ have been used as input.

Γ_{ee} (MeV)	83.5 ± 0.5
Γ_ν (MeV)	166.5 ± 1.0
Γ_{had} (MeV)	1737 ± 22

particles, such as the top quark; (ii) the value of the strong coupling constant α_s. The standard model predictions for the partial widths are given in table 4. The standard model predictions for σ^0 and Γ_Z assuming 2, 3, and 4 species of light neutrinos are shown in fig. 6. The value $N_\nu = 3$ is preferred. More precise information on N_ν is contained in the peak cross-section σ^0.

If the partial widths are taken from standard model predictions, a two parameter fit can be performed, leaving M_Z and N_ν as only free parameters. This fit, performed with the programs of refs. [12,15] on the two data samples, leaves the value of M_Z unchanged. The result for N_ν is

$$N_\nu = 3.27 \pm 0.24_{stat} \pm 0.16_{sys} \pm 0.05_{th} , \qquad (4)$$

where the errors coming from statistics, experimental systematics and theoretical uncertainty are shown separately. Theoretical uncertainties due to electroweak radiative effects consist mostly in a change of the overall scale of the partial widths, and largely cancel in σ^0. The uncertainty in Γ_{had}, related to the QCD correction, cancels in part in σ^0; the resulting uncertainty $\Delta\sigma^0/\sigma^0 = 0.4\Delta\Gamma_{had}/\Gamma_Z = 0.003$ is much smaller than the effect produced by a single neutrino family, $\Delta\sigma^0/\sigma^0 = -0.13$. Combining these errors in quadrature one finds

$$N_\nu = 3.27 \pm 0.30 . \qquad (5)$$

The hypothesis $N_\nu = 4$ is ruled out at 98% confidence level. This measurement improves in a decisive way upon previous determinations of the number of neutrino species from the UA1 [16] and UA2 [17] experiments, from PEP [18] and PETRA [19], from cosmological [20] or astrophysical [21] arguments, as well as from a similar determination at the Z peak [22].

The demonstration that there is a third neutrino confirms that the τ neutrino is distinct from the e and μ neutrinos. The absence of a fourth light neutrino indicates that the quark–lepton families are closed with the three which are already known, except for the possibility that higher order families have neutrinos with masses in excess of ~30 GeV.

Acknowledgement

We would like to express our gratitude and admiration to our colleagues of the LEP division for the timely and beautiful operation of the machine. We thank the Technical Coordinator, Pierre Lazeyras, and the technical staff of the ALEPH Collaboration for their excellent work. We would like to dedicate this paper to the memory of those who died during LEP construction: Luigi Barito, Miloud Ferras, Luigi Filippi, Frédéric Mouly, Noël Piccini and François Pierrus. Those of us from non-member countries thank CERN for its hospitality.

References

[1] UA1 Collab., G. Arnison et al., Phys. Lett. B 126 (1983) 398.
[2] UA2 Collab., P. Bagnaia et al., Phys. Lett. B 129 (1983) 130.

[3] S.L. Glashow, Nucl. Phys. 22 (1961) 579;
S. Weinberg, Phys. Rev. Lett. 19 (1967) 1264;
A. Salam, Elementary particle theory, ed. N. Svartholm (Almquist and Wiksell, Stockholm, 1968) p. 367.

[4] For references see D.Y. Bardin et al., Z-line-shape group, in: Proc. Workshop of Z physics at LEP, CERN report 89-08;
R.N. Cahn, Phys. Rev. D 36 (1987) 2666;
O. Nicrosini and L. Trentadue, Phys. Lett. B 196 (1987) 551;
F.A. Berends, G. Burgers, W. Hollik and W.L. van Neerven, Phys. Lett. B 203 (1988) 177;
G. Burgers, in: Polarization at LEP, preprint CERN 88-06 (1988);
D.C. Kennedy et al., Nucl. Phys. B 321 (1989) 83.

[5] F. Abe et al., Phys. Rev. Lett. 63 (1989) 720.

[6] G.S. Abrams et al., Phys. Rev. Lett. 63 (1989) 724.

[7] ALEPH - a detector for electron–positron annihilation at LEP, Nucl. Instrum. Methods, to be published.

[8] F.A. Berends and R. Kleiss, Nucl. Phys. B 228 (1983) 737;
M. Böhm, A. Denner and W. Hollik, Nucl. Phys. B 304 (1988) 687;
F.A. Berends, R. Kleiss and W. Hollik, Nucl. Phys. B 304 (1988) 712.

[9] H. Burkhardt, F. Jegerlehner, G. Penso and C. Verzegnassi, Z. Phys. C 43 (1989) 497.

[10] D.Y. Bardin et al., Monte Carlo working group, Proc. Workshop of Z physics at LEP, CERN report 89-08.

[11] Computer program ZAPP, courtesy of G. Burgers.

[12] A. Borelli, M. Consoli, L. Maiani and R. Sisto, preprint CERN-TH-5441 (1989).

[13] Computer program ZHADRO, courtesy of G. Burgers.

[14] S. Myers, private communication; and LEP note, to appear.

[15] Computer program ZAPPH, courtesy of G. Burgers.

[16] UA1 Collab., C. Albajar et al., Phys. Lett. B 185 (1987) 241; B 198 (1987) 271.

[17] UA2 Collab., R. Ansari et al., Phys. Lett. B 186 (1987) 440.

[18] MAC Collab., W.T. Ford et al., Phys. Rev. D 19 (1986) 3472.
ASP Collab., C. Hearty et al., Phys. Rev. Lett. 58 (1987) 1711.

[19] CELLO Collab., H.J. Behrend et al., Phys. Lett. B 215 (1988) 186.

[20] G. Steigman, K.A. Olive, D.N. Schramm and M.S. Turner, Phys. Lett. B 176 (1986) 33;
J. Ellis, K. Enqvist, D.V. Nanopoulos and S. Sarkar, Phys. Lett. B 167 (1986) 457.

[21] J. Ellis and K.A. Olive, Phys. Lett. B 193 (1987) 525;
R. Schaeffer, Y. Declais and S. Jullian, Nature 330 (1987) 142;
L.M. Krauss, Nature 329 (1987) 689.

[22] MarkII Collab., J.M. Dorfan, Intern. Europhysics Conf. on High energy physics (Madrid, Spain, September 1989).

Z. Phys. C – Particles and Fields 53, 1–20 (1992)

Zeitschrift für Physik C Particles and Fields

Improved measurements of electroweak parameters from Z decays into fermion pairs

ALEPH Collaboration*

D. Decamp, B. Deschizeaux, C. Goy, J.-P. Lees, M.-N. Minard
Laboratoire de Physique des Particules (LAPP), IN^2P^3-CNRS, F-74019 Annecy-le-Vieux Cedex, France

R. Alemany, J.M. Crespo, M. Delfino, E. Fernandez, V. Gaitan, Ll. Garrido, Ll.M. Mir, A. Pacheco
Laboratorio de Fisica de Altas Energias, Universidad Autonoma de Barcelona, E-08193 Bellaterra (Barcelona), Spain[8]

M.G. Catanesi, D. Creanza, M. de Palma, A. Farilla, G. Iaselli[1], G. Maggi, M. Maggi, S. Natali, S. Nuzzo, M. Quattromini, A. Ranieri, G. Raso, F. Romano, F. Ruggieri, G. Selvaggi, L. Silvestris, P. Tempesta, G. Zito
INFN Sezione di Bari e Dipartimento di Fisica dell' Università, I-70126 Bari, Italy

Y. Gao, H. Hu[21], D. Huang, X. Huang, J. Lin, J. Lou, C. Qiao[21], T. Ruan[21], T. Wang, Y. Xie, D. Xu, R. Xu, J. Zhang, W. Zhao
Institute of High-Energy Physics, Academia Sinica, Beijing, People's Republic of China[9]

W.B. Atwood[2], L.A.T. Bauerdick, F. Bird, E. Blucher, G. Bonvicini, F. Bossi, J. Boudreau, D. Brown, T.H. Burnett[3], H. Drevermann, R.W. Forty, C. Grab, R. Hagelberg, S. Haywood, J. Hilgart, B. Jost, M. Kasemann, J. Knobloch, A. Lacourt, E. Lançon, I. Lehraus, T. Lohse, A. Lusiani, A. Marchioro, M. Martinez, P. Mato, S. Menary, A. Minten, A. Miotto, R. Miquel, H.-G. Moser, J. Nash, P. Palazzi, F. Ranjard, G. Redlinger, A. Roth, J. Rothberg[3], H. Rotscheidt, M. Saich, R.St. Denis, D. Schlatter, M. Takashima, M. Talby[4], W. Tejessy, H. Wachsmuth, S. Wasserbaech, S. Wheeler, W. Wiedenmann, W. Witzeling, J. Wotschack
European Laboratory for Particle Physics (CERN), CH-1211 Geneva 23, Switzerland

Z. Ajaltouni, M. Bardadin-Otwinowska, R. El Fellous, A. Falvard, P. Gay, J. Harvey, P. Henrard, J. Jousset, B. Michel, J.-C. Montret, D. Pallin, P. Perret, J. Proriol, F. Prulhière, G. Stimpfl
Laboratoire de Physique Corpusculaire, Université Blaise Pascal, IN^2P^3-CNRS, Clermont-Ferrand, F-63177 Aubière, France

J.D. Hansen, J.R. Hansen, P.H. Hansen, R. Møllerud, B.S. Nilsson
Niels Bohr Institute, DK-2100 Copenhagen, Denmark[10]

I. Efthymiopoulos, E. Simopoulou, A. Vayaki
Nuclear Research Center Demokritos (NRCD), Athens, Greece

J. Badier, A. Blondel, G. Bonneaud, J. Bourotte, F. Braems, J.C. Brient, G. Fouque, A. Gamess, R. Guirlet, S. Orteu, A. Rosowsky, A. Rougé, M. Rumpf, R. Tanaka, H. Videau
Laboratoire de Physique Nucléaire et des Hautes Energies, Ecole Polytechnique, IN^2P^3-CNRS, F-91128 Palaiseau Cedex, France

D.J. Candlin, E. Veitch
Department of Physics, University of Edinburgh, Edinburgh EH9 3JZ, UK

G. Parrini
Dipartimento di Fisica, Università di Firenze, INFN Sezione di Firenze, I-50125 Firenze, Italy

M. Corden, C. Georgiopoulos, M. Ikeda, J. Lannutti, D. Levinthal[16], M. Mermikides, L. Sawyer
Supercomputer Computations Research Institute and Department of Physics, Florida State University, Tallahassee, FL 32306, USA[13,14,15]

A. Antonelli, R. Baldini, G. Bencivenni, G. Bologna[5], P. Campana, G. Capon, F. Cerutti, V. Chiarella, B. D'Ettorre-Piazzoli[6], G. Felici, P. Laurelli, G. Mannocchi[6], F. Murtas, G.P. Murtas, G. Nicoletti, L. Passalacqua, M. Pepe-Altarelli, P. Picchi[5], P. Zografou
Laboratori Nazionali dell'INFN (LNF-INFN), I-00044 Frascati, Italy

B. Altoon, O. Boyle, A.W. Halley, I. ten Have, J.L. Hearns, J.G. Lynch, W.T. Morton, C. Raine, J.M. Scarr, K. Smith, A.S. Thompson, R.M. Turnbull
Department of Physics and Astronomy, University of Glasgow, Glasgow G128QQ, UK [11]

B. Brandl, O. Braun, R. Geiges, C. Geweniger, P. Hanke, V. Hepp, E.E. Kluge, Y. Maumary, A. Putzer, B. Rensch, A. Stahl, K. Tittel, M. Wunsch
Institut für Hochenergiephysik, Universität Heidelberg, W-6900 Heidelberg, Federal Republic of Germany [17]

A.T. Belk, R. Beuselinck, D.M. Binnie, W. Cameron, M. Cattaneo, P.J. Dornan [1], S. Dugeay, A.M. Greene, J.F. Hassard, N.M. Lieske, S.J. Patton, D.G. Payne, M.J. Phillips, J.K. Sedgbeer, G. Taylor, I.R. Tomalin, A.G. Wright
Department of Physics, Imperial College, London SW72BZ, UK [11]

P. Girtler, D. Kuhn, G. Rudolph
Institut für Experimentalphysik, Universität Innsbruck, A-6020 Innbruck, Austria [19]

C.K. Bowdery [1], T.J. Brodbeck, A.J. Finch, F. Foster, G. Hughes, N.R. Keemer, M. Nuttall, A. Patel, B.S. Rowlingson, T. Sloan, S.W. Snow, E.P. Whelan
Department of Physics, University of Lancaster, Lancaster LA14YB, UK [11]

T. Barczewski, K. Kleinknecht, J. Raab, B. Renk, S. Roehn, H.-G. Sander, M. Schmelling, H. Schmidt, F. Steeg, S.M. Walther, B. Wolf
Institut für Physik, Universität Mainz, W-6500 Mainz, Federal Republic of Germany [17]

J-P. Albanese, J-J. Aubert, C. Benchouk, V. Bernard, A. Bonissent, D. Courvoisier, F. Etienne, S. Papalexiou, P. Payre, B. Pietrzyk, Z. Qian
Centre de Physique des Particules, Faculté des Sciences de Luminy, IN^{2}P^3-CNRS, F-13288 Marseille, France

H. Becker, W. Blum, P. Cattaneo, G. Cowan, B. Dehning, H. Dietl, F. Dydak [26], M. Fernandez-Bosman, T. Hansl-Kozanecka [22], A. Jahn, W. Kozanecki [2,23], E. Lange, J. Lauber, G. Lütjens, G. Lutz, W. Männer, Y. Pan, R. Richter, J. Schröder, A.S. Schwarz, R. Settles, U. Stierlin, J. Thomas, G. Wolf
Max-Planck-Institut für Physik und Astrophysik, Werner-Heisenberg-Institut für Physik, W-8000 München, Federal Republic of Germany [17]

V. Bertin, J. Boucrot, O. Callot, X. Chen, A. Cordier, M. Davier, G. Ganis, J.-F. Grivaz, Ph. Heusse, P. Janot, D.W. Kim [20], F. Le Diberder, J. Lefrançois [1], A.-M. Lutz, J.-J. Veillet, I. Videau, Z. Zhang, F. Zomer
Laboratoire de l'Accélérateur Linéaire, Université de Paris-Sud, IN^{2}P^3-CNRS, F-91405 Orsay Cedex, France

D. Abbaneo, S.R. Amendolia, G. Bagliesi, G. Batignani, L. Bosisio, U. Bottigli, C. Bradaschia, M. Carpinelli, M.A. Ciocci, R. Dell'Orso, I. Ferrante, F. Fidecaro, L. Foà, E. Focardi, F. Forti, C. Gatto, A. Giassi, M.A. Giorgi, F. Ligabue, E.B. Mannelli, P.S. Marrocchesi, A. Messineo, L. Moneta, F. Palla, G. Sanguinetti, J. Steinberger, R. Tenchini, G. Tonelli, G. Triggiani, C. Vannini, A. Venturi, P.G. Verdini, J. Walsh
Dipartimento di Fisica dell'Università, INFN Sezione di Pisa, e Scuola Normale Superiore, I-56010 Pisa, Italy

J.M. Carter, M.G. Green [1], P.V. March, T. Medcalf, I.S. Quazi, J.A. Strong, R.M. Thomas, L.R. West, T. Wildish
Department of Physics, Royal Holloway and Bedford New College, University of London, Surrey TW20OEX, UK [11]

D.R. Botterill, R.W. Clifft, T.R. Edgecock, M. Edwards, S.M. Fisher, T.J. Jones, P.R. Norton, D.P. Salmon, J.C. Thompson
Particle Physics Department, Rutherford Appleton Laboratory, Chilton, Didcot, OXON OX11 OQX, UK [11]

B. Bloch-Devaux, P. Colas, C. Klopfenstein, E. Locci, S. Loucatos, E. Monnier, P. Perez, J.A. Perlas, F. Perrier, J. Rander, J.-F. Renardy, A. Roussarie, J.-P. Schuller, J. Schwindling, B. Vallage
Département de Physique des Particules Elémentaires, CEN-Saclay, F-91191 Gif-sur-Yvette Cedex, France [18]

J.G. Ashman, C.N. Booth, C. Buttar, R. Carney, S. Cartwright, F. Combley, M. Dinsdale, M. Dogru, F. Hatfield, J. Martin, D. Parker, P. Reeves, L.F. Thompson
Department of Physics, University of Sheffield, Sheffield S37RH, UK [11]

E. Barberio, S. Brandt, H. Burkhardt [1], C. Grupen, H. Meinhard, L. Mirabito, U. Schäfer, H. Seywerd
Fachbereich Physik, Universität Siegen, W-5900 Siegen, Federal Republic of Germany [17]

G. Apollinari, G. Giannini, B. Gobbo, F. Liello, F. Ragusa [25], L. Rolandi, U. Stiegler
Dipartimento di Fisica, Università di Trieste e INFN Sezione di Trieste, I-34127 Trieste, Italy

L. Bellantoni, X. Chen, D. Cinabro, J.S. Conway, D.F. Cowen [24], Z. Feng, D.P.S. Ferguson, Y.S. Gao, J. Grahl, J.L. Harton, J.E. Jacobsen, R.C. Jared [7], R.P. Johnson, B.W. Le Claire, Y.B. Pan, J.R. Pater, Y. Saadi, V. Sharma, Z.H. Shi, Y.H. Tang, A.M. Walsh, J.A. Wear [27], F.V. Weber, M.H. Whitney, Sau Lan Wu, G. Zobernig
Department of Physics, University of Wisconsin, Madison, WI 53706, USA [12]

Received 5 July 1991

Abstract. The properties of the Z resonance are measured on the basis of 190000 Z decays into fermion pairs collected with the ALEPH detector at LEP. Assuming lepton universality, $M_Z = (91.182 \pm 0.009_{\mathrm{exp}} \pm 0.020_{\mathrm{LEP}})$ GeV, $\Gamma_Z = (2484 \pm 17)$ MeV, $\sigma^0_{\mathrm{had}} = (41.44 \pm 0.36)$ nb, and $\Gamma_{\mathrm{had}}/\Gamma_{\ell\ell} = 21.00 \pm 0.20$. The corresponding number of light neutrino species is 2.97 ± 0.07. The forward-backward asymmetry in leptonic decays is used to determine the ratio of vector to axial-vector coupling constants of leptons: $g_V^2(M_Z^2)/g_A^2(M_Z^2) = 0.0072 \pm 0.0027$. Combining these results with ALEPH results on quark charge and $b\bar{b}$ asymmetries, and τ polarization, $\sin^2\theta_W(M_Z^2) = 0.2312 \pm 0.0018$. In the context of the Minimal Standard Model, limits are placed on the top-quark mass.

* This paper is dedicated to Emilio Picasso for his contributions as LEP Project Leader
[1] Now at CERN
[2] Permanent address: SLAC, Stanford, CA 94309, USA
[3] Permanent address: University of Washington, Seattle, WA 98195, USA
[4] Also Centre de Physique des Particules, Faculté des Sciences, Marseille, France
[5] Also Istituto di Fisica Generale, Università di Torino, Torino, Italy
[6] Also Istituto de Cosmo-Geofisica del C.N.R., Torino, Italy
[7] Permanent address: LBL, Berkeley, CA 94720, USA
[8] Supported by CAICYT, Spain
[9] Supported by the National Science Foundation of China
[10] Supported by the Danish Natural Science Research Council
[11] Supported by the UK Science and Engineering Research Council
[12] Supported by the US Department of Energy, contract DE-AC02-76ER00881
[13] Supported by the US Department of Energy, contract DE-FG05-87ER40319
[14] Supported by the NSF, contract PHY-8451274
[15] Supported by the US Department of Energy, contract DE-FC0S-85ER250000
[16] Supported by SLOAN fellowship, contract BR2703
[17] Supported by the Bundesministerium für Forschung und Technologie, FRG
[18] Supported by the Institut de Recherche Fondamentale du C.E.A.
[19] Supported by Fonds zur Förderung der wissenschaftlichen Forschung, Austria
[20] Supported by the Korean Science and Engineering Foundation and Ministry of Education
[21] Supported by the World Laboratory
[22] On leave of absence from MIT, Cambridge, MA 02139, USA
[23] Supported by Alexander von Humboldt Fellowship, FRG
[24] Now at California Institute of Technology, Pasadena, CA 91125, USA
[25] Now at Dipartimento di Fisica, Università di Milano, Milano, Italy
[26] Also at CERN, PPE Division, 1211 Geneva 23, Switzerland
[27] Now at University of California, Santa Cruz, CA 95064, USA

1 Introduction

Measurements of cross sections and angular distributions for Z decay to fermion pairs place significant constraints on parameters of the Standard Model and allow tests of the consistency of the model.

In this paper, improved measurements of the parameters of the Z resonance based on 8.0 pb^{-1} of data collected with the ALEPH detector at LEP during 1989 and 1990 at and near the Z mass are presented. The data include 165000 hadronic and 25000 leptonic Z decays, and correspond to a sample three times larger than that of our previous analysis [1]. In addition to reduced statistical errors, improved analysis methods have led to substantially reduced systematic errors.

The ALEPH detector [2] and the techniques used to select different types of events [1] have been described elsewhere. Here, significant changes since the previous work and the additional systematic studies which have been done are discussed. The precision of the luminosity measurement is described in a separate paper [3], and is not addressed here.

2 Hadronic event selection

Hadronic events are selected with two independent methods, as described in [1]. One of these methods is based entirely on charged tracks, while the other depends on energy deposition in the electromagnetic (ECAL) and hadronic (HCAL) calorimeters. The track-based selection requires at least 5 charged tracks in the time-projection chamber (TPC), ALEPH's main charged-particle tracking chamber, and a charged-track energy sum (assuming pion masses) greater than 10% of the centre-of-mass energy. The calorimeter-based selection requires that the combined ECAL and HCAL energy exceeds 20% of the centre-of-mass energy. In addition, the minimum energy deposit in the ECAL must be either 7 GeV in the barrel or 1.5 GeV in each endcap. A time requirement of ± 100 ns with respect to the beam crossing, measured on ECAL signals, removes the bulk of cosmic-ray background. Additional requirements are imposed to suppress the background from lepton pairs [1].

For both selections, the background from leptonic decays of the Z is estimated from Monte Carlo. The two-photon background is measured with the data by exploiting the different centre-of-mass energy dependence of the two-photon background and hadronic decays of the Z. The trigger inefficiency is measured to

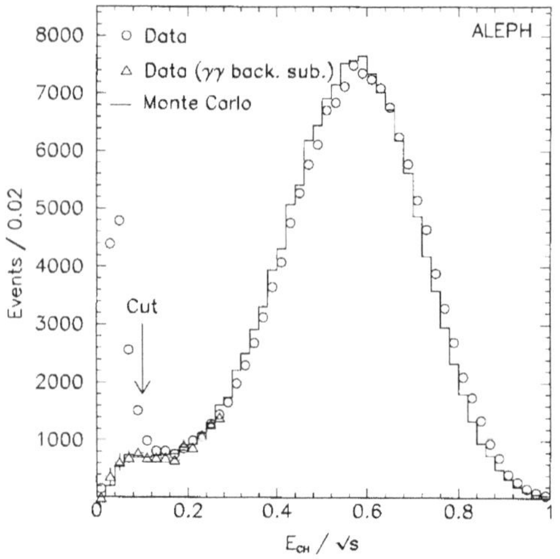

Fig. 1. Charged-track energy sum per event for data (circles) and Monte Carlo (solid histogram). The triangles represent measured numbers of resonant events after the subtraction of two-photon background (see text). The Monte Carlo is normalized to the data for $E_{CH}/\sqrt{s}>0.1$

be less than 0.03% by comparing two independent triggers, one based on ECAL energy, and the other based on hits in the Inner Tracking Chamber (ITC) in conjunction with hits in the HCAL.

The dominant systematic error in the track-based selection is a result of the requirement that the sum of the charged-track energies exceeds 10% of the centre-of-mass energy. The size of this systematic error is estimated from the effect of a shift of $0.01\sqrt{s}$ in the charged-energy distribution, corresponding to the observed shift between the data and the Monte Carlo simulation (see Fig. 1). The resulting uncertainty in the selection efficiency is 0.24%. As a check on this systematic error, the measured resonant cross section for events failing the charged energy requirement is compared with the Monte Carlo prediction. Resonant and nonresonant cross sections are separated with the technique already used to extract the two-photon background ($\sigma_{\gamma\gamma}=(118\pm23)$ pb corresponding to (0.39 ± 0.07)% of the peak hadronic cross section). In Fig. 1, the results of this measurement are plotted with the observed charged-energy distribution and the Monte Carlo prediction for $q\bar{q}$ events; the Monte Carlo distribution is normalized to the data for events with $E_{CH}>0.1\sqrt{s}$. The difference between the measured resonant cross section and the Monte Carlo prediction for events with $E_{CH}<0.1\sqrt{s}$ is (0.15 ± 0.17)%, which is consistent with the quoted systematic error. Combining all other systematic errors, including uncertainties in background contributions, the systematic error in the track-based selection is 0.26% at $\sqrt{s}=M_Z$.

The calorimeter-based selection has a larger efficiency than the track-based selection, and has a correspondingly smaller systematic uncertainty related to the selection requirements. The dominant systematic error comes instead from background in the $q\bar{q}$ sample. In particular, $\tau^+\tau^-$ events represent (0.62 ± 0.15)% of the selected event sample. The remaining background comes from e^+e^- events, two-photon interactions, and cosmic rays, corresponding to (33 ± 12) pb, (70 ± 25) pb, and (26 ± 5) events/pb^{-1}, respectively. These numbers are given as percentages of the peak cross section in Table 1. The total systematic error of this selection is about 0.2% at $\sqrt{s}=M_Z$.

Details of the two selections are summarized in Table 1. The error quoted for selection efficiency includes the statistical error from the Monte Carlo; the uncertainties in background subtractions are listed separately.

All of the events not common to both samples (about 4% of the events) have been scanned visually for evidence of detector or data-acquisition problems. An additional

Table 1. Hadronic event selections. The number of events corresponds to data at all centre-of-mass energies, while efficiencies, backgrounds, and systematic errors are quoted at the peak

	Charged track selection	Calorimeter selection
Events	166158	169993
Selection efficiency (%)	97.4 ± 0.24	99.1 ± 0.09
Backgrounds:		
e^+e^- (%)	0	0.11 ± 0.04
$\tau^+\tau^-$ (%)	0.26 ± 0.03	0.62 ± 0.15
Cosmic-ray (%)	0	0.08 ± 0.02
Two-photon (%)	0.39 ± 0.07	0.23 ± 0.08
Total syst. error ($\sqrt{s}=M_Z$)(%)	0.26	0.20

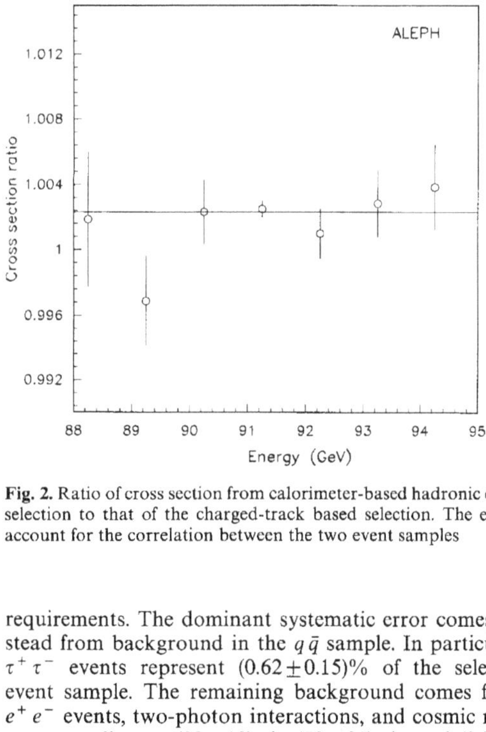

Fig. 2. Ratio of cross section from calorimeter-based hadronic event selection to that of the charged-track based selection. The errors account for the correlation between the two event samples

systematic error of 0.04% is assigned to account for the few problems which have been observed; these problems were related to a malfunction in the readout electronics of the TPC.

Since the systematic errors of the two selections are largely independent, the measured cross sections are averaged and an overall systematic error of 0.2% at $\sqrt{s} = M_Z$ is quoted. As a check on the systematic error, the ratio of cross sections measured with the two selections is plotted as a function of energy (Fig. 2); the large correlation between the event samples is taken into account in the errors. A fit to these ratios yields $\sigma_{cal}/\sigma_{track} = 1.0023 \pm 0.0004$, which is consistent with the quoted systematic errors.

3 Leptonic event selection

The leptonic branching ratios have been determined using procedures similar to those described in [1]. The leptonic branching fractions are measured separately for each type of lepton pair, and for all lepton pairs in common, without any attempt to distinguish among them. The comparison of this common-lepton sample with the separate lepton samples allows checks on the systematic biases of the event selection requirements. As will be discussed, the e^+e^- and $\mu^+\mu^-$ selections have been modified with respect to [1] to increase the acceptance for events with a radiated photon, thereby reducing systematic uncertainties in event selection efficiency. Improved understanding of backgrounds and efficiency also results in reduced systematic errors for the $\tau^+\tau^-$ and common-lepton $\ell^+\ell^-$ channels.

The theoretical treatment of the t-channel contribution to the e^+e^- and common-lepton samples is also improved with respect to [1]. A new program called ALIBABA [4], which allows a calculation of Bhabba scattering including $\mathcal{O}(\alpha^2)$ initial-state radiation, is now available. The program can calculate the full s-channel, t-channel, and interference term, the s-channel alone, or the t-channel alone. Throughout this paper, the sum of the t-channel and the interference term is referred to as the t-channel contribution to the cross section. The results of ALIBABA have a reported accuracy of approximately 0.5% of the total cross section [4]. Corrections are necessary to make the results of ALIBABA, a semi-analytical program which gives results as functions of the polar angles and energies of the outgoing electron and positron, applicable to the experimental acceptance. Including the error in these corrections, the total systematic error in the t-channel subtraction is 2% of the pure t-channel part (not including $s-t$ interference) of the subtraction. At the peak ($\sqrt{s} = M_Z$), the t-channel contribution is 12.2% of the total cross section in the angular range in which $Z \to e^+e^-(\gamma)$ events are selected. In addition, the sensitivity of the t-channel contribution to uncertainties in the Z mass and width is about 1% of the pure t-channel part of the cross section at the peak and smaller away from the peak. As a check of the t-channel subtraction procedure, a fit to the complete cross section (including t-channel and interference) has been done using the program MIBA [5]. This program, like ALIBABA, includes complete $\mathcal{O}(\alpha)$ and leading log $\mathcal{O}(\alpha^2)$ QED corrections with soft-photon exponentiation. The hadron and electron lineshapes have been fit simultaneously with and without t-channel subtraction. The results of the two methods agree to better than 0.1%; the results presented here are based on the t-channel subtraction method.

The forward-backward asymmetries $A_{FB}(s)$ at each centre-of-mass energy are extracted from a fit to the angular distribution with the function [6]

$$\frac{d\sigma}{d\cos\theta^*} = C(1 + \cos^2\theta^* + \tfrac{8}{3} A_{FB} \cos\theta^*)\, F(\cos\theta^*), \qquad (1)$$

using a maximum-likelihood method. C is a normalization constant and $F(\cos\theta^*)$ describes the effect of the t-channel exchange which is relevant only for the e^+e^-

Table 2. Lepton-pair selections. The number of selected events corresponds to data at all centre-of-mass energies, while selection efficiencies, background, and systematic errors are quoted at the peak

	e^+e^-	$\mu^+\mu^-$	$\tau^+\tau^-$	$\ell^+\ell^-$
Selected events	6947	6691	6260	24757
Angular range ($\cos\theta^*$)	(−0.9, 0.7)	(−0.9, 0.9)	(−0.9, 0.9)	(−0.9, 0.9)
Selection efficiency (%)[a]	98.8 ± 0.3	98.4 ± 0.5	86.4 ± 0.8	98.3 ± 0.2
Overall efficiency (%)	71.4 ± 0.4	83.1 ± 0.5	72.9 ± 0.8	83.0 ± 0.2
t-channel contribution (%)	12.2 ± 0.3	0	0	13.1 ± 0.3
Backgrounds:				
e^+e^- (%)	–	<0.03	1.1 ± 0.1	–
$\mu^+\mu^-$ (%)	<0.11	–	0.12 ± 0.08	–
$\tau^+\tau^-$ (%)	1.18 ± 0.07	0.12 ± 0.08	–	–
Hadrons (%)	0	0	1.12 ± 0.31	0.8 ± 0.07
Two-photon (%)	0	0	0.66 ± 0.13	0.22 ± 0.04
Cosmic rays (%)	0	0.23 ± 0.05	0.07 ± 0.02	0.02 ± 0.01
Total syst. error $\sqrt{s} = M_Z$ (%)	0.5	0.5	0.9	0.4

[a] In this Table, the selection efficiency is defined as the efficiency within the accepted angular range

channel. The scattering angle, θ^*, between the incoming e^- and the outgoing fermion is defined as

$$\cos\theta^* = \sin\tfrac{1}{2}(\theta^+ + \theta^-)/\sin\tfrac{1}{2}(\theta^+ - \theta^-), \qquad (2)$$

where θ^- and θ^+ are the polar angles of the vector sum of the track momenta* in the hemispheres corresponding to the outgoing fermion and antifermion, respectively. This variable preserves the true angular distribution in the e^+e^- centre-of-mass when hard collinear radiation takes place from the initial state.

Details of the lepton-pair selections are summarized in Table 2. A first series of requirements, which is common to all of the lepton-pair channels, is applied to eliminate hadronic events, as well as to suppress background from beam-gas, two-photon, and cosmic-ray events. These initial requirements are the following:

- Only tracks with 4 or more points measured in the TPC, which originate from the beam crossing within ± 10 cm along the beam direction and $\pm d_0$ in the transverse plane, are considered; these tracks are referred to as good. Two d_0 requirements are used: a more restrictive one ($d_0 = 2$ cm) and a less restrictive one ($d_0 = 5$ cm).
- The event is required to have either 2 and only 2 good tracks which pass the loose d_0 requirement or 3 to 8 good tracks which satisfy the restrictive d_0 requirement.
- The event is divided into two hemispheres by a plane perpendicular to the thrust axis. There must be at least one good track in each hemisphere.
- At least one good track with $d_0 < 2$ cm must have a reconstructed momentum greater than 3 GeV.
- Events with more than 4 tracks are rejected if any track has an angle greater than 31.8° with respect to the vector sum of the track momenta in the same hemisphere.
- The acollinearity, η_{acol}, defined as 180° minus the angle between the vector sum of the track momenta** in each hemisphere, must be smaller than 20°.

The multiplicity requirement makes the acceptance sensitive to inefficiencies in charged-track finding. This effect has been studied with a sample of e^+e^- pairs which were selected solely on the basis of calorimeter information. An inefficiency of $(0.25 \pm 0.10)\%$ in addition to the $(0.69 \pm 0.06)\%$ predicted by the Monte Carlo has been observed, and is corrected for in the analyses below; this inefficiency is the result of the TPC readout problem mentioned in the previous section.

In 1990, an additional charged electromagnetic trigger with a threshold of 0.2 GeV was implemented, improving the redundancy of the trigger. The trigger inefficiency is less than 0.15% for all lepton channels, and is known with a precision better than 0.05%.

* For the $e^+e^-(\gamma)$ final states, $\cos\theta^*$ is calculated from the two highest momentum tracks only

** For the $e^+e^-(\gamma)$ final states, the acollinearity is calculated from the two highest momentum tracks only

3.1 $Z \to \ell^+\ell^-$

The selection of leptons without distinguishing the lepton flavour is done in a manner similar to that described in [1] except that the angular acceptance of the selection is increased to $|\cos\theta^*| < 0.9$. To further reduce the background from cosmic rays and two-photon interactions, some criteria are added to the requirements described above.

Cosmic-ray background is reduced by requiring that for two track events, at least one track originate from the beam crossing within 1 cm in the transverse direction and 5 cm in the beam direction. Also, $\ell^+\ell^-$ candidates are required to have at least 4 ITC points associated to a track; this requirement rejects most cosmic-ray events because of the more restrictive timing of the ITC.

Two-photon background is suppressed by the following requirements:

- The transverse momentum relative to the beam of the vector sum of the tracks in each hemisphere (the jet transverse momentum) must be larger than 2.5 GeV in at least one hemisphere.
- For two-track events with both momenta less than 6 GeV, the transverse momenta of the two tracks with respect to the beam must differ from each other by more than 15%.
- The visible invariant mass of the event must be greater than 4.5 GeV.

Using the above criteria, 24757 events are selected as lepton-pair candidates with an efficiency of $(98.3 \pm 0.2)\%$ inside the angular range $|\cos\theta^*| < 0.9$. The remaining two-photon background of (9.9 ± 1.9) pb (corresponding to $(0.22 \pm 0.04)\%$ of the selected event sample at the peak) has been subtracted.

The main systematic errors in the common-lepton cross section result from uncertainties in the angular acceptance and acollinearity cut (0.14%), the t-channel contribution (0.3%), and the track parameters, such as momentum errors (0.16%, which is included in the quoted systematic error in selection efficiency). Combining errors in quadrature, the total systematic error is less than 0.4%.

3.2 $Z \to e^+e^-(\gamma)$

The selection of electron pairs is based on the sum of the momenta of the two most energetic tracks in an event and on the sum of energies of the ECAL clusters associated with these tracks [1]. The current analysis includes energy which escapes through cracks in the ECAL but is detected by the HCAL, and includes the energy of radiated photons. As shown in Fig. 3, much of the energy lost in ECAL cracks is recovered in the HCAL. When ECAL clusters are close to an ECAL crack, the HCAL energy which is associated to the ECAL cluster is included in the energy sum. This procedure is applied only to two-track events; the background from $\tau^+\tau^-$ pairs would be increased by about a factor

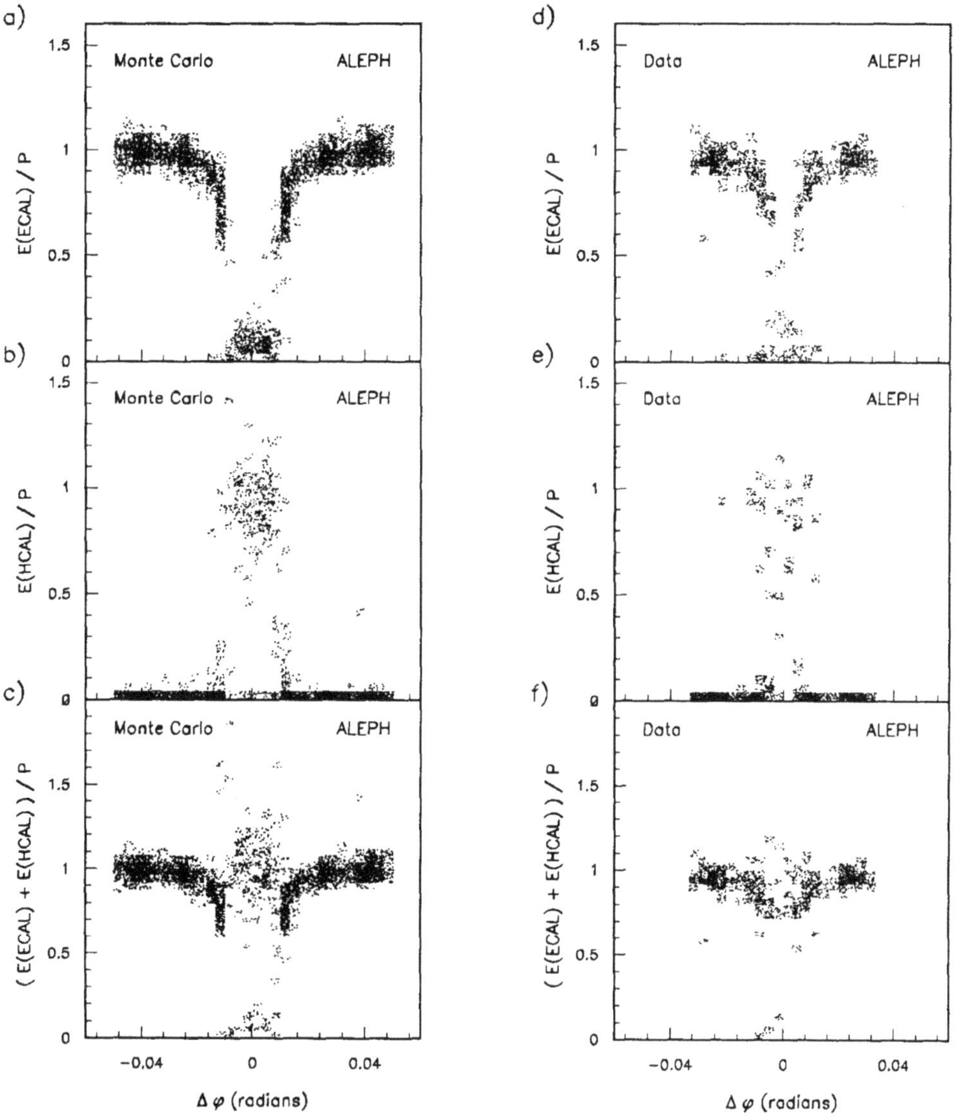

Fig. 3a–f. a ECAL barrel energy versus azimuth $\Delta\varphi$, the distance to the crack, for Bhabha candidates, **b** HCAL barrel energy versus azimuth $\Delta\varphi$ for Bhabha candidates, **c** the sum of a and b, for Monte Carlo; **d**, **e** and **f** are the equivalent plots for data. The 12 modules in φ have been plotted together and only the part of the φ range around the crack regions is shown

of two if the procedure were applied to higher multiplicity events.

The $e^+e^-(\gamma)$ candidates must satisfy the following requirements:

$$\sum E > 0.20\sqrt{s},$$

$$\sum p > 0.05\sqrt{s},$$

$$\sum p + \sum \mathcal{E} > 1.2\sqrt{s}.$$

Here, $\sum p$ refers to the sum of the momenta of the two most energetic tracks and $\sum E(\sum \mathcal{E})$ refers to the energy sum of the clusters associated with these tracks before (after) the inclusion of energy from a radiated photon and of the associated HCAL energy.

This selection yields a sample of 6947 events in the angular range $-0.9 < \cos\theta^* < +0.7$. The efficiency of the selection within the geometrical acceptance is calculated to be $(98.8 \pm 0.3)\%$ from Monte Carlo [7]. The dominant

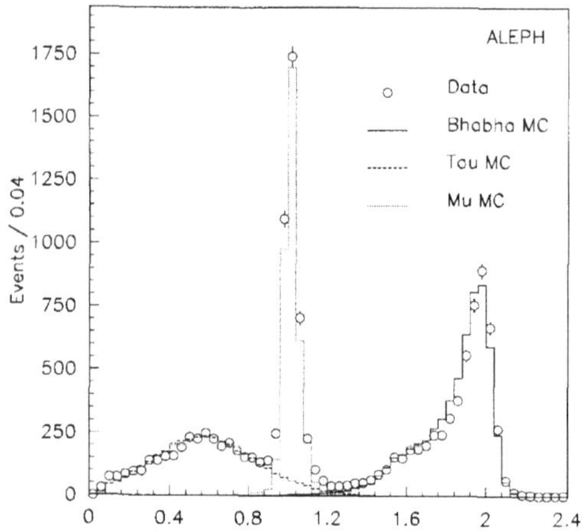

Fig. 4. $\sum p + \sum \mathscr{E}$ for lepton pairs, together with Monte Carlo predictions from the different lepton species

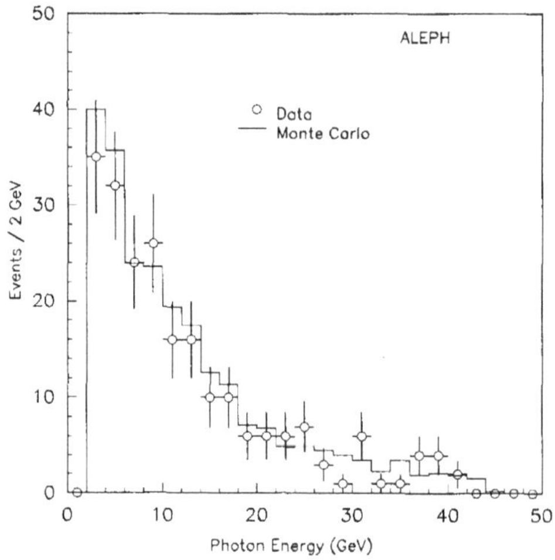

Fig. 5. Photon-energy distribution for $\mu^+\mu^-\gamma$ events from data and Monte Carlo

background in the electron sample comes from $\tau^+\tau^-$ pairs and is estimated to be $(1.18 \pm 0.07)\%$ from Monte Carlo simulation [8].

The systematic error in the selection efficiency and background has been estimated from the variation of the number of accepted events for data and Monte Carlo [7, 8] as a function of the most sensitive cut, $\sum p + \sum \mathscr{E}$, and has been found to be 0.3% inside the acceptance (Fig. 4).

The other important systematic error comes from the t-channel subtraction. At the peak energy, uncertainties in ALIBABA result in a systematic error of 2% of the pure t-channel part of the subtracted cross section; the uncertainty in the value of $\sqrt{s} - M_Z$ (10 MeV) gives an additional contribution to the systematic error of 1.6% of the subtracted cross section in the angular range $-0.9 < \cos\theta^* < +0.7$. This contribution to the systematic error would be larger, i.e. 1.2% of the subtracted cross section (about 30% of the total cross section), in the range $|\cos\theta^*| < 0.9$. Away from the peak, the latter error is negligible and the total systematic error is estimated to be 2% of the pure t-channel part of the cross section. Since this relative error in the t-channel subtraction varies slowly across the Z peak, the value at the peak has been used. The 0.7% systematic error in the absolute luminosity [3] introduces a further systematic uncertainty in the t-channel subtraction.

The total systematic error in the cross section is about 60% of the statistical error at the peak and less than 20% away from the peak. The systematic uncertainty in the forward-backward asymmetry is about 10% of the statistical error.

3.3 $Z \to \mu^+\mu^-(\gamma)$

The muon-pair selection has been extended in two respects since the previous analysis [1]. The muon chambers* are now used in conjunction with the HCAL to identify muons, and the kinematic requirements of the selection have been changed to improve the efficiency for events with a radiated photon. If the momenta of the two fastest tracks both exceed $35\sqrt{s}/M_Z$ GeV, the event is accepted as a muon-pair candidate. If only one track exceeds this momentum cut, the event is accepted if a photon can be found which is consistent in energy and position with the $\mu^+\mu^-\gamma$ hypothesis. Figure 5 shows the photon-energy distribution for the selected events; the Monte Carlo [8] distribution is consistent with the data. In the current selection one of the two highest momentum tracks is required to have an HCAL hit pattern consistent with that expected for a muon (as described in [1]), or to be matched to a hit in the muon chambers.

A total of 6691 $\mu^+\mu^-$ candidates has been selected in the angular range $|\cos\theta^*| < 0.9$. The efficiency of the selection in this angular range is $(98.4 \pm 0.5)\%$. The backgrounds in the sample are $\tau^+\tau^-$ pairs and cosmic-ray events. The $\tau^+\tau^-$ background has been calculated from Monte Carlo [8] to be $(0.12 \pm 0.08)\%$. The cosmic-ray background has been estimated to be (3.3 ± 0.7) pb (corresponding to $(0.23 \pm 0.05)\%$ of the selected event sample at the peak) by loosening vertex requirements and using the number of additional events found to estimate the number of background events accepted by the standard vertex requirements.

In order to estimate the systematic error of the event selection based on HCAL, the cross sections obtained with this selection are compared with those obtained

* One double-layer of muon chambers surrounding the hadronic calorimeter has been installed for the 1990 period. The muon chambers, which are constructed of streamer tubes with a 1 cm pitch, provide a single 3-dimensional space point for charged tracks which penetrate the 7.5 interaction lengths of material between the chambers and the interaction point at normal incidence

by a selection based on the presence of minimum ionizing particle signals in ECAL and HCAL. A systematic error of 0.2% is assigned to the cross sections to account for the observed difference of $(0.1\pm0.2)\%$.

The Monte Carlo generator used to measure the efficiency of the kinematic cuts does not include all higher-order radiative corrections, and therefore underestimates the number of events in which two hard photons are produced by final state radiation. The comparison of the number of $\mu^+\mu^-\gamma$ events found in data and in Monte Carlo shows a disagreement of about 10% on 3.5% of the selected event sample; the resulting systematic uncertainty is estimated to be 0.35%. Combining this uncertainty with the statistical error in the determination of the efficiency of the kinematic cuts gives a total systematic error of 0.5% in this efficiency. To check this value, the cross sections have been compared with those obtained when relaxing the momentum cut on the second track from $35\sqrt{s}/M_Z$ GeV to $22\sqrt{s}/M_Z$ GeV. The average difference is $(0.3\pm0.2)\%$, which is consistent with the quoted systematic error.

Including the 0.2% uncertainty in backgrounds, a 0.6% systematic error is assigned to the $\mu^+\mu^-$ cross section, in addition to the systematic uncertainty in the luminosity measurement.

To estimate the systematic error in the forward-backward asymmetry, results obtained with two sets of kinematic cuts and two sets of selection cuts (as described above) have been compared. The average difference in the asymmetry between the two selection methods is 0.002 ± 0.004; the average difference between the two sets of kinematic cuts is 0.002 ± 0.003. The resulting systematic error in the forward-backward asymmetry is about 30% of the statistical error at the peak and less than 10% away from the peak.

3.4 $Z\rightarrow\tau^+\tau^-$

The $\tau^+\tau^-$ selection procedure used for the cross-section measurement is similar to that described in [1]. The requirements are those of the common-lepton selection together with additional cuts to separate $\tau^+\tau^-$ candidates from e^+e^- and $\mu^+\mu^-$ events. In particular, the $\tau^+\tau^-$ candidates are required to satisfy the criteria described below.

To suppress e^+e^- and $\mu^+\mu^-$ events, the square of the missing mass calculated from the tracks (assuming pion masses) and $\sqrt{s}$ is required to exceed $400\sqrt{s}/M_Z$ GeV^2. The e^+e^- background is reduced further by requiring that the energy measured in ECAL be less than $55\sqrt{s}/M_Z$ GeV.

This selection yields a sample of 6260 events with $|\cos\theta^*|<0.9$. The efficiency of the selection within the geometrical acceptance is $(86.4\pm0.8)\%$; the inefficiency results from the missing mass and ECAL energy cuts in equal proportions. The main backgrounds are $Z\rightarrow q\bar{q}$, $Z\rightarrow e^+e^-$, and two-photon events. The $q\bar{q}$ and two-photon background has been determined from Monte Carlo [9, 10]. A background of $(1.12\pm0.31)\%$ for $q\bar{q}$ and (9.9 ± 1.9) pb (i.e., $(0.66\pm0.13)\%$ of the selected event sample at the peak) for two-photon events has been subtracted from the cross sections. The remaining background results mainly from e^+e^- pairs where two photons are radiated, producing a large missing mass, and at the same time part of the electromagnetic energy is lost either in inactive parts of ECAL or because the photons are collinear with the initial electrons. This background was measured by applying the $\tau^+\tau^-$ selection to the e^+e^- sample of events obtained using the described $Z\rightarrow e^+e^-(\gamma)$ selection, and taking the efficiencies of the two selections into account. The small $\tau^+\tau^-$ contamination in the e^+e^- data has been estimated by Monte Carlo. The e^+e^- background was found to be $(1.1\pm0.5)\%$ and has been subtracted.

The systematic error in the selection efficiency has been evaluated with data. The following procedure is used to produce an unbiased sample of $\tau^+\tau^-$ events by combining taus from different events. Only events with less than 7 charged tracks and at least one track per hemisphere are considered. The energy sum of the charged tracks is required to be larger than 8 GeV and the acollinearity less than 30°. A τ selection procedure is applied to each event hemisphere. If there is only one charged track in a hemisphere, the sum of energy of the charged track and ECAL energy must be less than 35 GeV; if there are 2 or 3 charged tracks, in addition, the invariant mass of the tracks must be less than 2 GeV. If such a τ candidate is found in an event, only the τ

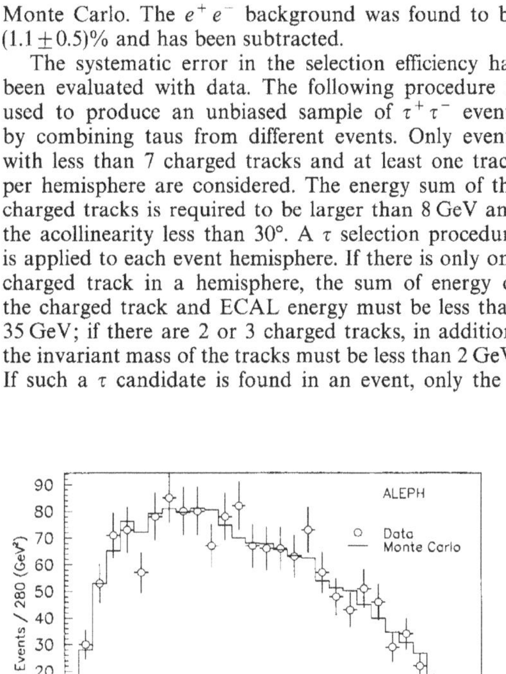

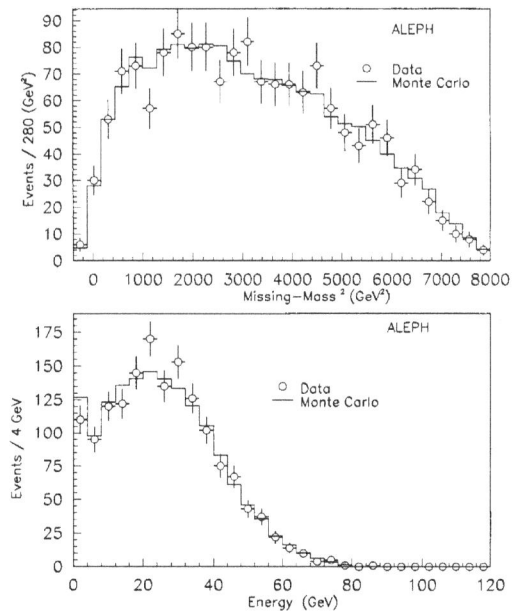

Fig. 6a, b. a Missing-mass squared distribution for $\tau^+\tau^-$ Monte Carlo (histogram) and paired half events from data (solid circles); **b** ECAL energy distribution for $\tau^+\tau^-$ Monte Carlo and paired half events from data in the opposite hemisphere is used in the study. Half events having opposite charge and nearly opposite polar and azimuthal angles are paired to form complete events with unbiased ECAL energy and missing mass distributions. Figure 6 shows these distributions compared to those of $\tau^+\tau^-$ Monte Carlo events. The efficiency observed in the mixed-event sample confirms that determined with the $\tau^+\tau^-$ Monte Carlo within the 0.6% statistical precision of the comparison. The event-mixing method destroys the intrinsic correlation of helicities between two taus of a real event. This effect has been studied with Monte Carlo and found to be negligible.

To measure the forward-backward asymmetry, a more selective procedure has been adopted since a charge-asymmetric background can produce a bias in the measurement. To reduce the already small contamination of e^+e^- events, a selection procedure which positively identifies τ decays based on muon and pion identification is used. For each track with momentum larger than 3 GeV, the same muon identification as used in the $\mu^+\mu^-$ selection is applied. If the track cannot be identified as a muon, the charged-pion hypothesis is tested. A pion is distinguished from an electron on the basis of either its ECAL energy deposit, if its momentum is larger than 3 GeV, or its dE/dx measurement in the TPC.

In addition, reconstructed π^0s with the energy of each photon greater than 250 MeV are used; details of the π^0 identification are given in [11]. An event is accepted as a $\tau^+\tau^-$ candidate if either at least one muon, or at least two pions (including π^0s), or one pion with an ECAL energy (E_{ECAL}) less than $0.8\,E_{\mathrm{beam}}$ in each hemisphere of the event are identified. 6095 $\tau^+\tau^-$ candidates are selected with an efficiency of 80.8% inside the defined acceptance. The remaining e^+e^- background is (0.23 ± 0.12)%; it has been measured directly from data by tagging an electron (or positron) in one hemisphere of the event and counting the number of identified pions in the other hemisphere. Figure 7 shows the probability of identifying either one muon or one or more pions in one hemisphere once an electron has been tagged in the other hemisphere; the same probability for an unbiased sample of $\tau^+\tau^-$ Monte Carlo events is shown also.

For the measurement of A_{FB} in the $\tau^+\tau^-$ channel, the $\mu^+\mu^-$ and $q\bar{q}$ background can be neglected while the small e^+e^- background has been subtracted. The systematic error in the forward-backward asymmetry, resulting mainly from uncertainty in the e^+e^- background subtraction, is about 2% of the statistical error at all centre-of-mass energies.

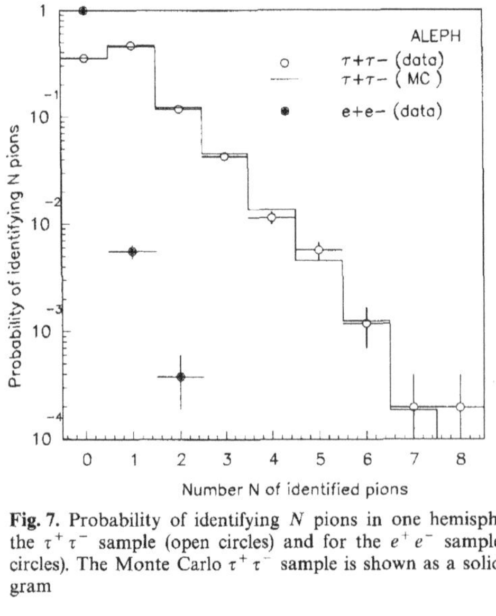

Fig. 7. Probability of identifying N pions in one hemisphere for the $\tau^+\tau^-$ sample (open circles) and for the e^+e^- sample (solid circles). The Monte Carlo $\tau^+\tau^-$ sample is shown as a solid histogram

4 Comparison among lepton channels

As discussed earlier, a substantial background in the separate lepton analyses comes from other lepton channels. To check the estimated efficiency and background for each lepton channel, the selections are compared among themselves and with the common-lepton selection. Figure 8 shows the number of events in the regions of overlap among the different selections.

Only data from the 1990 running period and with $|\cos\theta^*| < 0.9$ are used for this comparison. 18 events out of 22031 events found in either the individual or the common-lepton selections do not appear in the common-lepton sample. These 18 events consist of 5 e^+e^- and 13 $\mu^+\mu^-$ candidates. In this comparison, one must consider the different methods of cosmic-ray rejection used by the $Z \to \mu^+\mu^-$ and $Z \to \ell^+\ell^-$ analyses. The 7 events rejected by the ITC hit requirement in the common-lepton selection are consistent with the higher cosmic-ray background in the $Z \to \mu^+\mu^-$ selection. Subtracting these 7 events leaves a discrepancy of 11 events out of 22031, i.e., 0.05%.

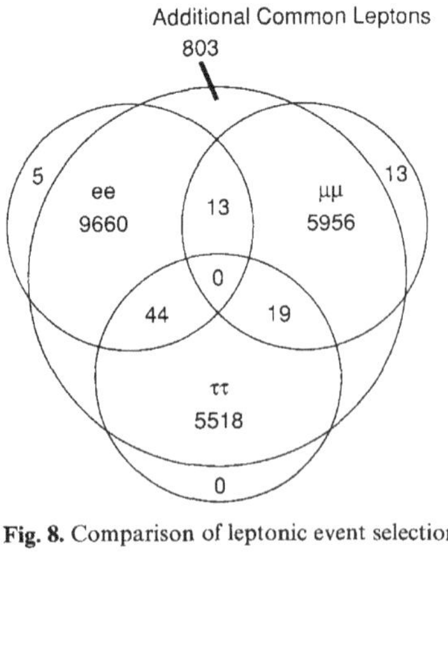

Fig. 8. Comparison of leptonic event selections for 1990 period

A total of 803 events are found in the common-lepton sample but not in the separate lepton samples. The events are consistent with being leptonic Z decays and represent less than 4% of the events found by the individual selections. Although these events are difficult to identify individually, one can estimate the expected contribution of each lepton species by taking into account the relative efficiencies of selecting e^+e^-, $\mu^+\mu^-$ or $\tau^+\tau^-$ inside the common-lepton sample and the relevant backgrounds. The total expected contribution is 760 ± 23 events (110 ± 10 e^+e^-, 82 ± 16 $\mu^+\mu^-$, 568 ± 13 $\tau^+\tau^-$), and is statistically consistent with the 803 events observed. Thus, the separate and common-lepton selections agree at the level of (0.2 ± 0.1)%.

Events which are present in more than one lepton selection have been studied also. There are no events which pass the selection requirements for all three lepton species, and there are no events passing two selections which are not also present in the common-lepton sample.

Taking into account the estimated background in each sample and the "conditional" efficiencies of selecting events as candidates of one type when they have already been selected as candidates of another type, the number of events expected to be common to different pairs of selections can be calculated. For the e^+e^- and $\tau^+\tau^-$ species, where the overlap between the two species is large, the "conditional" efficiencies have been obtained from Monte Carlo [7, 8]. For all other "conditional" efficiencies, it has been assumed that these efficiencies were equivalent to the background of one species inside the other (see Table 2). A total of 50 ± 6 events are expected to be common to the e^+e^- and $\tau^+\tau^-$ samples, statistically consistent with the 44 events found. Less than 9 events are expected to be common to e^+e^- and $\mu^+\mu^-$ selections and 13 events have been found. Less than 17 events are expected to be common to the $\mu^+\mu^-$ and $\tau^+\tau^-$ selections and 19 events have been found. All the events appearing in more than one individual lepton selection have been scanned visually, and the observed numbers of events in each category are consistent with these estimates. The largest discrepancy between the measured and predicted numbers of events in any overlap region is 0.1%.

5 Cross sections and forward-backward asymmetries

Data have been collected at 16 energies at and near the peak of the Z resonance. Table 3 summarizes the cross sections for hadrons and lepton pairs as a function of centre-of-mass energy; the seven highest luminosity points correspond to data taken during 1990. Only statistical errors are given; the systematic errors resulting from the event selection requirements are summarized in Tables 1 and 2. The cross sections given in Table 3 have been corrected for the effect of the center-of-mass energy spread of LEP, as described in the following section.

The luminosity, which is measured with small-angle Bhabhas [3], has an experimental error of 0.6%. The theoretical calculation for small angle Bhabha scattering includes corrections of second order in α [12, 13]. The uncertainty in this calculation within the acceptance is 0.3% [3], giving a total luminosity error of 0.7%.

Table 4 summarizes the forward-backward asymmetries for lepton pairs as a function of centre-of-mass energy. The systematic errors are negligible compared to the statistical errors. The determination of the cross section and of the forward-backward asymmetry from the angular distribution is not strictly model independent because of the subtraction of the t-channel contribution; this effect, however, is negligible with respect to the quoted systematic error. The systematic uncertainty in the measurement of the forward-backward asymmetry resulting from the LEP beam-energy uncertainty [14] is negligible compared to the statistical error.

Table 3. Hadron and lepton cross sections. Only statistical errors are given

$\sqrt{s}$ (GeV)	$\mathscr{L}_{int}$ (nb^{-1})	σ_{had} (nb)	σ_{ee} (nb)	$\sigma_{\mu\mu}$ (nb)	$\sigma_{\tau\tau}$ (nb)	$\sigma_{\ell\ell}$ (nb)
88.224	482.2 ± 4.1	4.596 ± 0.105	0.233 ± 0.039	0.247 ± 0.025	0.171 ± 0.024	0.599 ± 0.063
88.277	108.5 ± 2.0	4.534 ± 0.220	0.324 ± 0.088	0.228 ± 0.050	0.352 ± 0.072	1.027 ± 0.152
89.220	520.3 ± 4.3	8.409 ± 0.146	0.334 ± 0.042	0.502 ± 0.034	0.387 ± 0.034	1.279 ± 0.074
89.277	46.5 ± 1.3	8.865 ± 0.503	0.439 ± 0.150	0.288 ± 0.089	0.480 ± 0.127	1.260 ± 0.246
90.222	447.1 ± 4.1	18.464 ± 0.265	0.924 ± 0.063	0.902 ± 0.050	0.882 ± 0.054	2.751 ± 0.104
90.277	72.6 ± 1.6	19.823 ± 0.691	1.135 ± 0.170	0.953 ± 0.109	0.790 ± 0.127	2.811 ± 0.260
91.030	144.3 ± 2.3	29.254 ± 0.657	1.527 ± 0.133	1.312 ± 0.099	1.384 ± 0.118	4.262 ± 0.217
91.222	3655.7 ± 11.8	30.502 ± 0.135	1.489 ± 0.026	1.432 ± 0.022	1.494 ± 0.024	4.432 ± 0.044
91.277	137.6 ± 2.3	30.357 ± 0.692	1.249 ± 0.123	1.412 ± 0.109	1.319 ± 0.117	4.151 ± 0.218
91.529	142.8 ± 2.5	30.918 ± 0.691	1.539 ± 0.131	1.510 ± 0.109	1.406 ± 0.119	4.419 ± 0.217
92.216	555.6 ± 4.7	21.762 ± 0.271	1.093 ± 0.055	1.003 ± 0.047	1.060 ± 0.053	3.146 ± 0.091
92.277	112.4 ± 2.1	21.346 ± 0.594	1.055 ± 0.119	1.072 ± 0.109	1.075 ± 0.118	3.259 ± 0.207
93.220	597.5 ± 4.9	12.410 ± 0.177	0.634 ± 0.040	0.634 ± 0.036	0.552 ± 0.037	1.861 ± 0.069
93.277	42.7 ± 1.3	12.623 ± 0.670	0.760 ± 0.164	0.685 ± 0.139	0.514 ± 0.138	2.175 ± 0.276
94.215	641.7 ± 5.1	7.989 ± 0.128	0.406 ± 0.032	0.437 ± 0.029	0.409 ± 0.031	1.230 ± 0.056
94.278	66.9 ± 1.6	7.989 ± 0.399	0.452 ± 0.104	0.347 ± 0.079	0.389 ± 0.096	1.025 ± 0.161

Table 4. Forward-backward asymmetries as a function of the centre-of-mass energy for $Z \to$ lepton pairs

$\sqrt{s}$(GeV)	$A_{FB}^{e^+e^-}$	$A_{FB}^{\mu^+\mu^-}$	$A_{FB}^{\tau^+\tau^-}$	$A_{FB}^{\ell^+\ell^-}$
88.224	-0.389 ± 0.228	-0.132 ± 0.103	-0.268 ± 0.123	-0.319 ± 0.098
88.277	-0.195 ± 0.357	-0.341 ± 0.196	-0.582 ± 0.140	-0.356 ± 0.167
89.220	-0.512 ± 0.187	-0.296 ± 0.062	-0.129 ± 0.082	-0.259 ± 0.060
89.277	-1.168 ± 0.856	-0.304 ± 0.203	-0.330 ± 0.200	-0.903 ± 0.281
90.222	-0.170 ± 0.082	-0.159 ± 0.052	-0.084 ± 0.062	-0.123 ± 0.037
90.277	-0.123 ± 0.172	-0.142 ± 0.129	-0.020 ± 0.130	-0.147 ± 0.093
91.030	$+0.065 \pm 0.090$	$+0.016 \pm 0.074$	-0.049 ± 0.082	$+0.027 \pm 0.049$
91.222	-0.009 ± 0.018	-0.002 ± 0.015	$+0.021 \pm 0.016$	-0.002 ± 0.009
91.277	-0.035 ± 0.110	-0.058 ± 0.076	$+0.019 \pm 0.084$	-0.040 ± 0.050
91.529	-0.102 ± 0.092	-0.011 ± 0.076	$+0.018 \pm 0.079$	-0.029 ± 0.047
92.216	$+0.145 \pm 0.049$	$+0.108 \pm 0.046$	$+0.130 \pm 0.050$	$+0.130 \pm 0.027$
92.277	$+0.030 \pm 0.120$	$+0.106 \pm 0.091$	$+0.184 \pm 0.101$	$+0.085 \pm 0.061$
93.220	$+0.237 \pm 0.058$	$+0.149 \pm 0.054$	$+0.270 \pm 0.059$	$+0.197 \pm 0.035$
93.277	$+0.200 \pm 0.218$	$+0.287 \pm 0.179$	$+0.415 \pm 0.192$	$+0.320 \pm 0.114$
94.215	$+0.130 \pm 0.078$	$+0.180 \pm 0.065$	$+0.326 \pm 0.072$	$+0.165 \pm 0.041$
94.278	$+0.242 \pm 0.250$	$+0.101 \pm 0.204$	-0.232 ± 0.220	-0.045 ± 0.147

6 Z resonance parameters

The cross section $\sigma_{e^+e^- \to f\bar{f}}$ for fermion-pair production in e^+e^- annihilation, after correction for initial-state radiation, can be expressed in a model-independent formulation [15–18] as a function of the physical parameters of the Z resonance. It contains three terms: the Z-exchange contribution which is represented by a Breit-Wigner function, the photon-exchange contribution, and the interference term. In a completely model-independent description, the precise value of the interference term is unknown. The dependence of the results on this term is less than 0.1% [19], so the Standard Model value has been assumed. The effect of initial-state radiation is large, of the order of 30% at the peak, but is known to better than 0.2% [16] of the cross section, and is taken into account. The peak cross section from Z exchange, when unfolded from initial-state radiation, is

$$\sigma_{f\bar{f}}^0 = \frac{12\pi}{M_Z^2} \frac{\Gamma_{ee}\Gamma_{f\bar{f}}}{\Gamma_Z^2}.$$

The resonance parameters have been determined by fitting the model-independent lineshape as implemented in the computer program MIZA [19] to the cross sections presented in Table 3. In the χ^2 minimization, the various correlations in the data which result from the common luminosity determination and common event selection criteria are taken into account. For the electron-pair sample and the common-lepton sample, points with $|\sqrt{s} - M_Z| > 1.5$ GeV are omitted to minimize the contribution of the interference term since it depends on Γ_{ee}, and to minimize the uncertainty resulting from the t-channel subtraction.

Fitting the hadronic and three lepton-pair cross sections simultaneously, one can determine six parameters: the Z mass, M_Z, the total width, Γ_Z, the peak hadronic cross section, σ_{had}^0, and the three ratios of hadron to lepton partial widths, $\Gamma_{\text{had}}/\Gamma_{ee}$, $\Gamma_{\text{had}}/\Gamma_{\mu\mu}$, and $\Gamma_{\text{had}}/\Gamma_{\tau\tau}$. A more precise determination of $R = \Gamma_{\text{had}}/\Gamma_{\ell\ell}$ is obtained from a four-parameter fit, assuming lepton universality. The results are summarized in Table 5. Additional errors resulting from uncertainty in the absolute scale of the LEP energy, the beam energy spread, nonreproducibility of energy from fill to fill, and systematic point-to-point energy uncertainties among the scan points are not shown in Table 5. The absolute scale of the centre-of-mass energy is estimated to be 2.2×10^{-4} (± 20 MeV) [14], which is the main uncertainty in the Z mass. The beam energy spread distorts the Z line shape. This effect has been included by convoluting the line shape with a Gaussian centre-of-mass energy spread with $\sigma = 50$ MeV at the Z peak [14]. The energy spread in-

Table 5. Fit results for hadron and lepton cross sections. Errors resulting from uncertainty in the LEP energy scale are not included

Parameter	No lepton univ.	lepton univ.	Common-leptons	S.M. Pred. [20]
M_Z [GeV]	91.182 ± 0.009	91.182 ± 0.009	91.182 ± 0.010	
Γ_Z [MeV]	2485 ± 17	2484 ± 17	2482 ± 18	2489 ± 23
σ_{had}^0 [nb]	41.44 ± 0.36	41.44 ± 0.36	41.45 ± 0.36	41.42 ± 0.07
$\Gamma_{\text{had}}/\Gamma_{ee}$	20.66 ± 0.33	–		
$\Gamma_{\text{had}}/\Gamma_{\mu\mu}$	21.26 ± 0.29	–		
$\Gamma_{\text{had}}/\Gamma_{\tau\tau}$	21.00 ± 0.36	–		
$\Gamma_{\text{had}}/\Gamma_{\ell\ell}$	–	21.00 ± 0.20	20.91 ± 0.20	20.80 ± 0.08
$\Gamma_{\text{inv}}/\Gamma_{\ell\ell}$	–	5.91 ± 0.15	5.93 ± 0.15	5.91 ± 0.01
χ^2	47.2/51 d.o.f.	49.1/53 d.o.f.	11.1/21 d.o.f.	

Table 6. Correlation coefficients for 4 parameter fit to the Z line shape assuming lepton universality

	M_Z	Γ_Z	σ^0_{had}	R
M_Z	1	+0.06	+0.01	0.00
Γ_Z	+0.06	1	−0.31	−0.01
σ^0_{had}	+0.01	−0.31	1	+0.13
$\Gamma_{\text{had}}/\Gamma_{\ell\ell}$	0.00	−0.01	+0.13	1

Table 7. Z resonance parameters derived from the measured hadron and lepton cross sections

Final state	Partial width (MeV)	Branching ratio	Peak cross section (nb)
	Six parameter fit		
Hadrons	1730 ±18	0.696 ±0.006	41.44 ±0.36
e^+e^-	83.8±0.9	0.0337±0.0003	2.006±0.035
$\mu^+\mu^-$	81.4±1.4	0.0328±0.0005	1.949±0.030
$\tau^+\tau^-$	82.4±1.6	0.0332±0.0006	1.974±0.037
Invisible	507 ±18		
	Four parameter fit		
Hadrons	1744 ±15	0.702 ±0.005	41.44 ±0.36
Lepton	83.1±0.7	0.0334±0.0002	1.974±0.023
Invisible	491 ±13	–	–

creases the measured total width by 4 MeV and decreases the peak cross section by 0.13%. The cross sections in Table 3 are corrected for this effect. The uncertainty in the reproducibility of machine conditions from fill to fill is estimated to be 10 MeV [14]. With 7 successive 7-point scans in 1990, however, this effect contributes a negligible uncertainty of the order of 2 MeV in both the Z mass and total width. Finally, the systematic point-to-point relative energy uncertainty of 10 MeV [14] contributes an additional systematic uncertainty of 5 MeV in both the Z mass and total width.

The parameters given in Table 5 are largely independent of each other; the largest correlation is −31% between σ^0_{had} and Γ_Z, as shown in Table 6.

The four-parameter fit has been repeated with the cross sections from the common-lepton sample. The results are consistent with those found using the separate lepton samples (see Table 5) if one takes into account the angular acceptance of the common-lepton selection, which yields 20% more events.

Other parameters, including the partial widths and branching ratios, can be derived from these results (Table 7). The errors in these parameters, however, have large correlation coefficients; in particular, Γ_{ee} is strongly anticorrelated with $\Gamma_{\mu\mu}$ and $\Gamma_{\tau\tau}$. The three leptonic widths agree with each other as expected from lepton universality.

7 The determination of vector and axial-vector couplings

The leptonic width, $\Gamma_{\ell\ell}$, and the lepton forward-backward asymmetry depend on the leptonic vector and axial-vector coupling constants. In the Born approximation, the QED-corrected partial width is given by

$$\Gamma_{\ell\ell}=\frac{G_F M_Z^3}{6\sqrt{2}\pi}(g_{V\ell}^2+g_{A\ell}^2)\left(1+\frac{3}{4}\frac{\alpha}{\pi}\right). \tag{3}$$

The corresponding forward-backward asymmetry at the Z peak is

$$A_{\text{FB}}=\frac{3}{4}\frac{2g_{Ve}g_{Ae}}{g_{Ve}^2+g_{Ae}^2}\frac{2g_{Vf}g_{Af}}{g_{Vf}^2+g_{Af}^2}\sim 3\frac{g_{V\ell}^2}{g_{A\ell}^2}. \tag{4}$$

The variation of A_{FB} away from the Z peak depends mainly on $g_{Ae}g_{Af}$.

The statistical precision of the data requires that higher-order electroweak corrections be included. In the Improved Born Approximation [21–23], which includes the bulk of the $\mathcal{O}(\alpha)$ electroweak corrections, the above expressions are still valid, but the coupling constants become effective running coupling constants evaluated at $q^2=M_Z^2$: $g_V(M_Z^2)$ and $g_A(M_Z^2)$. These effective constants are determined from the measured partial width and asymmetry; in the limit of the Improved Born Approximation, this determination is therefore independent of the Higgs and top masses.

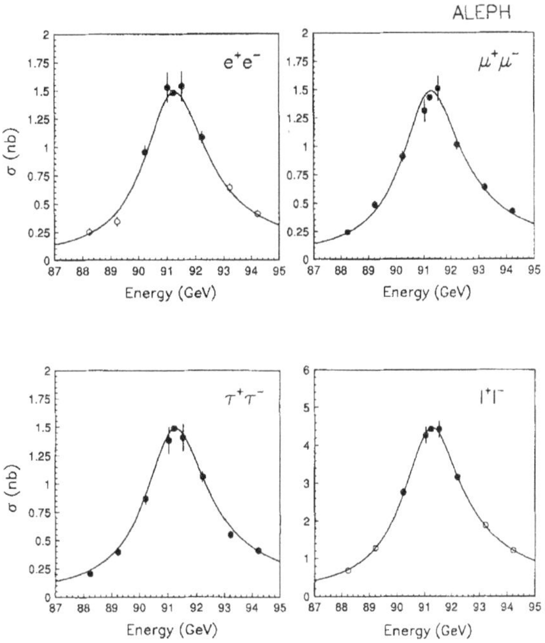

Fig. 9a–d. Cross sections for $e^+e^-\to$lepton pairs as a function of the centre-of-mass energy: **a** $e^+e^-\to e^+e^-$, **b** $e^+e^-\to\mu^+\mu^-$, **c** $e^+e^-\to\tau^+\tau^-$, and **d** $e^+e^-\to\ell^+\ell^-$. For points where the energy difference is less than 100 MeV the average cross section is plotted. The lines are the Standard Model predictions. Points with open circles are not used in the fit

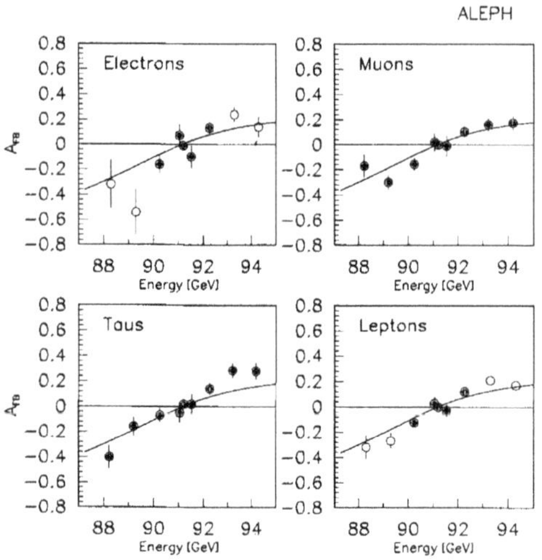

Fig. 10. Forward-backward asymmetry for e^-, μ^-, τ^-, and ℓ^- in lepton-pair events as a function of the centre-of-mass energy. For points where the energy difference is less than 100 MeV, the average asymmetry is plotted. The lines are the results of the fit. Points with open circles are not used in the fit

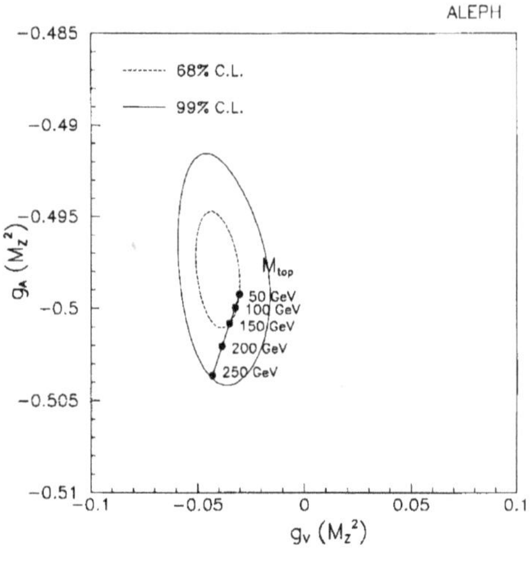

Fig. 11. Probability contours for $g_V(M_Z^2)$ and $g_A(M_Z^2)$ from leptonic forward-backward asymmetries and τ polarization. The points are the expectations of the Standard Model for different top masses, assuming $\alpha_s = 0.12$ and $M_{\text{Higgs}} = 200$ GeV

The measured asymmetries are fit as a function of $\sqrt{s}$, assuming lepton universality and using the lineshape measurements of M_Z, Γ_Z, and $\Gamma_{\ell\ell}$ as constraints (see the Appendix for a detailed description of the fitting formulae). In the e^+e^- channel, only the five points closest to the Z peak have been used to minimize the uncertainty resulting from t-channel subtraction.

The measured forward-backward asymmetries and the results of the fit are shown in Fig. 10. The corresponding coupling constants for leptons are

$$g_V^2(M_Z^2)/g_A^2(M_Z^2) = 0.0072 \pm 0.0027,$$

$$g_V(M_Z^2) = -0.042^{+0.009}_{-0.007} \quad \text{and}$$

$$g_A(M_Z^2) = -0.498 \pm 0.002.$$

Since the leptonic width and asymmetry depend on $g_V^2(M_Z^2)$ and $g_A^2(M_Z^2)$, the signs of $g_V(M_Z^2)$ and $g_A(M_Z^2)$ have been inferred from other ALEPH measurements [11, 24], and from neutrino-electron scattering experiments [25–28].

The fit can be repeated without assuming lepton universality, by replacing the constraint from $\Gamma_{\ell\ell}$ by Γ_{ee}, $\Gamma_{\mu\mu}$, and $\Gamma_{\tau\tau}$. In addition, the measurement of the τ polarization [11] is used to constrain the τ coupling further. The coupling constants as given in Table 8 are consistent with lepton universality. The smaller error in $g_V(M_Z^2)$ for the τ is a result of the inclusion of the τ polarization measurement. The probability contours for $g_A(M_Z^2)$ and $g_V(M_Z^2)$ are shown in Fig. 11.

8 Standard model interpretation

8.1 The number of light neutrino species

The number of light neutrino species can be obtained from the measured value of $\Gamma_{\text{inv}}/\Gamma_{\ell\ell}$, under the assumption that $\Gamma_{\text{inv}} = N_\nu \Gamma_\nu$. In the Standard Model, $\Gamma_{\ell\ell}/\Gamma_\nu$ is related to the lepton coupling constants by

$$\frac{\Gamma_{\ell\ell}}{\Gamma_\nu} = \frac{1}{2}\left(1 + \left(\frac{g_V(M_Z^2)}{g_A(M_Z^2)}\right)^2\right)\left(1 + \frac{3}{4}\frac{\alpha}{\pi}\right)(1 + \delta_V), \qquad (5)$$

Table 8. Effective vector and axial-vector coupling constants for e, μ, and τ from forward-backward asymmetries without/with τ polarization [11]

	A_{FB} only		$A_{FB} + P_\tau$	
	$g_V(M_Z^2)$	$g_A(M_Z^2)$	$g_V(M_Z^2)$	$g_A(M_Z^2)$
e	$-0.035^{+0.013}_{-0.014}$	$-0.501^{+0.003}_{-0.003}$	$-0.045^{+0.013}_{-0.011}$	$-0.500^{+0.003}_{-0.003}$
μ	$-0.023^{+0.029}_{-0.037}$	$-0.494^{+0.005}_{-0.004}$	$-0.018^{+0.023}_{-0.026}$	$-0.494^{+0.004}_{-0.004}$
τ	$-0.104^{+0.040}_{-0.066}$	$-0.486^{+0.019}_{-0.009}$	$-0.045^{+0.010}_{-0.011}$	$-0.495^{+0.005}_{-0.005}$
Lepton	$-0.042^{+0.009}_{-0.007}$	$-0.498^{+0.002}_{-0.002}$	$-0.041^{+0.007}_{-0.006}$	$-0.498^{+0.002}_{-0.002}$

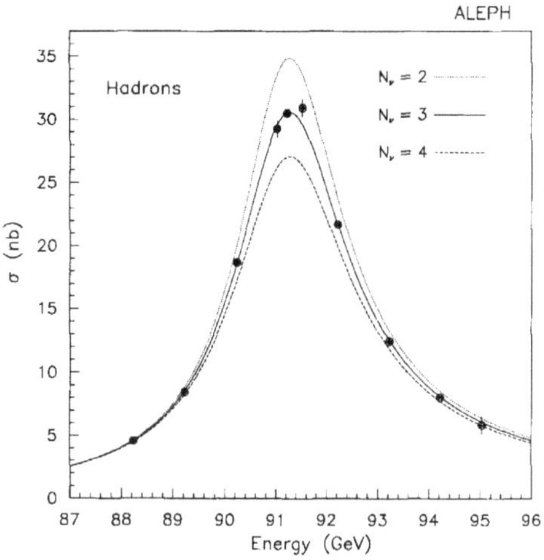

Fig. 12. Cross sections for $e^+e^- \to$hadrons as a function of the centre-of-mass energy. For points where the energy difference is less than 100 MeV the average cross section is plotted. The Standard Model predictions for $N_\nu = 2$, 3, and 4 are shown

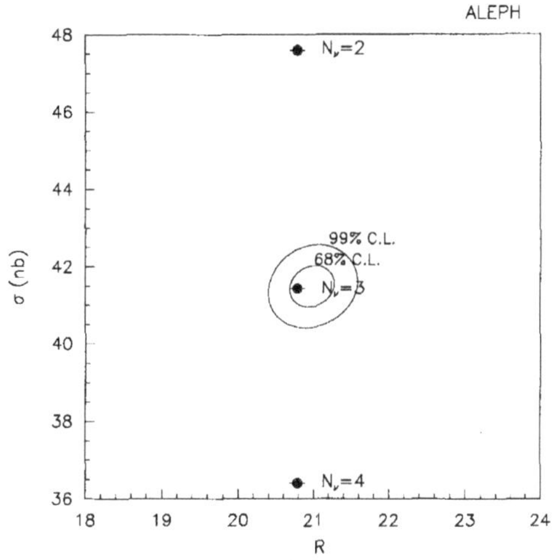

Fig. 13. Contours of constant χ^2 for the hadronic peak cross section σ^0_{had} as a function of $\Gamma_{\text{had}}/\Gamma_{\ell\ell}$ together with the Standard Model prediction. The uncertainty in the QCD correction is indicated by the error bar and corresponds to $\Delta\alpha_s = \pm 0.01$

and is equal to 0.5023 ± 0.0010 using the measured value of $g_{V\ell}/g_{A\ell}$ given in Sect. 7.* $\delta_V = -0.003 \pm 0.0003$ is the electroweak vertex correction.** From the present measurement

$$\Gamma_{\text{inv}}/\Gamma_{\ell\ell} = \sqrt{\frac{12\pi R}{M_Z^2 \sigma^0_{\text{had}}}} - R - 3 = 5.91 \pm 0.15,$$

and the above value of $\Gamma_{\ell\ell}/\Gamma_\nu$,

$$N_\nu = 2.97 \pm 0.07.$$

The only Standard Model assumption required to obtain this result is the expression for $\Gamma_{\ell\ell}/\Gamma_\nu$; therefore, it is still valid if unexpected states yielding hadrons are present in Z decay. The expected line shapes for hadronic Z decay for 2, 3, and 4 neutrino species are shown with the measured cross sections in Fig. 12.

Fixing $N_\nu = 3$, the neutrino partial width can be determined:

$$\Gamma_\nu = (163.7 \pm 4.3)\ \text{MeV}.$$

8.2 Peak cross sections, ratios of partial widths, and α_s

The measurement of the Z mass, together with α and G_F, constrains Standard Model predictions, leaving only a small dependence on the unknown Higgs and top masses, which enter through higher-order corrections. The ratio of partial widths such as $R = \Gamma_{\text{had}}/\Gamma_{\ell\ell}$ as well

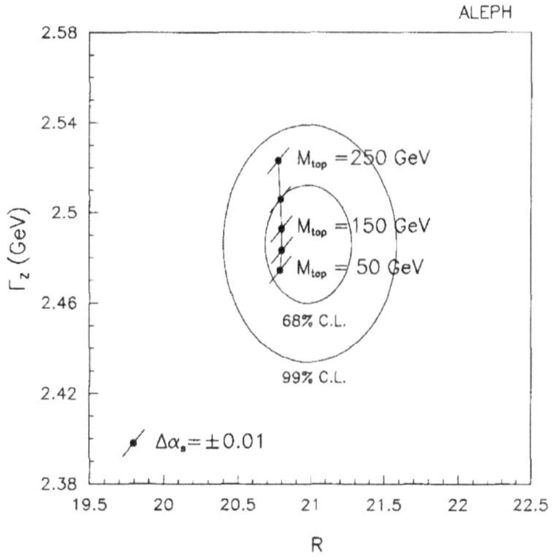

Fig. 14. Contours of constant χ^2 for Γ_Z as a function of $\Gamma_{\text{had}}/\Gamma_{\ell\ell}$ together with the Standard Model predictions for various top masses (assuming $M_{\text{Higgs}} = 200$ GeV). The uncertainty resulting from the QCD correction is indicated by the error bars

as the cross sections at the peak depend only weakly on M_{top} and M_{Higgs}, and are therefore particularly suitable for a comparison between experiment and theory. The partial widths themselves, however, are sensitive to $\sin^2\theta_W(M_Z^2)$ and therefore to the top and Higgs masses.

In Fig. 13, the correlation between the peak cross section for Z decay into hadrons and R is compared with the Standard Model prediction for 2, 3, and 4 neutrino

* The Standard Model prediction [20] for $\Gamma_{\ell\ell}/\Gamma_\nu$ is 0.5022 ± 0.0008

** δ_V has been obtained from [22] and [29]; the uncertainty in its value has been obtained by varying M_{top} from 80 GeV to 250 GeV, and M_{Higgs} from 50 GeV to 1 TeV species; the agreement for $N_\nu=3$ is good. The main uncertainty in the prediction is the uncertainty in the strong coupling constant α_s [30]. Each of the two measurements checks the Standard Model with a precision of $\sim 1\%$.

In Fig. 14, the results for Γ_Z and R are compared with the Standard Model. The variation of the theoretical prediction with M_{top} and the uncertainty resulting from the α_s correction are indicated. The agreement is good for top masses in the range 50–200 GeV.

The dependence of R on α_s, assuming a second-order expansion in α_s in the $\overline{MS}$ scheme, is [31]

$$R=R^0\left(1+\frac{\alpha_s}{\pi}+1.41\left(\frac{\alpha_s}{\pi}\right)^2\right),$$

where $R^0=19.98\pm 0.03$ [20] is the Standard Model prediction for R in the absence of strong interactions. Refitting the data imposing $N_\nu=3$, one obtains $R=R^0$ (1.049 ± 0.008), which gives

$$\alpha_s(M_Z^2)=0.144\pm 0.024.$$

8.3 The electroweak mixing angle and radiative effects

In the Standard Model, the weak vector and axial-vector couplings are defined in terms of the weak mixing angle $\sin^2\theta_W$. The measurements of the Z lineshape and the forward-backward asymmetries may be interpreted as different measurements of $\sin^2\theta_W(M_Z^2)$. The comparison of these measurements provides a fundamental test of the Standard Model and is sensitive to physics outside of the Standard Model.

The effective vector and axial-vector couplings may be written as

$$g_A(M_Z^2)=-1/2\sqrt{1+\Delta\rho_\ell}, \quad (6)$$

$$g_V(M_Z^2)=g_A(M_Z^2)(1-4(\sin^2\theta_W(M_Z^2)+C_\ell)). \quad (7)$$

The effective parameters $\Delta\rho_\ell$ and $\sin^2\theta_W(M_Z^2)$ [21–23] absorb, by definition, any deviation from the tree-level couplings not explicitly included in the fitting formula (see Appendix). $C_\ell=0.0007$ is the flavor-dependent electroweak vertex correction [32]; $\Delta\rho_\ell$, as defined in (6) [33], includes the corresponding electroweak vertex correction to $\Delta\rho$.*

The lepton forward-backward asymmetry provides a direct measurement of $\sin^2\theta_W(M_Z^2)$ since it determines the ratio of vector to axial-vector coupling constants of the leptons ((4) and (7)). The measurement of the forward-backward asymmetry for leptons presented here corresponds to

$$\sin^2\theta_W(M_Z^2)=0.2281\pm 0.0040.$$

Other ALEPH asymmetry measurements similarly determine $\sin^2\theta_W(M_Z^2)$:

- the quark charge asymmetry [24],
- the tau polarization [11],
- the $b\bar{b}$ and $c\bar{c}$ forward-backward asymmetries [35].

The different measurements of $\sin^2\theta_W(M_Z^2)$ are summarized in Table 9, showing good agreement in the values of $\sin^2\theta_W(M_Z^2)$, with an average of

$$\sin^2\theta_W(M_Z^2)=0.2285\pm 0.0025.$$

The leptonic width can also be related to $\sin^2\theta_W(M_Z^2)$:

$$\Gamma_{\ell\ell}=(1+\kappa)\frac{\alpha(M_Z^2)M_Z}{48\sin^2\theta_W(M_Z^2)\cos^2\theta_W(M_Z^2)} \cdot[(1-4(\sin^2\theta_W(M_Z^2)+C_\ell))^2+1]\left(1+\frac{3}{4}\frac{\alpha}{\pi}\right).$$

Comparing this expression with the one obtained by substituting (6) and (7) into (3), M_Z^2 can be related to $\sin^2\theta_W(M_Z^2)$:

$$M_Z^2=\frac{\pi\alpha(M_Z^2)(1+\kappa)}{\sqrt{2}G_F(1+\Delta\rho_\ell)\sin^2\theta_W(M_Z^2)\cos^2\theta_W(M_Z^2)}. \quad (8)$$

The factor κ absorbs the running of the Z self-energy across the Z resonance, as well as residual vertex corrections. κ, which is related to the variable S discussed in [36] ($S=\kappa\times 4\sin^2\theta_W(M_Z^2)\cos^2\theta_W(M_Z^2)/\alpha$), depends logarithmically on both M_{top} and M_{Higgs}, and is therefore relatively more sensitive to M_{Higgs} than $\Delta\rho_\ell$. There is no explicit $\Delta\rho_\ell$ dependence in the expression for $\Gamma_{\ell\ell}$, while the $\Delta\rho_\ell$ dependence remains in the relationship between $\sin^2\theta_W(M_Z^2)$ and M_Z.

If one assumes the Minimal Standard Model value* of $\kappa=0.0033\pm 0.0010_{top}\pm 0.0015_{Higgs}$, the measurement of $\Gamma_{\ell\ell}/M_Z$ corresponds to

$$\sin^2\theta_W(M_Z^2)=0.2340\pm 0.0025.$$

It is interesting to note that the precision of $\sin^2\theta_W(M_Z^2)$ as determined from the asymmetry measurements is comparable to the one obtained from $\Gamma_{\ell\ell}/M_Z$ (see Table 9). The combined result is $\sin^2\theta_W(M_Z^2)=0.2312$ ± 0.0018.

The correlation between $\Gamma_{\ell\ell}$ and $\sin^2\theta_W(M_Z^2)$ measured from the asymmetries is shown in Fig. 15, along with the Standard Model predictions. The comparison of these measurements may also be interpreted as a measurement of κ:

$$\kappa=-0.016\pm 0.012.$$

---

* $\Delta\rho_\ell$ is related to $\Delta\rho$ by an electroweak vertex correction [34]:

$$\Delta\rho_\ell=\Delta\rho+C.$$

For large values of M_{top}, the leading behaviour of $\Delta\rho$ is given by

$$\Delta\rho\simeq\frac{\alpha}{\pi}\frac{M_{top}^2}{M_Z^2}-\frac{\alpha}{4\pi}\ln\frac{M_{Higgs}^2}{M_Z^2}.$$

In a similar approximation, $\sin^2\theta_W(M_Z^2)$ may be written as

$$\sin^2\theta_W(M_Z^2)\simeq 1-\frac{M_W^2}{(1+\Delta\rho)M_Z^2}$$

* The quoted errors correspond to the M_{top} and M_{Higgs} ranges given in [20]

Table 9. Different measurements of $\sin^2\theta_W(M_Z^2)$. $b\bar{b}$ and $c\bar{c}$ asymmetries have been combined using a 20% correlation

Measurement	Measured quantity	value	$\sin^2\theta_W(M_Z^2)$
Lepton F-B asymmetry	$g_V^2(M_Z^2)/g_A^2(M_Z^2)$	0.0072 ± 0.0027	0.2281 ± 0.0040
Quark charge asymmetry	$\langle Q_{FB} \rangle$	-0.0084 ± 0.0016	0.2300 ± 0.0052
Tau polarization	P_τ^0	-0.152 ± 0.045	0.2302 ± 0.0057
$b\bar{b}$ asymmetry	A_{FB}^b	0.126 ± 0.030	0.2262 ± 0.0054
$c\bar{c}$ asymmetry	A_{FB}^c	0.064 ± 0.049	0.2310 ± 0.0120
Asymmetry average			0.2285 ± 0.0025
Line shape	$\Gamma_{\ell\ell}$	83.05 ± 0.67 MeV	0.2340 ± 0.0025
Overall average			0.2312 ± 0.0018

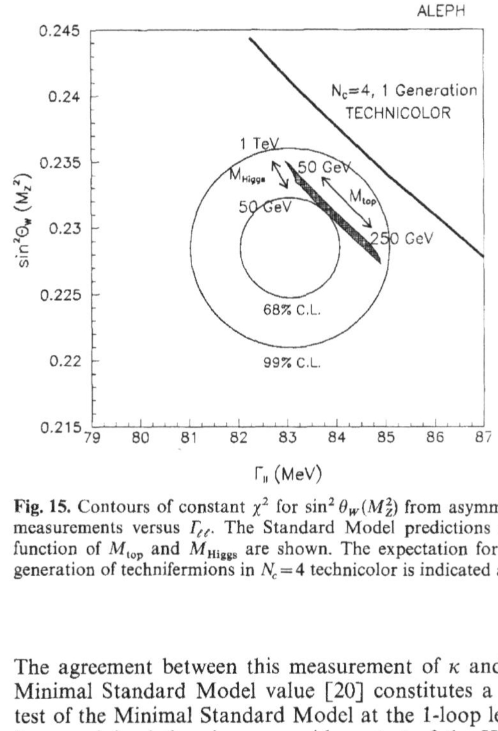

Fig. 15. Contours of constant χ^2 for $\sin^2\theta_W(M_Z^2)$ from asymmetry measurements versus $\Gamma_{\ell\ell}$. The Standard Model predictions as a function of M_{top} and M_{Higgs} are shown. The expectation for one generation of technifermions in $N_c = 4$ technicolor is indicated also

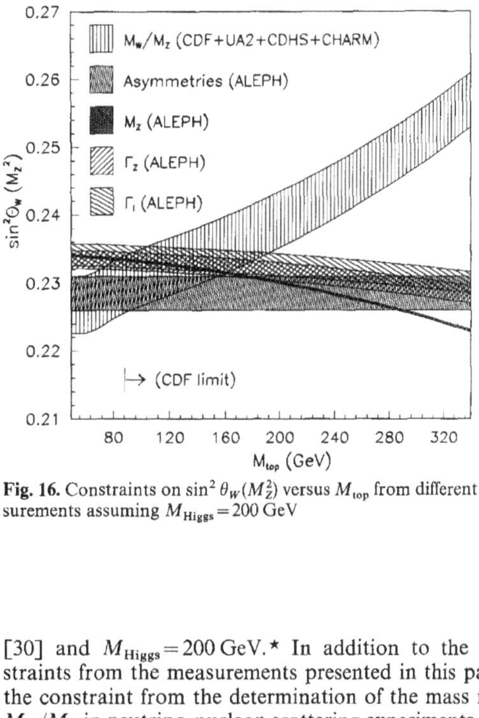

Fig. 16. Constraints on $\sin^2\theta_W(M_Z^2)$ versus M_{top} from different measurements assuming $M_{Higgs} = 200$ GeV

The agreement between this measurement of κ and its Minimal Standard Model value [20] constitutes a 1% test of the Minimal Standard Model at the 1-loop level. In nonminimal theories, κ provides a test of the Higgs sector [37]. Such a scenario is discussed in [36] where the quantity S is expected to be increased by 2.1 for one generation of technifermions in $N_c = 4$ technicolor; this change corresponds to an increase of 0.024 in κ. The measured value of κ excludes this possibility at more than a 95% confidence level (see Fig. 15).

8.4 $\sin^2\theta_W(M_Z^2)$ and limits on the top mass

In the Minimal Standard Model with three neutrino species, any observable can be computed as a function of $\sin^2\theta_W(M_Z^2)$, M_{top}, M_{Higgs}, and α_s. Therefore, for given values of α_s and M_{Higgs}, each observable defines a relationship between $\sin^2\theta_W(M_Z^2)$ and M_{top}. These relationships are shown in Fig. 16, assuming $\alpha_s = 0.121 \pm 0.008$ [30] and $M_{Higgs} = 200$ GeV.* In addition to the constraints from the measurements presented in this paper, the constraint from the determination of the mass ratio M_W/M_Z in neutrino-nucleon scattering experiments [38] and from the direct measurement of M_W in $p\bar{p}$ colliders [39, 40] is shown. The width of the band for each observable corresponds to the experimental uncertainty in the measurement. As shown in the previous section, the asymmetries depend on $\sin^2\theta_W(M_Z^2)$, but have no explicit top-mass dependence. Therefore, the constraint from the asymmetry measurements appears as a horizontal band. The other observables depend on M_{top} as well as $\sin^2\theta_W(M_Z^2)$, and appear as curved bands in Fig. 16.**

The data displayed in Fig. 16 may be combined to find best values of $\sin^2\theta_W(M_Z^2)$ and M_{top}. The following

* The theoretical calculations used include M_{top}^4 two-loop contributions as well as QCD corrections to the top-mass dependent terms

** Note that the constraints derived from $\Gamma_{\ell\ell}$ and M_Z are displayed separately. The constraint from the ratio $\Gamma_{\ell\ell}/M_Z$, as discussed in the previous section, would appear as a horizontal band measurements are considered in a combined fit:

- from the lineshape:

$M_Z = (91.182 \pm 0.009)$ GeV

$\Gamma_Z = (2484 \pm 17)$ MeV

$\Gamma_{\text{had}} = (1744 \pm 15)$ MeV

$\Gamma_{\ell\ell} = (83.1 \pm 0.7)$ MeV

- from the asymmetries:

$\sin^2 \theta_W(M_Z^2) = 0.2285 \pm 0.0025$

- from measurements of the vector-boson masses ratio [38–40]:

$$\sin^2 \theta_W = 1 - \frac{M_W^2}{M_Z^2} = 0.2292 \pm 0.0042.$$

A fit to the ALEPH results alone yields

$\sin^2 \theta_W(M_Z^2) = 0.2312 \pm 0.0016 \pm 0.0002_{\text{Higgs}}$

and

$M_{\text{top}} = (170 \pm^{42}_{55} \pm^{21}_{14\,\text{Higgs}})$ GeV,

with $\chi^2 = 2.8$ for 2 degrees of freedom. The second error shows the uncertainty corresponding to a change in the Higgs mass from 50 GeV [41] to 1000 GeV. These results may be interpreted as a measurement of the W boson mass:

$M_W = (80.33^{+0.30}_{-0.32})$ GeV.

Combining the ALEPH results with the determination of the M_W/M_Z ratio from other experiments [39, 40], one obtains a more precise determination of $\sin^2 \theta_W(M_Z^2)$ and M_{top}:

$\sin^2 \theta_W(M_Z^2) = 0.2322 \pm 0.0009 \pm 0.0003_{\text{Higgs}}$,

$M_{\text{top}} = (139 \pm^{30}_{35} \pm^{22}_{15\,\text{Higgs}})$ GeV,

with $\chi^2 = 4.4$ for 5 degrees of freedom.

An independent analysis has been done in which the data are tested for consistency with the Minimal Standard Model using EXPOSTAR [21, 29, 42]. The only free parameters in this analysis are M_Z, M_{top}, M_{Higgs}, and α_s. Imposing $M_{\text{Higgs}} = 200$ GeV and using the ALEPH measurement of α_s [30] as a constraint, the data are found to agree with the Minimal Standard Model, and result* in $M_Z = (91.182 \pm 0.009)$ GeV and $M_{\text{top}} = (157^{+49}_{-71})$ GeV, consistent with the values given above.

9 Conclusion

On the basis of 190000 Z decays collected with the ALEPH detector, the following parameters of the Z resonance have been measured:

$M_Z = (91.182 \pm 0.009_{\text{exp}} \pm 0.020_{\text{L.P}})$ GeV,

$\Gamma_Z = (2484 \pm 17)$ MeV,

$\sigma^0_{\text{had}} = (41.44 \pm 0.36)$ nb,

$R = 21.00 \pm 0.20.$

The corresponding number of light neutrino species is 2.97 ± 0.07. From the forward-backward asymmetry, the ratio of vector to axial vector-couplings is found to be

$g_V(M_Z^2)^2 / g_A(M_Z^2)^2 = 0.0072 \pm 0.0027.$

Using ALEPH's asymmetry measurements together with $\Gamma_{\ell\ell}/M_Z$ yields

$\sin^2 \theta_W(M_Z^2) = 0.2312 \pm 0.0018.$

A fit to the ALEPH data assuming $M_{\text{Higgs}} = 200$ GeV gives

$M_{\text{top}} = (170 \pm^{42}_{55} \pm^{21}_{14\,\text{Higgs}})$ GeV,

where the second error corresponds to a change in the Higgs mass from 50 GeV to 1 TeV.

All of the results presented here are consistent with the Minimal Standard Model at a 1% level, as well as with previous measurements [1, 43].

Acknowledgements. It is a pleasure to thank our colleagues from the SL division for the operation of LEP. We are indebted to the engineers and technicians at CERN and our home institutes for their contributions to ALEPH's success. Those of us not from member states thank CERN for its hospitality.

Appendix

In this Appendix, the parametrizations used for fitting the lineshape and forward-backward asymmetry are described. Some small contributions, such as initial-final state interference, are not easily treated in the model-independent scheme used, and have been neglected.

The cross section $\sigma_{e^+e^- \to f\bar{f}}$ can be expressed in a model-independent way in terms of three contributions:

- the Z exchange that is represented by a Breit-Wigner function;
- the photon exchange;
- the $Z-\gamma$ interference term I_f.

The following expression is used for the fits:

$$\sigma(s)_{ff} = \frac{s}{(s - M_Z^2)^2 + s^2 \Gamma_Z^2 / M_Z^2} \left(\frac{12\pi \Gamma_{ff}}{M_Z^2} \frac{\Gamma_{ee}}{1 + \frac{3}{4}\frac{\alpha}{\pi}} + I_f \frac{N_c(s - M_Z^2)}{s} \right) + \frac{4}{3} \pi N_c Q_f^2 \frac{\alpha^2(s)}{s}, \tag{9}$$

where Γ_{ee}, Γ_{ff} are the partial widths for Z decay into e^+e^- or any fermion pair $f\bar{f}$; Q_f is the charge and N_c is the colour factor of the fermion. Γ_Z is the total width

---

* The quark-charge and $b\bar{b}$ asymmetry measurements have not been included in this analysis

and M_Z the mass of the Z boson. The invisible width is defined by $\Gamma_{inv}=\Gamma_Z-\Gamma_{had}-\Gamma_{ee}-\Gamma_{\mu\mu}-\Gamma_{\tau\tau}$, where Γ_{had} denotes the hadronic decay width. In this formula, Γ_{ee} is divided by the factor $\left(1+\frac{3}{4}\frac{\alpha}{\pi}\right)$ to allow for a separate treatment of the initial state radiation [19], as described below. Initial state real and virtual pair production has been studied but is not yet taken into account.

The interference term, I_f, cannot be expressed in a model-independent way in terms of the Z partial widths. Since I_f is small, the Standard Model value is assumed.

To include the effect of initial-state radiation, the expression for the cross section is convoluted with a radiator function [17, 44].

The QED-corrected forward-backward asymmetry in the angular distribution can be well approximated as [45]:

$$A_{FB}(s)=\frac{\int_0^{x_{max}} H(s,x)\,\sigma_{NR}^{FB}(s(1-x))\,\mathrm{d}x}{\int_0^{x_{max}} H(s,x)\,\sigma_{NR}(s(1-x))\,\mathrm{d}x},$$

where $H(s, x)$ is the same radiator function used in the total cross section. In the numerator, $\sigma_{NR}^{FB}=\sigma_{NR}^{F}-\sigma_{NR}^{B}$ is the integrated forward minus backward non-radiatively-corrected cross section, which in terms of the effective coupling constants $g_A(M_Z^2)$ and $g_V(M_Z^2)$ is

$$\sigma_{NR}^{FB}(s)=\frac{\pi\alpha^2(s)}{s}\left(2g_{Ae}g_{Af}F_G(s)\cdot\frac{s(s-M_Z^2)+s^2\frac{\Gamma_Z}{M_Z}\mathrm{Im}(\Delta\alpha)}{(s-M_Z^2)^2+s^2\frac{\Gamma_Z^2}{M_Z^2}}+4g_{Ve}g_{Ae}g_{Vf}g_{Af}F_G^2(s)\frac{s^2}{(s-M_Z^2)^2+s^2\frac{\Gamma_Z^2}{M_Z^2}}\right),$$

where $\mathrm{Im}(\Delta\alpha)$ is the imaginary part of the photon vacuum polarization. The denominator is the total cross section, which can be expressed as

$$\sigma_{NR}(s)=\frac{4}{3}\frac{\pi\alpha^2(s)}{s}\left(1+2g_{Ve}g_{Vf}F_G(s)\frac{s(s-M_Z^2)}{(s-M_Z^2)^2+s^2\frac{\Gamma_Z^2}{M_Z^2}}+((g_{Ve})^2+(g_{Ae})^2)\cdot((g_{Vf})^2+(g_{Af})^2)\cdot F_G^2(s)\frac{s^2}{(s-M_Z^2)^2+s^2\frac{\Gamma_Z^2}{M_Z^2}}\right),$$

where

$$F_G(s)=\frac{G_F M_Z^2}{2\sqrt{2}\pi\alpha(s)}. \qquad (10)$$

The parametrizations used for the lineshape and the forward-backward asymmetry are good approximations of the most accurate Standard Model calculations available.

References

1. D. Decamp et al. (ALEPH Coll.): Z. Phys. C – Particles and Fields 48 (1990) 365
2. D. Decamp et al. (ALEPH Coll.): Nucl. Instrum. Methods A294 (1990) 121
3. D. Decamp et al. (ALEPH Coll.): Measurement of the Absolute Luminosity with the ALEPH detector. CERN-PPE 91-129 to be submitted to Z. Phys. C – Particles and Fields
4. W. Beenakker, F.A. Berends, S.C. van der Marck: Nucl. Phys. B349 (1991) 323
5. M. Martinez, R. Miquel: Z. Phys. C – Particles and Fields 53 (1992) 115
6. S. Jadach, Z. Wąs: Phys. Rev. D41 (1990) 1425
7. F.A. Berends, R. Kleiss: Nucl. Phys. B228 (1983) 737; M. Böhm, A. Denner, W. Hollik: Nucl. Phys. B304 (1988) 687; F.A. Berends, R. Kleiss, W. Hollik: Nucl. Phys. B304 (1988) 712; Computer program BABAMC, courtesy of R. Kleiss
8. S. Jadach, Z. Wąs: Comput. Phys. Commun. 36 (1985) 191; Monte Carlo Group: in: Proceedings of the Workshop on Z Physics at LEP, CERN Report 89-08 (1989) Vol. III; S. Jadach, B.F.L. Ward, Z. Wąs: CERN-TH961-90, to be published in Comput. Phys. Commun. Computer program KORALZ, courtesy of S. Jadach, B.F.L. Ward, Z. Wąs
9. M. Bengtsson, T. Sjostrand: Phys. Lett. B185 (1987) 435; G. Marchesini, B.R. Weber: Nucl. Phys. B310 (1988) 461
10. S. Kawabata: Comput. Phys. Commun. 41 (1986) 127
11. D. Decamp et al. (ALEPH Coll.): Measurement of the Polarization of τ Leptons Produced in Z Decays. CERN-PPE91-94, submitted to Phys. Lett. B
12. S. Jadach, E. Richter-Wąs, Z. Wąs, B.F.L. Ward: Phys. Lett. B260 (1991) 438
13. W. Beenakker, F.A. Berends, S.C. van der Marck: Nucl. Phys. B355 (1991) 281
14. V. Hatton et al.: LEP absolute energy in 1990, LEP Performance R. Baily et al.: LEP energy calibration, CERN-SL90-95 J.M. Jowett: Luminosity and energy spread in LEP, CERN-LEP-TH85-04 and private communication for updated values
15. A. Borelli et al.: Nucl. Phys. B333 (1990) 357
16. F.A. Berends et al.: Z Line Shape group in: Proceedings of the Workshop of Z Physics at LEP, CERN Report 89-08 Vol. I, 89
17. F.A. Berends, G. Burgers, W.L. van Neerven: Nucl. Phys. B297 (1988) 429; Nucl. Phys. B304 (1988) 921
18. D. Bardin et al.: Z. Phys. C – Particles and Fields 44 (1989) 493
19. M. Martinez et al.: Z. Phys. C – Particles and Fields 49 (1991) 645
20. The Standard model predictions correspond to the following values: $M_{top}=(130^{+120}_{-50})$ GeV, $M_{Higgs}=(200^{+800}_{-150})$ GeV, and $\alpha_s=0.12\pm0.01$
21. D.C. Kennedy, B.W. Lynn: Nucl. Phys. B322 (1989) 1
22. M. Consoli, W. Hollik: in: Proceedings of the Workshop of Z Physics at LEP, CERN Report 89-08 Vol. I, 7
23. W.J. Marciano, A. Sirlin: Phys. Rev. Lett. 46 (1981) 163
24. D. Decamp et al. (ALEPH Coll.): Phys. Lett. B259 (1991) 377
25. K. Abe et al.: Phys. Rev. Lett. 62 (1989) 1709
26. D. Geiregat et al. (CHARM II Coll.): Phys. Lett. B259 (1991) 499
27. J. Dorenbosch et al. (CHARM Coll.): Z. Phys. C – Particles and Fields 41 (1989) 567
28. R.C. Allen et al.: Phys. Rev. Lett. 64 (1990) 1330
29. D.C. Kennedy, B.W. Lynn, C.J.-C. Im, R.G. Stuart: Nucl. Phys. 321B (1989) 83; B.W. Lynn, S. Selypsky, R.G. Stuart, D. Levinthal, in preparation
30. D. Decamp et al. (ALEPH Coll.): Phys. Lett. B255 (1990) 623; D. Decamp et al. (ALEPH Coll.): Phys. Lett. B257 (1991) 479

31. K.G. Cheterkyn, A.L. Kateev, F.V. Tkachov: Phys. Lett. B85 (1979) 277; M. Dine, J. Sarpinstein: Phys. Rev. Lett. 43 (1979) 668; W. Celemaster, R.J. Gonsalves: Phys. Rev. Lett. 44 (1979) 560; Phys. Rev. D21 (1980) 3112
32. D. Bardin, W. Hollik, T. Riemann: Z. Phys. C – Particles and Fields 49 (1991) 485
33. This definition of $\Delta\rho_\ell$ follows the suggestion of G. Altarelli, R. Barbieri: Phys. Lett. B230 (1991) 161
34. W. Hollik: private communication
35. D. Decamp et al. (ALEPH Coll.): Phys. Lett. B 263 (1991) 325
36. M.E. Peskin, T. Takeuchi: Phys. Rev. Lett. 65 (1990) 964
37. B.W. Lynn, M.E. Peskin, R.G. Stuart: in: Physics at LEP, CERN Report 86-02 Vol. I, 90
38. H. Abramowicz et al. (CDHS Coll.): Phys. Rev. Lett. 57 (1986) 298; A. Blondel et al.: Z. Phys. C – Particles and Fields 45 (1990) 361; J.V. Allaby et al. (CHARM Coll.): Phys. Lett. B177 (1986) 446; Z. Phys. C – Particles and Fields 36 (1987) 611
39. J. Alitti et al. (UA2 Coll.): Phys. Lett. B241 (1990) 150
40. F. Abe et al. (CDF Coll.): Phys. Rev. Lett. 65 (1990) 2243
41. D. Decamp et al. (ALEPH Coll.): Searches for the Standard Higgs Boson Produced in the Reaction $e^+e^- \to H^0 Z^*$, CERN-PPE91-19
42. D. Levinthal, F. Bird, R.G. Stuart, B.W. Lynn: CERN-TH6094/91 submitted to Z. Phys. C – Particles and Fields
43. P. Abreu et al. (DELPHI Coll.): Determination of the Z^0 Resonance Parameters and Couplings from its Hadronic and Leptonic Decays, CERN-PPE91-95; B. Adeva et al. (L3 Coll.): Z. Phys. C – Particles and Fields 51 (1991) 179; G. Alexander et al. (OPAL Coll.): Z. Phys. C – Particles and Fields 52 (1991) 175
44. E.A. Kuraev, V.S. Fadin: Sov. J. Nucl. Phys. 41 (1985) 466; G. Altarelli, G. Martinelli: in: Physics at LEP, CERN86-02 (1986) Vol. I, 47; O. Nicrosini, L. Trentadue: Phys. Lett. B196 (1987) 551
45. D. Bardin et al.: in: Proceedings of the Workshop of Z Physics at LEP, CERN Report 89-08, 1989, Vol. I, 203

ELSEVIER

22 May 1997

PHYSICS LETTERS B

Physics Letters B 401 (1997) 163–175

A measurement of R_b using mutually exclusive tags

ALEPH Collaboration

R. Barate [a], D. Buskulic [a], D. Decamp [a], P. Ghez [a], C. Goy [a], J.-P. Lees [a], A. Lucotte [a], M.-N. Minard [a], J.-Y. Nief [a], B. Pietrzyk [a], M.P. Casado [b], M. Chmeissani [b], P. Comas [b], J.M. Crespo [b], M. Delfino [b], E. Fernandez [b], M. Fernandez-Bosman [b], Ll. Garrido [b,15], A. Juste [b], M. Martinez [b], R. Miquel [b], Ll.M. Mir [b], S. Orteu [b], C. Padilla [b], I.C. Park [b], A. Pascual [b], J.A. Perlas [b], I. Riu [b], F. Sanchez [b], F. Teubert [b], A. Colaleo [c], D. Creanza [c], M. de Palma [c], G. Gelao [c], G. Iaselli [c], G. Maggi [c], M. Maggi [c], N. Marinelli [c], S. Nuzzo [c], A. Ranieri [c], G. Raso [c], F. Ruggieri [c], G. Selvaggi [c], L. Silvestris [c], P. Tempesta [c], A. Tricomi [c,3], G. Zito [c], X. Huang [d], J. Lin [d], Q. Ouyang [d], T. Wang [d], Y. Xie [d], R. Xu [d], S. Xue [d], J. Zhang [d], L. Zhang [d], W. Zhao [d], D. Abbaneo [e], R. Alemany [e], U. Becker [e], A.O. Bazarko [e,20], P. Bright-Thomas [e], M. Cattaneo [e], F. Cerutti [e], H. Drevermann [e], R.W. Forty [e], M. Frank [e], R. Hagelberg [e], J. Harvey [e], P. Janot [e], B. Jost [e], E. Kneringer [e], J. Knobloch [e], I. Lehraus [e], G. Lutters [e], P. Mato [e], A. Minten [e], L. Moneta [e], A. Pacheco [e], J.-F. Pusztaszeri [e], F. Ranjard [e], P. Rensing [e,12], G. Rizzo [e], L. Rolandi [e], D. Schlatter [e], M. Schmitt [e], O. Schneider [e], W. Tejessy [e], I.R. Tomalin [e], H. Wachsmuth [e], A. Wagner [e], Z. Ajaltouni [f], A. Barrès [f], C. Boyer [f], A. Falvard [f], C. Ferdi [f], P. Gay [f], C. Guicheney [f], P. Henrard [f], J. Jousset [f], B. Michel [f], S. Monteil [f], J-C. Montret [f], D. Pallin [f], P. Perret [f], F. Podlyski [f], J. Proriol [f], P. Rosnet [f], J.-M. Rossignol [f], T. Fearnley [g], J.B. Hansen [g], J.D. Hansen [g], J.R. Hansen [g], P.H. Hansen [g], B.S. Nilsson [g], B. Rensch [g], A. Wäänänen [g], G. Daskalakis [h], A. Kyriakis [h], C. Markou [h], E. Simopoulou [h], I. Siotis [h], A. Vayaki [h], A. Blondel [i], G. Bonneaud [i], J.C. Brient [i], P. Bourdon [i], A. Rougé [i], M. Rumpf [i], A. Valassi [i,6], M. Verderi [i], H. Videau [i], D.J. Candlin [j], M.I. Parsons [j], E. Focardi [k], G. Parrini [k], K. Zachariadou [k], M. Corden [ℓ], C. Georgiopoulos [ℓ], D.E. Jaffe [ℓ], A. Antonelli [m], G. Bencivenni [m], G. Bologna [m,4], F. Bossi [m], P. Campana [m], G. Capon [m], D. Casper [m], V. Chiarella [m], G. Felici [m], P. Laurelli [m], G. Mannocchi [m,5], F. Murtas [m], G.P. Murtas [m], L. Passalacqua [m], M. Pepe-Altarelli [m], L. Curtis [n], S.J. Dorris [n], A.W. Halley [n], I.G. Knowles [n], J.G. Lynch [n], V. O'Shea [n], C. Raine [n], J.M. Scarr [n], K. Smith [n], P. Teixeira-Dias [n], A.S. Thompson [n], E. Thomson [n], F. Thomson [n], R.M. Turnbull [n], C. Geweniger [o], G. Graefe [o], P. Hanke [o], G. Hansper [o], V. Hepp [o], E.E. Kluge [o], A. Putzer [o], M. Schmidt [o], J. Sommer [o], K. Tittel [o], S. Werner [o], M. Wunsch [o], R. Beuselinck [p], D.M. Binnie [p], W. Cameron [p], P.J. Dornan [p], M. Girone [p], S. Goodsir [p], E.B. Martin [p],

PII S0370-2693(97)00407-3

A. Moutoussi[p], J. Nash[p], J.K. Sedgbeer[p], A.M. Stacey[p], M.D. Williams[p], G. Dissertori[q], V.M. Ghete[q], P. Girtler[q], D. Kuhn[q], G. Rudolph[q], A.P. Betteridge[r], C.K. Bowdery[r], P. Colrain[r], G. Crawford[r], A.J. Finch[r], F. Foster[r], G. Hughes[r], T. Sloan[r], M.I. Williams[r], A. Galla[s], I. Giehl[s], A.M. Greene[s], C. Hoffmann[s], K. Jakobs[s], K. Kleinknecht[s], G. Quast[s], B. Renk[s], E. Rohne[s], H.-G. Sander[s], P. van Gemmeren[s], C. Zeitnitz[s], J.J. Aubert[t], C. Benchouk[t], A. Bonissent[t], G. Bujosa[t], D. Calvet[t], J. Carr[t], P. Coyle[t], C. Diaconu[t], F. Etienne[t], N. Konstantinidis[t], O. Leroy[t], F. Motsch[t], P. Payre[t], D. Rousseau[t], M. Talby[t], A. Sadouki[t], M. Thulasidas[t], K. Trabelsi[t], M. Aleppo[u], F. Ragusa[u,2], R. Berlich[v], W. Blum[v], D. Brown[v], V. Büscher[v], H. Dietl[v], F. Dydak[v,2], G. Ganis[v], C. Gotzhein[v], H. Kroha[v], G. Lütjens[v], G. Lutz[v], W. Männer[v], H.-G. Moser[v], R. Richter[v], A. Rosado-Schlosser[v], S. Schael[v], R. Settles[v], H. Seywerd[v], R. St. Denis[v], H. Stenzel[v], W. Wiedenmann[v], G. Wolf[v], J. Boucrot[w], O. Callot[w,2], S. Chen[w], Y. Choi[w,21], A. Cordier[w], M. Davier[w], L. Duflot[w], J.-F. Grivaz[w], Ph. Heusse[w], A. Höcker[w], A. Jacholkowska[w], M. Jacquet[w], D.W. Kim[w,24], F. Le Diberder[w], J. Lefrançois[w], A.-M. Lutz[w], I. Nikolic[w], M.-H. Schune[w], S. Simion[w], E. Tournefier[w], J.-J. Veillet[w], I. Videau[w], D. Zerwas[w], P. Azzurri[x], G. Bagliesi[x], G. Batignani[x], S. Bettarini[x], C. Bozzi[x], G. Calderini[x], M. Carpinelli[x], M.A. Ciocci[x], V. Ciulli[x], R. Dell'Orso[x], R. Fantechi[x], I. Ferrante[x], L. Foà[x,1], F. Forti[x], A. Giassi[x], M.A. Giorgi[x], A. Gregorio[x], F. Ligabue[x], A. Lusiani[x], P.S. Marrocchesi[x], A. Messineo[x], F. Palla[x], G. Sanguinetti[x], A. Sciabà[x], P. Spagnolo[x], J. Steinberger[x], R. Tenchini[x], G. Tonelli[x,19], C. Vannini[x], A. Venturi[x], P.G. Verdini[x], G.A. Blair[y], L.M. Bryant[y], J.T. Chambers[y], Y. Gao[y], M.G. Green[y], T. Medcalf[y], P. Perrodo[y], J.A. Strong[y], J.H. von Wimmersperg-Toeller[y], D.R. Botterill[z], R.W. Clifft[z], T.R. Edgecock[z], S. Haywood[z], P. Maley[z], P.R. Norton[z], J.C. Thompson[z], A.E. Wright[z], B. Bloch-Devaux[aa], P. Colas[aa], S. Emery[aa], W. Kozanecki[aa], E. Lançon[aa], M.C. Lemaire[aa], E. Locci[aa], P. Perez[aa], J. Rander[aa], J.-F. Renardy[aa], A. Roussarie[aa], J.-P. Schuller[aa], J. Schwindling[aa], A. Trabelsi[aa], B. Vallage[aa], S.N. Black[ab], J.H. Dann[ab], R.P. Johnson[ab], H.Y. Kim[ab], A.M. Litke[ab], M.A. McNeil[ab], G. Taylor[ab], C.N. Booth[ac], R. Boswell[ac], C.A.J. Brew[ac], S. Cartwright[ac], F. Combley[ac], M.S. Kelly[ac], M. Lehto[ac], W.M. Newton[ac], J. Reeve[ac], L.F. Thompson[ac], A. Böhrer[ad], S. Brandt[ad], G. Cowan[ad], C. Grupen[ad], P. Saraiva[ad], L. Smolik[ad], F. Stephan[ad], M. Apollonio[ae], L. Bosisio[ae], R. Della Marina[ae], G. Giannini[ae], B. Gobbo[ae], G. Musolino[ae], J. Rothberg[af], S. Wasserbaech[af], S.R. Armstrong[ag], E. Charles[ag], P. Elmer[ag], D.P.S. Ferguson[ag], Y.S. Gao[ag,23], S. González[ag], T.C. Greening[ag], O.J. Hayes[ag], H. Hu[ag], S. Jin[ag], P.A. McNamara III[ag], J.M. Nachtman[ag], J. Nielsen[ag], W. Orejudos[ag], Y.B. Pan[ag], Y. Saadi[ag], I.J. Scott[ag], J. Walsh[ag], Sau Lan Wu[ag], X. Wu[ag], J.M. Yamartino[ag], G. Zobernig[ag]

[a] *Laboratoire de Physique des Particules (LAPP), IN2P3-CNRS, 74019 Annecy-le-Vieux Cedex, France*
[b] *Institut de Fisica d'Altes Energies, Universitat Autonoma de Barcelona, 08193 Bellaterra (Barcelona), Spain*[7]
[c] *Dipartimento di Fisica, INFN Sezione di Bari, 70126 Bari, Italy*
[d] *Institute of High-Energy Physics, Academia Sinica, Beijing, People's Republic of China*[8]
[e] *European Laboratory for Particle Physics (CERN), 1211 Geneva 23, Switzerland*
[f] *Laboratoire de Physique Corpusculaire, Université Blaise Pascal, IN2P3-CNRS, Clermont-Ferrand, 63177 Aubière, France*

[g] *Niels Bohr Institute, 2100 Copenhagen, Denmark*[9]
[h] *Nuclear Research Center Demokritos (NRCD), Athens, Greece*
[i] *Laboratoire de Physique Nucléaire et des Hautes Energies, Ecole Polytechnique, IN2P3-CNRS, 91128 Palaiseau Cedex, France*
[j] *Department of Physics, University of Edinburgh, Edinburgh EH9 3JZ, United Kingdom*[10]
[k] *Dipartimento di Fisica, Università di Firenze, INFN Sezione di Firenze, 50125 Firenze, Italy*
[ℓ] *Supercomputer Computations Research Institute, Florida State University, Tallahassee, FL 32306-4052, USA*[13,14]
[m] *Laboratori Nazionali dell'INFN (LNF-INFN), 00044 Frascati, Italy*
[n] *Department of Physics and Astronomy, University of Glasgow, Glasgow G12 8QQ, United Kingdom*[10]
[o] *Institut für Hochenergiephysik, Universität Heidelberg, 69120 Heidelberg, Germany*[16]
[p] *Department of Physics, Imperial College, London SW7 2BZ, United Kingdom*[10]
[q] *Institut für Experimentalphysik, Universität Innsbruck, 6020 Innsbruck, Austria*[18]
[r] *Department of Physics, University of Lancaster, Lancaster LA1 4YB, United Kingdom*[10]
[s] *Institut für Physik, Universität Mainz, 55099 Mainz, Germany*[16]
[t] *Centre de Physique des Particules, Faculté des Sciences de Luminy, IN2P3-CNRS, 13288 Marseille, France*
[u] *Dipartimento di Fisica, Università di Milano e INFN Sezione di Milano, 20133 Milano, Italy*
[v] *Max-Planck-Institut für Physik, Werner-Heisenberg-Institut, 80805 München, Germany*[16]
[w] *Laboratoire de l'Accélérateur Linéaire, Université de Paris-Sud, IN2P3-CNRS, 91405 Orsay Cedex, France*
[x] *Dipartimento di Fisica dell'Università, INFN Sezione di Pisa, e Scuola Normale Superiore, 56010 Pisa, Italy*
[y] *Department of Physics, Royal Holloway & Bedford New College, University of London, Surrey TW20 OEX, United Kingdom*[10]
[z] *Particle Physics Dept., Rutherford Appleton Laboratory, Chilton, Didcot, Oxon OX11 OQX, United Kingdom*[10]
[aa] *CEA, DAPNIA/Service de Physique des Particules, CE-Saclay, 91191 Gif-sur-Yvette Cedex, France*[17]
[ab] *Institute for Particle Physics, University of California at Santa Cruz, Santa Cruz, CA 95064, USA*[22]
[ac] *Department of Physics, University of Sheffield, Sheffield S3 7RH, United Kingdom*[10]
[ad] *Fachbereich Physik, Universität Siegen, 57068 Siegen, Germany*[16]
[ae] *Dipartimento di Fisica, Università di Trieste e INFN Sezione di Trieste, 34127 Trieste, Italy*
[af] *Experimental Elementary Particle Physics, University of Washington, WA 98195 Seattle, USA*
[ag] *Department of Physics, University of Wisconsin, Madison, WI 53706, USA*[11]

Received 20 February 1997
Editor: K. Winter

Abstract

A measurement of R_b using five mutually exclusive hemisphere tags has been performed by ALEPH using the full LEP1 statistics. Three tags are designed to select the decay of the Z^0 to b quarks, while the remaining two select Z^0 decays to c and light quarks, and are used to measure the tagging efficiencies. The result, $R_b = 0.2159 \pm 0.0009(\text{stat}) \pm 0.0011(\text{syst})$, is in agreement with the electroweak theory prediction of 0.2158 ± 0.0003.

[1] Now at CERN, 1211 Geneva 23, Switzerland.
[2] Also at CERN, 1211 Geneva 23, Switzerland.
[3] Also at Dipartimento di Fisica, INFN, Sezione di Catania, Catania, Italy.
[4] Also Istituto di Fisica Generale, Università di Torino, Torino, Italy.
[5] Also Istituto di Cosmo-Geofisica del C.N.R., Torino, Italy.
[6] Supported by the Commission of the European Communities, contract ERBCHBICT941234.
[7] Supported by CICYT, Spain.
[8] Supported by the National Science Foundation of China.
[9] Supported by the Danish Natural Science Research Council.
[10] Supported by the UK Particle Physics and Astronomy Research Council.
[11] Supported by the US Department of Energy, grant DE-FG0295-ER40896.
[12] Now at Dragon Systems, Newton, MA 02160, USA.
[13] Supported by the US Department of Energy, contract DE-FG05-92ER40742.
[14] Supported by the US Department of Energy, contract DE-FC05-85ER250000.
[15] Permanent address: Universitat de Barcelona, 08208 Barcelona, Spain.
[16] Supported by the Bundesministerium für Bildung, Wissenschaft, Forschung und Technologie, Germany.
[17] Supported by the Direction des Sciences de la Matière, C.E.A.
[18] Supported by Fonds zur Förderung der wissenschaftlichen Forschung, Austria.
[19] Also at Istituto di Matematica e Fisica, Università di Sassari,

1. Introduction

The foregoing measurement [1] of R_b, the fraction of hadronic Z^0 decays to b quarks, is based on hemisphere b quark selection by means of one tag which utilises lifetime and mass information. In the present analysis, this lifetime-mass hemisphere tag is complemented by four other mutually exclusive tags, using event shape information as well as lifetime and leptons. Mutually exclusive here means that the tags are constructed such that a hemisphere will be tagged at most by one tag. In this way it has been possible to increase the statistical accuracy as well as to reduce the systematic uncertainty.

Three of the five tags are designed to tag b events, one is designed to select c events, and one designed to select the combination of the three lighter quarks, u, d and s, together. The b tags include the lifetime-mass tag of [1], a tag that uses both lifetime and event shape information, and a tag based on the identification of leptons with large momenta and transverse momenta. The lifetime-mass tag has the highest purity and the largest impact on the analysis.

The five mutually exclusive hemisphere tags result in 20 statistically independent measurements: 5 singly tagged fractions, 5 doubly tagged with the same tag, and 10 doubly tagged with different tags. These are used in the present analysis to determine 14 quantities: R_b and 13 of the 15 efficiencies of the 5 tags for b, for c and for the combination of u, d, s flavours. The two background efficiencies of the lifetime-mass tag cannot be determined experimentally with success. There remain six constraints, which serve as a check on the analysis.

Monte Carlo simulation is used to calculate the two lifetime-mass tag background efficiencies and the correlations in the tagging of the two hemispheres in the same event. These contribute the dominant systematic uncertainties of the analysis.

Sassari, Italy.

[20] Now at Princeton University, Princeton, NJ 08544, USA.

[21] Permanent address: Sung Kyun Kwan University, Suwon, Korea.

[22] Supported by the US Department of Energy, grant DE-FG03-92ER40689.

[23] Now at Harvard University, Cambridge, MA 02138, USA.

[24] Permanent address: Kangnung National University, Kangnung, Korea.

2. The method

Events are divided into hemispheres by the plane perpendicular to the thrust axis.

The fraction of tagged hemispheres f_s^I with tag I is

$$f_s^I = R_b\epsilon_b^I + R_c\epsilon_c^I + (1 - R_b - R_c)\epsilon_x^I$$

where R_b and R_c are the b and c branching fractions of hadronic Z^0 decays and ϵ_a^I are the hemisphere tagging efficiencies for flavour a using tag I. ϵ_x is the average efficiency for u, d and s flavours. For five tags, there are $3 \times 5 = 15$ efficiencies.

The fraction of doubly tagged events $f_d^{I,J}$ with tags I and J is

$$f_d^{I,J} = \left[R_b\epsilon_b^I\epsilon_b^J(1+\rho_b^{I,J}) + R_c\epsilon_c^I\epsilon_c^J(1+\rho_c^{I,J}) + (1 - R_b - R_c)\epsilon_x^I\epsilon_x^J(1+\rho_x^{I,J})\right](2-\delta_{I,J})$$

where $\rho_a^{I,J}$ are the hemisphere-hemisphere efficiency correlations for flavour a and tags I and J. The effects on the correlations due to the small differences in efficiencies between u, d and s flavours are taken into account in the calculation of $\rho_x^{I,J}$. There are $3 \times 15 = 45$ correlations.

The 20 independent measured quantities consist of the 5 numbers of singly tagged events, N_s^I

$$N_s^I = N\left(2f_s^I - \sum_{K=1}^{5} f_d^{I,K}\cdot(1+\delta_{I,K})\right)$$

where N is the total number of events, and the 15 numbers of doubly tagged events $N_d^{I,J}$

$$N_d^{I,J} = Nf_d^{I,J}.$$

These 20 measurements are described by 62 unknown parameters: R_b, R_c, the 15 efficiencies of the five tags for the three flavours, and the 45 correlations of the 15 pairs of tags for the three flavours. In the following analysis R_b and 13 efficiencies are fitted to the data. The remaining two efficiencies of the lifetime-mass tag, ϵ_c^Q and ϵ_x^Q, and the 45 correlations are calculated using Monte Carlo simulation. The systematic error reflects the uncertainties in these calculations. The result will be given as a function of the difference between R_c and its value of 0.172 in the electroweak theory.

3. Event selection

The data used for the analysis were obtained on and near the Z^0 resonance in the ALEPH detector [2] during the period 1992 to 1995, since the introduction of a double sided microstrip vertex detector, with strip readout in $r - \phi$ and z.

Events are selected as in [1] except for the following requirements:

a) $|\cos\theta_{\mathrm{THRUST}}| < 0.65$, where θ_{THRUST} is the angle between the beam and the thrust axes.

b) $y_3 < 0.2$, where y_3 is the value of y_{cut} that sets the transition from 2 to 3 jets using the JADE algorithm [3]. This cut eliminates the 3% of events with the largest gluon radiation, for which the correspondence between data and Monte Carlo is poorer.

There remain 2 057 618 events. The Monte Carlo simulation is used to determine a selection bias in favour of b quarks relative to the lighter quarks of $0.1 \pm 0.1\%$, where the error is dominated by statistics, and a contamination from tau events of $0.30 \pm 0.01\%$, where the error is dominated by systematics.

4. The five hemisphere tags

The tags are designed in a pragmatic attempt to isolate the desired quark flavour with high efficiency and purity while keeping the hemisphere-hemisphere correlations small. The latter is accomplished by deriving the tags from hemisphere quantities exclusively. In particular, the primary (Z^0 decay) vertex is reconstructed independently in the two hemispheres, as described in [1].

The five tags are constructed from the following eight derived hemisphere quantities, of which the first two are described in greater detail in [1]:

(i) $\mathcal{P}_H$, the confidence level that all of the hemisphere tracks originate from the primary hemisphere vertex.

(ii) μ_H, the variable related to the invariant mass of the tracks inconsistent with originating from the primary vertex. Tracks in a hemisphere are ordered inversely to their probability $\mathcal{P}_T$ to originate from the primary vertex. Tracks are combined, in this order, until the invariant mass of the combination exceeds 1.8 GeV/c^2. The quantity μ_H is the $\mathcal{P}_T$ of the last track added.

(iii) $\mathcal{N}_B$, the output of a neural network [4] trained to select b quark hemispheres. The input quantities to the neural network are 25 event shape quantities, of which none depend explicitly on b lifetime effects. These inputs are listed in the Appendix.

(iv) p, the momentum of an identified electron or muon. Lepton identification is described in [5]: electrons are primarily identified by their characteristic shower development in the calorimeter, and muons are primarily identified by their penetration pattern. If more than one lepton is found in a hemisphere the highest momentum lepton is used.

(v) $p_\perp$, the transverse momentum of the lepton with respect to the direction of its jet after removing the lepton from the jet.

(vi) $\mathcal{N}_C$, the output of a neural network trained to select c quark hemispheres. The neural net inputs are one lifetime and 19 event shape quantities, which are given in the Appendix.

(vii) $\mathcal{P}_+$, a variable used together with $\mathcal{P}_-$, to select c quark hemispheres. Hemisphere tracks are divided into two groups on the basis of rapidity with respect to their associated jet axes greater or less than 5.1, chosen so that b hemispheres find equal numbers of tracks in the two groups. $\mathcal{P}_+$ is the confidence level that the tracks in the higher rapidity group originate from the primary vertex.

(viii) $\mathcal{P}_-$, the confidence level that the tracks in the lower rapidity group originate from the primary vertex.

Distributions of these eight variables are shown in Figs. 1 and 2, where the Monte Carlo distributions are given for b, c and x, together with a comparison of the flavour-combined Monte Carlo distributions with the data. Disagreements exist between the data and the Monte Carlo simulation, particularly for $\mathcal{N}_C$. The analysis is insensitive to such disagreements as they affect the efficiencies. The inadequacies of the Monte Carlo simulation as concerns the correlations are discussed in Section 6.3.

The definitions of the five tags in terms of the eight variables are given below. In order to satisfy the exclusive tag requirement each tag is given a priority, and if a hemisphere satisfies more than one tag, it is assigned to the tag with the highest priority. The lifetime-mass

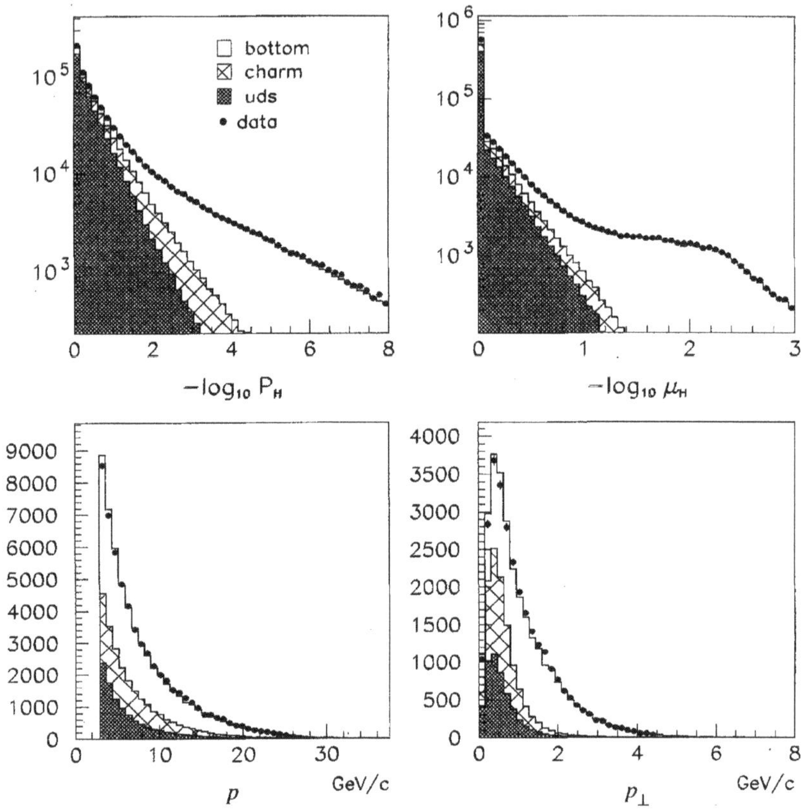

Fig. 1. Comparison of data (points) and Monte Carlo simulation (histogram) for tagging variables $\mathcal{P}_H$, μ_H, p, and $p_\perp$. The vertical axis is the number of hemispheres in the data per bin. The p, and $p_\perp$ spectra are not reweighted using the latest measurements of the semileptonic branching ratios. Quark flavour contributions in the simulation are indicated by the shaded regions.

tag variable is given by

$$\mathcal{B}_{\text{lm}} = -(0.7 \log_{10} \mu_H + 0.3 \log_{10} \mathcal{P}_H)$$

and the five tags are:

Q tag: $\mathcal{B}_{\text{lm}} > 2.2$
Priority 1

S tag: $0.85 < \mathcal{B}_{\text{lm}} < 2.2$ and $\mathcal{N}_B > (1.05 - 0.2\,\mathcal{B}_{\text{lm}})$
Priority 2

L tag: $p > 3$ GeV/c and $p_\perp > 1.4$ GeV/c
Priority 5

C tag: $\mathcal{P}_- > 0.07$ and $0.0003 < \mathcal{P}_+ < 0.3$
and $\mathcal{B}_{\text{lm}} < 1.5$ and $\mathcal{N}_C > 0.68$
Priority 3

X tag: $\log_{10} \mathcal{P}_H > -0.25$ and $\mathcal{N}_B < 0$
Priority 4

The Monte Carlo expectations for the 15 efficiencies are given in Table 1. These are a measure of the performance of the tags. In the R_b determination only the charm and light quark efficiencies of the Q tag are taken from Monte Carlo simulation. The tighter cut on the Q tag with respect to the foregoing paper [1] results in a reduction in the background charm and u, d, s efficiencies of nearly a factor of two.

The Monte Carlo simulation is used to determine

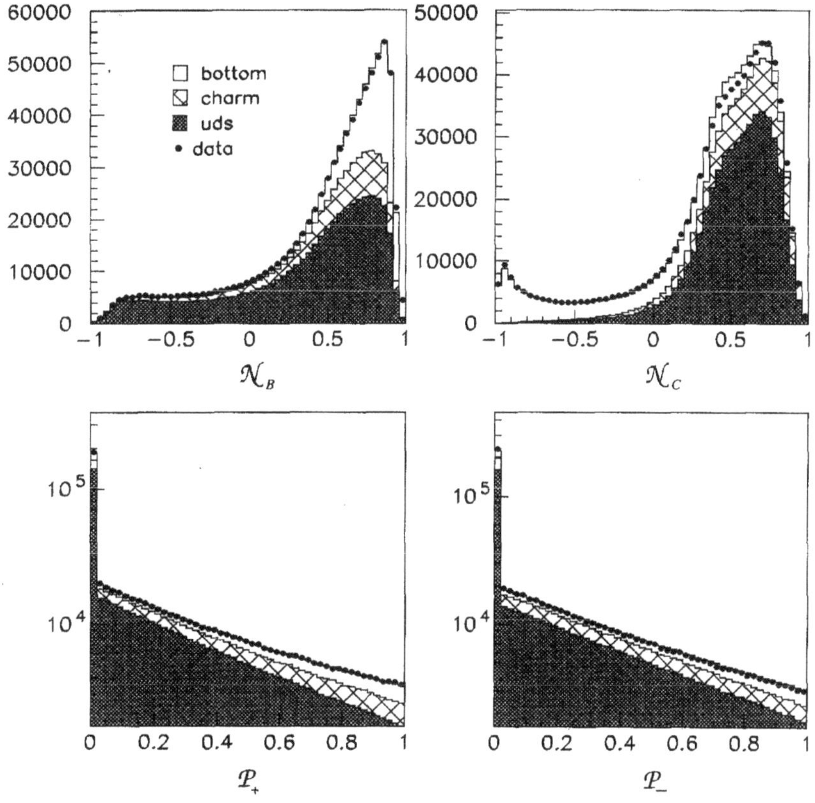

Fig. 2. Comparison of data (points) and Monte Carlo simulation (histogram) for tagging variables $\mathcal{N}_B$, $\mathcal{N}_C$, $\mathcal{P}_+$, and $\mathcal{P}_-$. The vertical axis is the number of hemispheres in the data per bin. Quark flavour contributions in the simulation are indicated by the shaded regions.

Table 1
Monte Carlo results for the tagging efficiencies.

Tag	ϵ_{uds}	ϵ_c	ϵ_b
Q	0.00043	0.00216	0.1957
S	0.00204	0.01402	0.1759
L	0.00158	0.00694	0.0425
C	0.07927	0.16193	0.0260
X	0.11686	0.03962	0.0022

the hemisphere-hemisphere correlations for all tags. The impact on R_b of a particular correlation is given by

$$\frac{\Delta R_b}{R_b \Delta \rho_a^{I,J}},$$

that is, the relative uncertainty in R_b is the impact times the uncertainty in the correlation. The impacts vary a great deal, so that only about 10 of the 45 correlations are significant to the analysis. The Monte Carlo expectations for the dominant correlations and their impacts are given in Table 2. For comparison, in the single tag analysis [1] the impact of the single correlation considered is unity and the impacts of the other two correlations are negligible.

5. Results

The basic experimental result consists of 20 measurements: the numbers of singly and doubly tagged events. These are given in Table 3. The fit of R_b and

Table 2
Monte Carlo results for the dominant correlations with statistical errors and their impacts on R_b.

Correlation	Value	Impact
$\rho_b^{Q,Q}$	0.0274 ± 0.0033	0.450
$\rho_b^{Q,S}$	-0.0022 ± 0.0025	0.420
$\rho_b^{Q,L}$	-0.0176 ± 0.0050	0.084
$\rho_b^{Q,C}$	0.0246 ± 0.0072	-0.084
$\rho_x^{X,C}$	0.0964 ± 0.0049	-0.070
$\rho_b^{S,C}$	0.0226 ± 0.0076	0.064
$\rho_x^{C,C}$	0.0718 ± 0.0085	-0.045
$\rho_c^{C,C}$	0.0872 ± 0.0075	-0.041
$\rho_b^{S,S}$	-0.0060 ± 0.0036	0.026
$\rho_c^{S,C}$	-0.0085 ± 0.0179	0.021
$\rho_x^{X,X}$	0.1335 ± 0.0056	-0.016
$\rho_c^{X,C}$	0.0984 ± 0.0120	-0.012
$\rho_b^{L,C}$	0.0076 ± 0.0168	0.010

Table 3
Measured numbers of singly and doubly tagged events.

$N_Q = 96504$	$N_{QQ} = 16715$
$N_S = 100266$	$N_{SS} = 13980$
$N_L = 33973$	$N_{LL} = 1097$
$N_X = 243002$	$N = 18947$
$N_C = 265994$	$N_{CC} = 20123$
$N_{QS} = 30138$	$N_{QL} = 8098$
$N_{QX} = 504$	$N_{QC} = 4331$
$N_{SL} = 7534$	$N_{SX} = 1480$
$N_{SC} = 5838$	$N_{LX} = 1079$
$N_{LC} = 2570$	$N_{XC} = 32293$

13 of the 15 efficiencies to these 20 data points gives the result

$$R_b = 0.21585 \pm 0.00087$$

with a χ^2 of 8.1 for the 6 degrees of freedom. The error is the statistical error. This result has been corrected for the event selection bias and tau contamination discussed in Section 3. In addition, a correction of $+0.0003$ has been applied to remove the contribution of the photon propagator.

The fitted efficiencies can be compared with the Monte Carlo simulation predictions and are given together in Table 4.

Table 4
Tagging efficiencies as predicted by Monte Carlo simulation and the fitted efficiencies with statistical errors. For a complete comparison of the fit results with the simulation predictions an estimate of the systematic error must be included.

Efficiency	Simulation prediction	Fit result
ϵ_b^Q	0.1957	0.1922 ± 0.0008
ϵ_b^S	0.1759	0.1769 ± 0.0006
ϵ_c^S	0.0140	0.0128 ± 0.0012
ϵ_x^S	0.0020	0.0028 ± 0.0002
ϵ_b^L	0.0425	0.0486 ± 0.0003
ϵ_c^L	0.0069	0.0083 ± 0.0007
ϵ_x^L	0.0016	0.0031 ± 0.0002
ϵ_b^C	0.0260	0.0233 ± 0.0004
ϵ_c^C	0.1619	0.1592 ± 0.0020
ϵ_x^C	0.0793	0.0866 ± 0.0005
ϵ_b^X	0.0022	0.0021 ± 0.0001
ϵ_c^X	0.0396	0.0418 ± 0.0005
ϵ_x^X	0.1169	0.1129 ± 0.0006

6. Systematic errors

The systematic errors arise only through the quantities calculated using Monte Carlo simulation: the uncertainty due to event selection, $\Delta R_b = 0.00017$, the correlations and the two Q tag background efficiencies. The impacts on R_b of the correlations are listed in Table 2, and the impacts of the two Q tag background efficiencies are given by

$$\frac{\Delta R_b}{R_b}\frac{\epsilon_b^Q}{\Delta\epsilon_c^Q} = -1.5, \quad \frac{\Delta R_b}{R_b}\frac{\epsilon_b^Q}{\Delta\epsilon_x^Q} = -5.4.$$

This section describes uncertainties entering via these Monte Carlo quantities. The total uncertainty due to the limited Monte Carlo statistics is $\Delta R_b = 0.00047$.

The task of estimating the systematic errors is essentially the same as described in [1], with the following observations:

(i) The two Q tag background efficiencies enter with the same impacts in the two analyses. However, because of the harder cut ($\mathcal{B}_{\text{lm}} > 2.2$ compared with $\mathcal{B}_{\text{lm}} > 1.9$) the background efficiencies relative to the b efficiency are reduced by almost a factor of two, with a reduction in systematic error.

(ii) The impact of $\rho_b^{Q,Q}$ is reduced by more than a factor of two, but other correlations also contribute substantially to the error in R_b.

Table 5
Uncertainties in Q tag efficiencies and in R_b due to uncertainties in detector simulation from the resolution on the impact parameter significance, S [1], and tracking efficiency.

	$\Delta\epsilon_c^Q$	$\Delta\epsilon_x^Q$	ΔR_b
θ dependence of S resolution	0.00002	< 0.00001	0.00003
p dependence of S resolution	0.00002	0.00001	0.00009
θ dependence of tracking efficiency	0.00001	< 0.00001	0.00002
p dependence of tracking efficiency	< 0.00001	< 0.00001	< 0.00001
Remaining inaccuracy of S resolution	0.00003	< 0.00001	0.00005
Track correlation in S resolution	0.00008	0.00005	0.00045
Total uncertainty	0.00009	0.00005	0.00046

6.1. Detector simulation uncertainty

Monte Carlo predictions for ϵ_c^Q and ϵ_x^Q depend on the assumed impact parameter resolution and efficiency for vertex detector hits to be associated to tracks. As described in [1], the Monte Carlo simulation is corrected to achieve better agreement with data in these tracking quantities, and a systematic uncertainty is assigned to the correction. The resulting uncertainties in ϵ_c^Q and ϵ_x^Q are given in Table 5.

6.2. Systematics from b and c physics uncertainties

Uncertainties in physical parameters that enter into the simulation result in uncertainties in ϵ_c^Q, ϵ_x^Q and the correlations.

These are calculated by varying the physics inputs to the Monte Carlo simulation within their allowed experimental ranges [6]. The procedure is described in Ref. [1].

Table 6 reports the errors in R_b due to these physical parameter variations. The overall error resulting from these uncertainties is $\Delta R_b = 0.00084$.

Table 6
Systematic errors in R_b due to uncertainties in charm and bottom physics.

Source	ΔR_b
Both b in the same hemisphere: $2.2 \pm 0.7\%$	$\mp$0.00011
$g \to c\bar{c}$: $(2.38 \pm 0.48)\%$ per event	$\mp$0.00043
Ratio $g \to b\bar{b}/g \to c\bar{c}$: 0.13 ± 0.04	$\mp$0.00054
$\langle x_E(g)\rangle = 0.76 \pm 0.03$	$\mp$0.00010
B fragmentation: 0.702 ± 0.008	$\mp$0.00010
B_s fraction: 0.112 ± 0.019	$\mp$0.00004
Λ_b fraction: 0.132 ± 0.041	$\mp$0.00007
B^+ lifetime: 1.62 ± 0.06 ps	$\mp$0.00005
B^0 lifetime: 1.56 ± 0.06 ps	$\mp$0.00005
B_s lifetime: 1.61 ± 0.10 ps	$\mp$0.00003
Λ_b lifetime: 1.14 ± 0.08 ps	$\mp$0.00003
B charged multiplicity: 5.73 ± 0.35	$\mp$0.00017
Charm fragmentation: 0.484 ± 0.008	$\pm$0.00011
D^+ lifetime: 1.057 ± 0.015 ps	$\mp$0.00001
D^0 lifetime: 0.415 ± 0.004 ps	$\mp$0.00001
D_s lifetime: 0.467 ± 0.017 ps	$\mp$0.00004
Λ_c lifetime: 0.206 ± 0.012 ps	$\mp$0.00001
$\mathcal{B}(D^+ \to K_S)$: 0.295 ± 0.035	$\pm$0.00009
$\mathcal{B}(D^0 \to K_S)$: 0.210 ± 0.025	$\pm$0.00009
$\mathcal{B}(D_s \to K_S)$: 0.195 ± 0.140	$\pm$0.00012
D^+ fraction: 0.233 ± 0.028	$\mp$0.00012
Λ_c fraction: 0.065 ± 0.029	$\pm$0.00007
D_s fraction: 0.102 ± 0.037	$\mp$0.00003
Charm charged multiplicity	$\mp$0.00025
Charm neutral multiplicity	$\mp$0.00016
V^0 rate ($\pm$10%) and eff. ($\pm$20%)	$\mp$0.00004
Total	0.00084

6.3. Correlation errors

Correlation errors are assigned following the same basic procedure as in [1]. Since the correlation now concerns events in which the two hemispheres may be tagged by different tags, I and J, the contribution to the correlation $\rho_f^{I,J}$ is

$$\rho_f^{I,J}(\mathrm{v}) = \frac{\int f_f(\mathrm{v}) \left[\epsilon_f^{I,\mathrm{same}}(\mathrm{v})\epsilon_f^{J,\mathrm{oppo}}(\mathrm{v}) + \epsilon_f^{J,\mathrm{same}}(\mathrm{v})\epsilon_f^{I,\mathrm{oppo}}(\mathrm{v})\right] d\mathrm{v}}{2\langle\epsilon_f^I\rangle\langle\epsilon_f^J\rangle} - 1$$

Table 7
Summary of differences in correlation contributions between data and Monte Carlo simulation. For each correlation, the table indicates the global value as given by simulation and the impact on R_b. For each of the four variables, the correlation contribution, the difference (ΔD) between the two data flavour isolation methods and the difference (D–MC) in contribution between the data and Monte Carlo simulation, are given for the 13 correlations of impact of 0. 1 or greater.

Corr.	MC Global Corr.	Impact	Jet momentum			y_3			$\cos\theta_{\text{THRUST}}$			ϕ_{THRUST}		
			contr.	ΔD	D–MC	contr.	ΔD	D–MC	contr.	ΔD	D–MC	contr.	ΔD	D–MC
						b correlations								
$\rho_b^{Q,Q}$	.0274	.450	.0103	.0012	−.0016	.0059	.0002	−.0005	.0132	.0001	−.0002	.0002	.0000	.0005
$\rho_b^{Q,S}$	−.0022	.420	.0040	.0009	.0003	.0037	.0002	.0003	.0012	.0000	−.0008	−.0002	.0000	.0000
$\rho_b^{Q,L}$	−.0176	.084	−.0095	−.0003	.0020	−.0028	−.0001	−.0001	−.0045	.0001	.0011	.0004	.0000	.0000
$\rho_b^{Q,C}$	.0246	.084	.0311	.0072	.0071	.0259	.0013	−.0006	−.0095	−.0007	−.0007	−.0002	−.0008	−.0012
$\rho_b^{S,C}$	.0226	.064	.0116	.0021	−.0002	.0166	.0015	.0029	−.0006	−.0003	.0003	.0001	.0001	.0001
$\rho_b^{S,S}$	−.0060	.026	.0004	.0007	.0013	.0024	.0002	.0005	.0002	.0000	−.0001	.0002	.0000	−.0001
$\rho_b^{L,C}$	.0076	.010	−.0435	−.0117	−.0213	−.0129	.0003	−.0010	.0029	.0001	.0004	.0000	−.0001	−.0003
						u, *d*, *s* correlations								
$\rho_x^{X,C}$	.0964	−.070	.0526	−.0017	−.0019	.0531	−.0007	−.0027	.0002	.0000	−.0007	−.0003	.0000	.0004
$\rho_x^{C,C}$	.0718	−.045	.0355	−.0018	−.0049	.0430	−.0008	−.0051	.0010	.0000	.0005	.0008	.0000	−.0007
$\rho_x^{X,X}$	.1335	−.016	.0775	−.0014	.0016	.0655	−.0005	.0014	.0001	.0000	.0001	.0001	.0000	−.0001
						c correlations								
$\rho_c^{C,C}$	.0872	−.041	.0540		.0031	.0654		−.0035	.0046		.0002	.0046		−.0006
$\rho_c^{S,C}$	−.0085	.021	.0057		.0047	.0021		−.0027	.0052		−.0009	.0045		.0000
$\rho_c^{X,C}$	.0984	−.012	.0701		.0033	.0738		−.0004	.0044		.0006	.0046		.0006
Impact weighted sum					**−.0010**			**.0006**			**−.0003**			**.0003**

where $f_f(\mathrm{v})$ is the fractional event distribution in variable v for flavour f, $\epsilon_f^{I,\text{same}}(\mathrm{v})$ is the corresponding efficiency for the tag I for the hemisphere in which v is measured, and $\epsilon_f^{J,\text{oppo}}(\mathrm{v})$ is the efficiency for the tag J on the side opposite to the one in which v is measured.

The errors are assigned as the differences between data and Monte Carlo multiplied by the relevant impacts on R_b, summed linearly for all correlations for a given variable, and in quadrature for the four chosen variables: jet momentum, y_3, $\cos\theta_{\text{THRUST}}$ and ϕ_{THRUST}. If, instead, the sum were made quadratically for each variable, taking the larger of the difference between data and Monte Carlo or its error, the resulting systematic error would be slightly smaller.

Comparison of data and Monte Carlo simulation for a particular flavour requires the isolation of this flavour in the data. For the flavour isolation the same procedure is used as in [1], resulting in two values for the data for the two different subtraction schemes (method 1 and method 2) which could be compared and whose average is used in the error assignment.

The isolation of the b flavour uses the requirement $\mathcal{B}_{\text{lm}} > 0.3$ for both hemispheres, with the resulting efficiencies 0.64, 0.18 and 0.03 for b, c and u, d, s flavours respectively.

For the isolation of the u, d, s flavours the requirements $-\log_{10}\mathcal{P}_H < 0.6$ and $\mathcal{P}_+ > 0.13$ are imposed on both hemispheres, with efficiencies 0.024, 0.22 and 0.54 for b, c and u, d, s flavours respectively.

No initial event selection requirement proved useful for purifying c flavour. As a consequence, the flavour isolation algorithm in which the unwanted flavours in the data are subtracted on the basis of the Monte Carlo (method 1) suffers from very large statistical errors and cannot be used. The errors for c flavour correlations are therefore assigned on the basis of the difference between the other method (method 2) and Monte Carlo.

Table 7 presents the results for the thirteen correlations with impact greater than 0.01. For each correlation, the overall value and impact are listed together with the contributions from the four variables studied. For each variable, the difference in background subtraction ΔD and the differences between data and Monte Carlo simulation are shown. The resultant uncertainty is $\Delta R_b = 0.00027$. The error in R_b due to the uncertainties in the remaining 32 correlations of very small impact is assigned by assuming $\pm 20\%$ uncertainties in each of these. The resultant error in R_b of ± 0.00005, combined with the errors due to the 13 larger impact correlations is finally $\Delta R_b = 0.00027$.

6.4. *Systematic error summary*

The systematic errors are summarised below:

$\Delta R_b = \pm 0.00047$ **Monte Carlo statistics**

± 0.00017 **Event selection**

± 0.00084 **Physics uncertainty**

± 0.00046 **Tracking uncertainty**

± 0.00027 **Hemisphere correlations uncertainty**

7. Discussion of the result

The systematic uncertainties evaluated in the previous section, added in quadrature, give a total systematic error of 0.00110 in R_b. The result presented in Section 5 becomes:

$$R_b = 0.2159 \pm 0.0009(\text{stat}) \pm 0.0011(\text{syst}) - 0.019 \times (R_c - 0.172)$$

where the explicit dependence on R_c is given.

Fig. 3 shows the stability of the result as a function of the cut on $\mathcal{N}_C$ of the C tag as well as a function of the $\mathcal{N}_B$ cut of the S tag. Fig. 4 shows the stability of the result with respect to the Q tag cut, together with the contributions to the error.

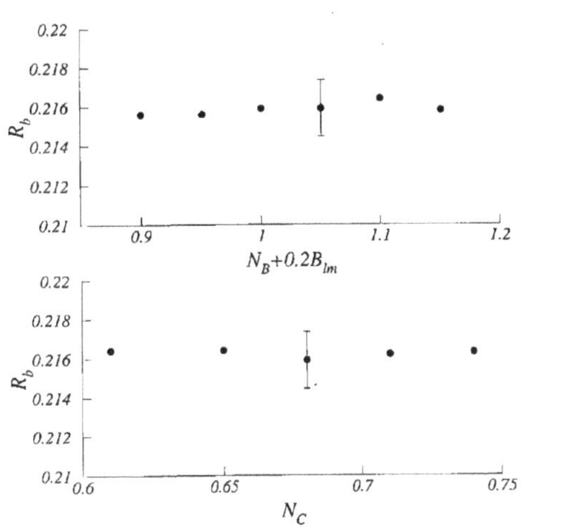

Fig. 3. R_b results as a function of the $\mathcal{N}_B$ cut of the S Tag (upper plot) and as a function of the $\mathcal{N}_C$ cut of the C Tag (lower plot).

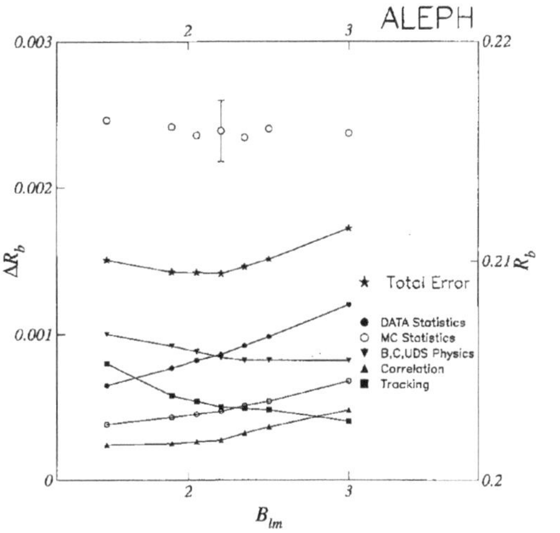

Fig. 4. R_b results as a function of the $\mathcal{B}_{lm}$ cut (upper plot) plotted against the right ordinate; the error bar is the total one. On the left ordinate the different sources of errors are plotted together with the total error as a function of the $\mathcal{B}_{lm}$ cut.

The present analysis relies heavily on the lifetime-mass tag which is the basis of the preceding paper [1]. The result of this single tag analysis is

$$R_b = 0.2167 \pm 0.0011(\text{stat}) \pm 0.0013(\text{syst}) - 0.037 \times (R_c - 0.172).$$

That result is highly correlated to the present one, both in the data used and in the systematic errors, so the two cannot be used independently. They are statistically consistent. The result presented here has the smaller error, and is therefore taken as the final result.

The result is in good agreement with the current expectation of Electroweak theory [7] as predicted for a top mass of 175 ± 6 GeV/c^2 [8]

$$R_b = 0.2158 \pm 0.0003 \quad \text{(Electroweak expectation)}.$$

If the radiative corrections, which are dominated by top quark effects, were left out of the electroweak calculation, the expected result would be

$$R_b = 0.2183 \pm 0.0001$$

(Electroweak expectation without rad. corr.).

This measurement is an indication of the top quark dominated radiative vertex correction.

Other measurements of R_b have been reported recently by SLD [9], DELPHI [10] and OPAL [11]. The more precise result, the one of OPAL, is $R_b = 0.2175 \pm 0.0014(\text{stat}) \pm 0.0017(\text{syst}) - 0.106 \times (R_c - 0.172)$.

Acknowledgements

We are indebted to our colleagues in the accelerator divisions for the good performance of the LEP storage ring. We thank also the engineers and technicians of our home institutes for their support in constructing ALEPH. Those of us from non-member countries thank CERN for its hospitality.

Appendix A

A.1. Neural network input variables for $\mathcal{N}_B$

- B(1): Longitudinal momentum $p_{\parallel}$ of the leading track of the most energetic jet J_{max} of the hemisphere.
- B(2): Fox-Wolfram moment of order 5 normalized to the 0th moment [12].
- B(3): Transverse mass with respect to the plane of the event

$$B(3) = \frac{\sum_i |p(i) \cos\theta_{i3}| \sqrt{s}}{E_{\text{Hemi}}}$$

where the subscript 3 refers to the minor axis (axis perpendicular to the plane of the event), i runs over all the energy flow tracks of the hemisphere (charged tracks, photons and neutral hadrons) and E_{Hemi} is the visible energy of the hemisphere.
- B(4): Invariant mass of the three most energetic tracks of J_{max}.
- B(5): "Forward Momentum": tracks of the most energetic jet of the hemisphere are boosted along its axis J_{max} assuming that the B hadrons carry on average 70% of the beam energy, and the total momentum of the tracks produced in a forward direction with respect to J_{max} is computed.
- B(6): Transverse momentum $p_{\perp}$ of the second leading track of J_{max}.
- B(7): Same as B(6) for the leading track of J_{max}.
- B(8): Multiplicity of the energy flow tracks of J_{max} with $p > 1$ GeV/c.
- B(9): Invariant mass of the charged tracks of J_{max} with $p > 1$ GeV/c.
- B(10): Invariant mass of the energy-flow tracks belonging to a cone of $\pm 40°$ around the axis of J_{max} (denoted C_{40}^{Jet} hereafter).
- B(11): Sum of the squared transverse momenta of the particles belonging to the cone C_{40}^{Jet}.
- B(12): Energy of C_{40}^{Jet} divided by E_{Hemi}.
- B(13): Energy of the leading track of the hemisphere divided by E_{Hemi}.
- B(14): Invariant mass of the energy-flow tracks belonging to a cone of $\pm 40°$ around the leading track of the hemisphere (denoted C_{40}^{track} hereafter).
- B(15): Energy of C_{40}^{track} divided by E_{Hemi}.
- B(16): "Directed Sphericity" [13] defined as $\sum_i p_{\perp}(i)^2 / \sum_i |P(i)|^2$, where i runs over a set of energy flow tracks of J_{max}, $p_{\perp}(i)$ is the transverse momentum of the track i w.r.t. its jet axis and $P(i)$ is the momentum of i estimated in the centre of mass of the set of considered tracks. This variable is an attempt to reflect the fact that the decay of a B-hadron in its rest frame is more isotropic than for light hadrons. Here the set of tracks used is the first, second and third most energetic energy flow tracks of J_{max}.

- B(17): Same as B(16) for the first, second and fourth tracks of J_{max}.
- B(18): Same as B(16) for the first and third tracks of J_{max}.
- B(19): Same as B(16) for the first and fourth tracks of J_{max}.
- B(20): Same as B(4) for the first and third tracks of J_{max}.
- B(21): Energy (in the laboratory frame) of the set of tracks used in B(16).
- B(22): Energy of the set of tracks used in B(17).
- B(23): Energy of the first and second most energetic tracks of J_{max}.
- B(24): Energy of the set of tracks used in B(19).
- B(25): Energy of the system of tracks obtained with a nucleated jet algorithm starting from the most energetic track of the hemisphere and stopping the nucleation when the invariant mass exceeds 2.1 GeV/c^2 [14]. This variable is intended to reproduce in an inclusive way the $\langle X_E \rangle$ of D mesons (which is different for D mesons produced in $Z \to c\bar{c}$ and $Z \to b\bar{b}$ events).

A.2. Neural network input variables for $\mathcal{N}_C$

- C(1): This variable is defined as

$$C(1) = \frac{\sum_j [\sum_{i \in j} p_\perp(i)]}{P_{\text{Hemi}}}$$

where j runs over all the jets of the hemispheres.
- C(2): Same as B(11) for the tracks of J_{max}.
- C(3): This variable is defined as:

$$C(3) = \frac{\sum_j [\sum_i p_\perp(i) p_\parallel(i)]}{P^2_{\text{Hemi}}}$$

where j runs over all jets of the hemisphere and i over all the tracks of each jet j.
- C(4): Same as B(1).
- C(5): Longitudinal momentum $p_\parallel$ of the second leading track of J_{max}.
- C(6): Same as B(7).
- C(7): Same as B(3).
- C(8): Visible energy of the hemisphere E_{Hemi}.
- C(9): "Directed Sphericity" calculated with the four most energetic tracks of J_{max}.
- C(10): Longitudinal momentum $p_\parallel$ of the third leading track of J_{max}.
- C(11): Same as B(5).
- C(12): Invariant mass of J_{max}.
- C(13): Sum of the masses of all the jets of the hemisphere.
- C(14): $\mathcal{P}_H$ as described in the text.
- C(15): Multiplicity of the charged tracks of J_{max} with $p > 0.25$ GeV/c.
- C(16): Multiplicity of the charged tracks of J_{max} with $p > 1.0$ GeV/c.
- C(17): Multiplicity of the energy flow tracks of J_{max} with $p > 0.25$ GeV/c.
- C(18): Same as B(8).
- C(19): Invariant mass of the energy flow tracks of J_{max} with $p > 1$ GeV/c.
- C(20): Number of identified leptons (electrons and muons) with $p > 3$ GeV/c in the hemisphere.

References

[1] D. Buskulic et al. (ALEPH Collab.), Phys. Lett. B 401 (1997) 150.
[2] D. Buskulic et al. (ALEPH Collab.), Nucl. Instr. Meth.A 360 (1995) 481.
[3] W. Bartel et al. (JADE Collab.), Z. Phys. C 33 (1986) 23.
[4] D. Buskulic et al. (ALEPH Collab.), Phys. Lett. B 313 (1993) 549.
[5] D. Buskulic et al. (ALEPH Collab.), Nucl. Instr. Meth. A 346 (1995) 461.
[6] Presentation of LEP Electroweak Heavy Flavour Results for Summer 1996 Conferences, LEPHF/96-01, ALEPH note 96-099.
[7] Reports of the Working group on Precision Calculations for the Z resonance, eds. D.Bardin, W.Hollik and G.Passarino, CERN 95-03 and references therein.
[8] J. Lys (CDF Collab.), FERMILAB-CONF-96/409-E. Proceedings ICHEP'96, Warsaw (1996);
E.W. Warnes (D0 Collab.), FERMILAB-CONF-96/243-E, Proceedings ICHEP'96, Warsaw (1996);
M. Demarteau, FERMILAB-CONF-96/354, Proceedings DPF'96, Minneapolis (1996).
[9] K. Abe et al. (SLD Collab.), Phys. Rev. D 53 (1996) 1023.
[10] P. Abreu et al. (DELPHI Collab.), Z. Phys. C 70 (1996) 531.
[11] K. Ackerstaff et al. (OPAL Collab.), CERN-PPE-96-167 (1996), submitted to Z. Phys. C.
[12] G.C. Fox, S. Wolfram, Phys. Rev. Lett., 41 (1978) 1581.
[13] L. Bellantoni et al., Nucl. Instr. Meth. A 310 (1991) 618.
[14] D. Buskulic et al. (ALEPH Collab.), Phys. Lett. B 295 (1992) 396.

www.ingramcontent.com/pod-product-compliance
Lightning Source LLC
Chambersburg PA
CBHW070156240326
41458CB00127B/5971
9789811264429